THE ENCYCLOPEDIA OF
SPACE TRAVEL AND ASTRONOMY

THE ENCYCLOPEDIA OF
SPACE TRAVEL AND ASTRONOMY

octopus

CONTENTS

First published 1979 by
Octopus Books Limited
59 Grosvenor Street
London W1

© 1979 Octopus Books Limited

ISBN 0 7064 0992 2

Produced by Mandarin Publishers Limited
22a Westlands Road,
Quarry Bay, Hong Kong

Printed in Italy

FOREWORD

By Carl Sagan
Director, Laboratory for Planetary Studies
David Duncan Professor of Astronomy and Space Sciences
Cornell University, Ithaca, New York, USA

In the early 17th century the great Italian scientist Galileo Galilei turned the first small astronomical telescope to the heavens and discovered wonders: that Venus underwent phases from new to crescent to full, like the Earth's moon, demonstrating that the Earth went around the Sun and not vice versa; that there were mountains and craters on the Moon. implying that it was a world, in some sense like ours, and not made of some special sky stuff, such as Aristotle's 'quintessence'; that there were dark spots on the Sun's visible disk, showing that it was not 'perfect', as 2000 years of religious mystics had maintained; and that there were four moons orbiting the planet Jupiter, shattering a theological ban on the existence of new worlds. Galileo's work, including his observations of the motions of the moons of Jupiter, was an important springboard for Isaac Newton's invention of mathematical physics, which has in turn led, in an important way, to our modern technological civilization.

Once Galileo had improved Dutch novelty lenses to make the first astronomical telescope, these breathtaking discoveries were inevitable. The astronomical phenomena were up there waiting to be discovered; it only required a fairly minor technological improvement over our human eyesight. Today a comparable revolution is being worked in astronomy. We are no longer restricted to ordinary visible light. Radio waves emitted by many astronomical objects are transmitted through the Earth's atmosphere and are now being studied by immense radio telescopes on the Earth's surface. The Earth's atmosphere is entirely opaque to gamma rays and x-rays, and partially opaque to ultraviolet and infrared light. Accordingly, we have launched instruments sensitive to these wavelengths to high altitudes, above much or all of the Earth's atmosphere – in airplanes, in balloons, in rockets and particularly in Earth-orbiting satellites. Until recently astronomy has been a kind of passive science in which the practitioners wistfully scan the heavens hoping something would happen. But now we have radar astronomy which is able to perform remote experiments on nearby objects; and, of the greatest significance, we have begun launching small instrumented space vehicles to fly by, orbit, and land on the nearest celestial objects. We have examined closely all of the planets known to the ancients. And we have sent four spacecraft –

Pioneers 10 and 11 and Voyagers 1 and 2 – on trajectories which will ultimately take them out of the solar system altogether, and into the realm of the stars.

With such remarkable instruments the pace of astronomical discovery has in the last decade or two become breathtaking. In many areas it has moved from myth and vague guesswork to precise knowledge. At the same time many more mysteries have been uncovered as old ones have been resolved. In recent years, the finding of deuterium between the stars with an orbiting ultraviolet observatory has cast light on the evolution of the universe: active volcanoes have been found on Io, the innermost of the four moons of Jupiter found by Galileo; distant astronomical objects have been found with apparent velocities greater than that of light, providing an interesting challenge to the theoreticians; the first serious searches in human history have been made for simple life on the planet Mars and for advanced civilizations on the planets of other stars; high energy cosmic rays discovered on the Earth have been identified as arising from colossal explosions in remote galaxies; ring systems have been identified around Jupiter and Uranus; a new worldlet has been discovered beyond the orbit of Saturn, and Pluto has been found to have a massive moon; radio observations reveal that two orbiting pulsars behave precisely as predicted by Einstein's theory of general relativity; and evidence has been presented suggesting that the clouds of Titan (the great moon of Saturn) and the grains and gas between the stars are all composed in part of organic molecules, related to those which four billion years ago, on our planet, led to the origin of life.

Like Galileo's findings, these recent discoveries, fragmentary data, and provocative first attempts have a significance ranging far beyond the perview of specialists in astronomy. We are in the midst of discovering nothing less than the nature of the universe and our place in it. In the long run some of these findings will have the most profound practical, as well as philosophical consequences. Modern astronomy and space sciences are providing answers to questions which we have asked for as long as there have been human beings. It evokes our sense of wonder; it speaks to us of who we are.

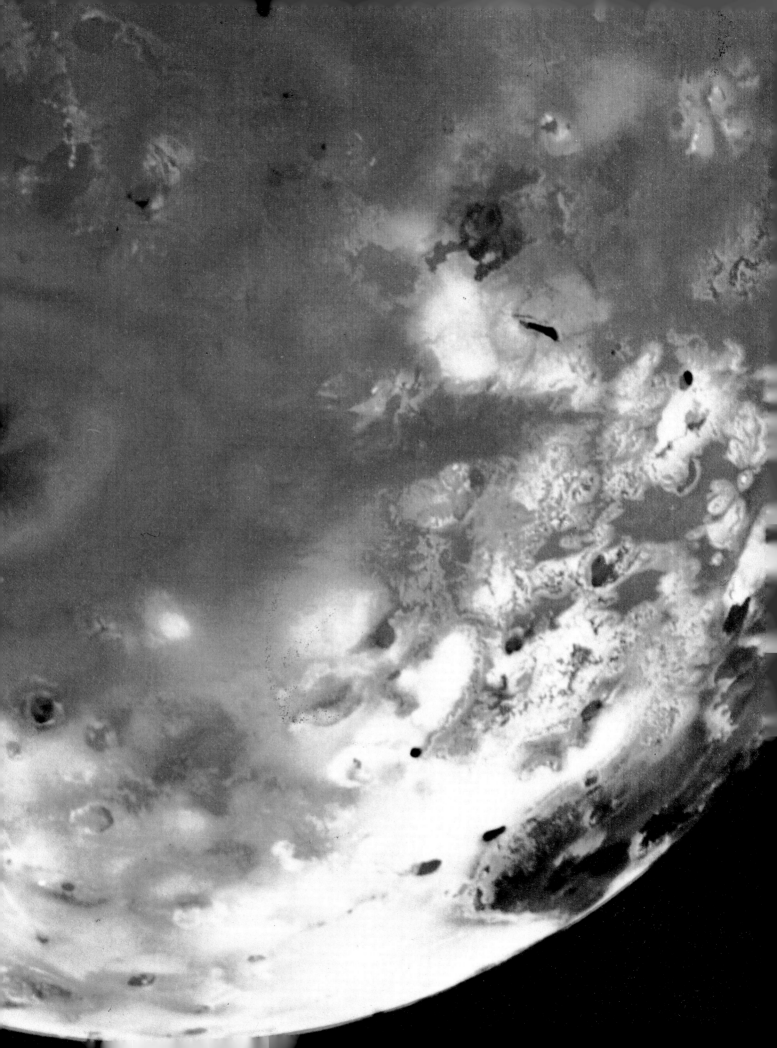

1/DAWN OF A NEW ERA

SIR BERNARD LOVELL

We are at present in the midst of a revolution in astronomy as profound in its way as that initiated by Copernicus when he showed that the Sun, not the Earth, is at the centre of our Solar System. At one level, the revolution is on the grandest of all themes – the beginning and end of the Universe. In addition, new space technologies will, within a few decades, allow mankind to live permanently in space. *Homo sapiens* may become a galactic, rather than simply a terrestrial species. The significance of these developments is assessed here by an elder statesman among astronomers, Sir Bernard Lovell, pioneer radio astronomer and founder of Britain's Jodrell Bank Radio Observatory (left), for years the world's largest fully steerable radio telescope. He talked to the editor, John Man, in the shadow of the 250-foot bowl that rears with surreal beauty over the Cheshire countryside.

I am often asked whether it is possible to comprehend the scale of the Universe. It's easy to describe the immensity of the Universe in conventional astronomical terms, but the question of *comprehending* it is an entirely different matter. I think that it is now a very long time since anyone has been able to comprehend the size of the Universe. 2,000 years ago, Posidonius calculated the distance of the Sun as 6,500 earth diameters. It wasn't a bad estimate: he was only out by a factor of two and it was a distance that could be comprehended if one had a feeling – as the ancient Greeks did – for the size of the Earth. Copernicus used a complicated argument to prove that the stars, which he assumed were all the same distance, were fixed to a sphere one and a half million earth radii distance – a long way, but still comprehensible. But in 1838, when Friedrich Bessel measured the distance to a nearby star, 61 Cygni, he obtained an answer of about ten light-years – a distance which light, travelling at 186,000 miles per second, takes ten years to cover. And today we know that the nearest stars are more than 5,000 million earth radii distances.

Well, these are immense distances that are entirely incomprehensible in everyday terms. If we try to think of such distances in terms of miles, the figures become meaningless. Even comparisons are not much use. If, in your mind, you reduce the Sun to the size of a six-foot ball, then Pluto, the most distant planet of our solar system, would be a marble five miles away. But the nearest star would be 34,000 miles away. And our own galaxy, the Milky Way – our own local collection of stars – would be 800 *million* miles across. So even on that scale, distances soon become meaningless.

'The best one can do is to think in terms of the speed of light, 186,000 miles per second. That's seven times round the world in a second. The light from the Sun takes about eight minutes reach us. It takes eleven hours to cross our Solar System. The light reaching us now from the nearest star set off four years ago. Our own Galaxy is about a hundred thousand light-years across. And our nearest galactic neighbour, the great spiral in Andromeda, is some two million light-years away. The most distant known objects are billions of light-years away. To talk in terms of light-years makes it possible for astronomers to deal with numbers they can handle. But in human terms, the distances are still unimaginably huge. At present rocket speeds, it would take more than a human life time to get to the nearest star and back again.

'One important point about these huge distances is that the further one goes, the less certain distances become. The very concept of distance begins to lose its meaning. We measure extreme distances in terms of what is known as the red-shift – distant objects are receding as the whole Universe expands and the light-waves from them are stretched so that when their light is analysed, it seems to be shifted towards the red end of the spectrum. When Edwin Hubble first published his measurements of the speed of recession in the late 1920's, he said that recession increased at 530 km. per second for every 3.25 million light-years of distance. He claimed an accuracy to within 15 per cent. When Allan Sandage estimated the same constant a few years ago, he gave the recession speed as 50 km. per second per 3.25 million light-years, just a tenth of what Hubble gave, and he was still claiming an accuracy of plus or minus 15 per cent. It's far more important, for the history of the Universe, to estimate not how far in terms of miles we can see into space, but how far back into time, bearing in mind that the light from these distant objects set off billions of years ago.

On the most significant recent discovery

'My answer is really quite an unambiguous one – the discovery in 1965 of low level background radiation which fills the whole Universe uniformly. This seems to be the relic of the radiation left over from the early stages of the expansion of our Universe – and the forces of that expansion led to the recession of the galaxies when they formed millions of years later. Of course the fact that the Universe is expanding implies that at some time in the past, all the objects in it were closer together. But we now seem to have very clear evidence that the Universe we know today began as a super-dense, super-hot "singularity".

'This is not only the strangest, but the most important of recent discoveries. There is no escape, apparently, from the fact that 10,000 million years ago – or at least a time scale of that order – the Universe was in a completely different state. For the first time we really have to face the problem of a Beginning. I find that for most of my professional life, I've been able to avoid having to face this problem, because there was always a possibility that the Universe was not like that, that it was balanced in a steady state. Whatever other problems there were to be faced, we didn't have to confront the possibility that the Universe was once entirely different.

'It's an extraordinary situation. The whole modern structure of mathematics and physics is built upon the laws of nature that we now see in operation. Yet we must now confront the possibility that these "laws" simply do not apply over an infinite period of time.

'Let's look at this problem a little more closely, for it seems to me to be of the very greatest importance, not only for astronomers, but also in relation to the prevalent attitude that scientific knowledge is valid for all time.

'There is some dispute as to the exact time of the origin of this background radiation, but the most likely theory is that it originated at a stage somewhat less than a million years after the beginning of the expansion of the Universe, an event commonly described as the "primordial fireball". If this is correct, we are receiving information that takes us 99.9 per cent of the way back in time to the beginning of the Universe (whereas investigations of remote objects give us nothing like this penetration – they take us back no more than about 75 per cent of the time to the beginning of the expansion). We can also provide an explanation of the events that probably led up to the formation of the background radiation, events which take us back to a minute fraction of a second (10^{-43} or one over ten followed by 42 zeros) after Time Zero. But the extraordinary thing is that if you ask what was the situation in the 10^{-43} seconds after the expansion began, no prediction is possible on the basis of modern

The 24 diagrams on this and the following three pages establish the scale of the Universe. Each picture represents a ten-fold increase in scale, from a height of 1 km. (0.6 miles) out to the limits of the known Universe at 10,000 million light-years. The 100,000-fold increase in distance on this page leaps from a bird's eye view of a launch site in the Kennedy Space Center, Cape Canaveral, Florida, to a view of the Earth familiar from Apollo space missions.

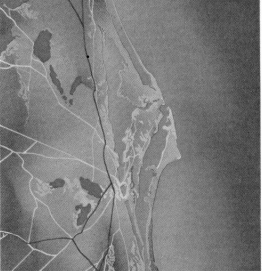
1   1 km: low altitude flight

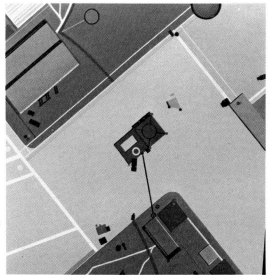
2   10 km: high altitude flight

3   100 km: low orbit

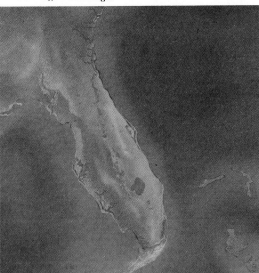
4   1000 km: mid-level orbit

5   10,000 km: high orbit

6   100,000 km: deep space

theory. We're concerned there with a Universe less than 10^{-33} centimetres across, of an incredibly high density. The theories of gravitation and atomic physics break down. We do not know whether this is a fundamental barrier to scientific description or whether there are other theories yet to be discovered.

On why the Universe is the way it is

'There are many other strange features when one searches for the meaning of this early history of the Universe. The primordial fireball contained

These drawings together represent a one million-fold increase in distance over the last picture on the previous page. The Earth-Moon system recedes rapidly into insignificance. In the final two pictures, the outer ring is the orbit of Pluto, which is shown at its average distance – it is in fact an irregular orbit that swings inside that of Neptune. At these scales the Sun becomes a mere pin-prick of light.

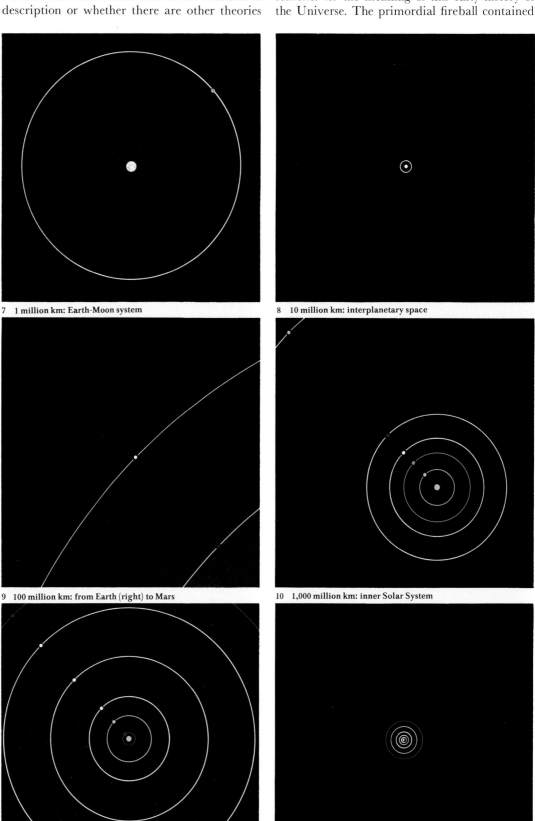

7 1 million km: Earth-Moon system

8 10 million km: interplanetary space

9 100 million km: from Earth (right) to Mars

10 1,000 million km: inner Solar System

11 10,000 million km: Solar System

12 100,000 million km: interstellar space

the reactions which led to the present distribution of hydrogen and helium – 75 per cent and 25 per cent respectively – a balance that explains the evolution of stars. Now, very small changes in the nature of the primordial fireball would have had an immense effect on the Universe. If certain atomic forces had been only slightly greater, then all the hydrogen would have become an isotope of helium and no long-lived stars could exist as they do at the present. They would have been explosive. Stars would have formed, but they would have used up all their energy in a very short

In the depths of interstellar space, the Sun remains a lonely point of light until a 100-fold increase in scale reveals the presence of its nearest neighbours. Distances can now not easily be measured in kilometres, and light-years become a more convenient scale. With further increases in scale, the Sun is lost among its neighbours, whose images begin to overlap as the larger structure that binds them, the Milky Way Galaxy with its characteristic Catherine Wheel spiral, begins to come into view.

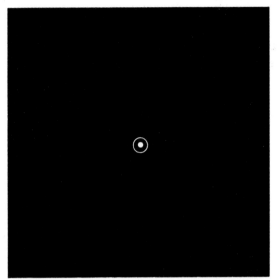

13   1 million million km: interstellar space

14   10 million million km (1 light year): interstellar space

15   10 light years: local stars

16   100 light years: local stars

17   1,000 light years: local region of galaxy

18   10,000 light years: arms of galactic spiral

The Galactic spiral with its 100 billion stars is just one of an uncounted number of galaxies, the nearest of which is two million light-years away. The galaxies themselves are grouped into clusters. These vast systems are all moving away from each other at a speed that is proportional to their distance apart. Light from the most distant known galaxies takes 10,000 million years to reach the Earth, and information about objects at such great distances is scanty, but astronomers have not yet seen any sign that the galaxies thin out.

time. There would have been no stars like the Sun, which give an output of energy for thousands of millions of years. It's only with stability on this time scale that life can evolve. If things had been just a little bit different at the beginning, therefore, there could have been no life, and the Universe would be unknowable.

'There's another extraordinary feature pointed out by Stephen Hawking of Cambridge in 1973: If, in the primordial fireball, the expansion of the Universe had differed by only one part in a million millionth from what it actually was, there

19   100,000 light years: Milky Way galaxy

20   1 million light years: intergalactic space

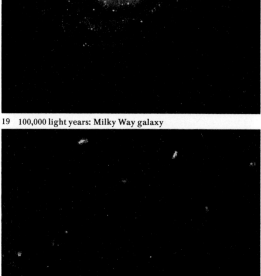

21   10 million light years: local group of galaxies

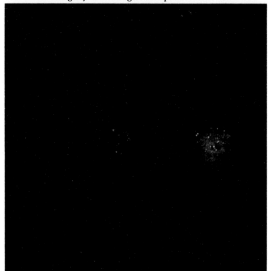

23   1,000 million light years: galactic clusters

22   100 million light years: galactic clusters

24   10,000 million light years: limits of known universe

would have been no possibility of the Universe existing as we know it now. If the Universe had expanded one million millionth part faster, then all the material in the Universe would have dispersed by now. There would have been no possibility of the gas being drawn together by gravity into stars. And if it had been a million millionth part slower, then gravitational forces would have caused the Universe to collapse within the first thousand million years or so of its existence. Again, there would have been no long-lived stars and no life.

'These things are to me immensely strange. Is it not extraordinary that the possibility of talking here this afternoon depends on events which were very narrowly determined over 10,000 million years ago in the very earliest moments of the Universe?'

On the significance of these discoveries for modern thought

'We live in a most peculiar time. Astronomers have extended the borders of science to a point where science is no longer capable of answering the questions raised. If at some point in the past, the Universe was once close to a singular state of infinitely small size and infinite density, we have to ask what was there before and what was outside the Universe. The answer at this moment is that such questions have no meaning. Yet they are questions that have absorbed man for over 2,000 years. When in the fourth century, Saint Augustine was asked why the world was not created sooner, his response was that there was no 'sooner' and that time was created when the world was created. Philosophically, we haven't made much progress beyond that. Similarly if you ask a modern astronomer what was outside the Universe before the Time Zero, the only possible answer is: there was no 'outside'. These answers do not seem to me to be answers with which scientists and philosophers will be happy in the future.

To get back to an earlier point, we have to face the problem of a Beginning, that is of a remote epoch when time and space, and the Universe, as we understand those terms today, did not exist. The American astronomer Robert Jastrow recently made the comment that astronomers have been climbing a great mountain and have at last hoisted themselves to the final rocky pinnacle only to find that the theologians have been sitting there for centuries. In the 13th century, Thomas Aquinas evolved a great synthesis and reconciliation of the science of Aristotle and theological dogma which survived and guided man for centuries. It seems to me that we need a modern Thomas Aquinas, who can synthesize the world of theology and science, to revolutionize, once again, the way we see and comprehend the Universe and man's relation to it.

On the major problems of astronomy

'There are two great scientific problems the solution to which may help in providing us with a way forward. They are the future of the Universe and the nature of gravitation.

'Will the Universe expand for ever, or will it collapse again? The answer to this question depends critically on the amount of matter in the Universe. If the density is at present more than 10^{-29} grammes per cc, then we calculate that there is enough matter in the Universe to overcome, by gravitational attraction, the present expansion, and the Universe will eventually collapse again to another state of super-density. If the density is less than that value the Universe will continue to expand for ever. It will be a once-for-all Universe.

'We do not know the answer yet, but I think that one can be optimistic that the solution will be found as a result of current attempts to measure the "deceleration parameter" – the search for evidence that there may be a departure from the linear relationship between the rate of expansion and distance. The issue is of immense interest. If this is a once-and-for-all Universe, the problems of the Beginning may be insoluble. If it is a cyclical series of many Universes, then it is possible to say, as some cosmologists do, that we are living in one of an infinite cycle of Universes. If so, we are in a privileged position, because this particular Universe has the combination of atomic and gravitational constants such that long-lived stars can exist, which, in turn, make possible the development of life.

'The second problem is that of gravity. It is a most remarkable fact that the major force which controls our lives and which determines the large-scale structure of the Universe is not understood. Although Newton discovered the laws of motion and the gravitational law, he evaded the problem of the nature of the gravitational force. He regarded gravity as a force of unknown type acting at a distance and transmitted instantaneously. Since the publication of Einstein's general theory of relativity in 1915 we have been forced to abandon Newton's concept of an absolute space and time and to regard the gravitational forces as a function of the geometry of space. These forces can no longer be envisaged as acting instantaneously at a distance but are propagated with the speed of light as gravitational waves. But no one has yet proved unequivocally the existence of gravitational waves. We do not understand the relationship between gravitational forces and the other forces which govern the structure and behaviour of atoms.

'So you see that although we think we have an accurate description of the Universe, there are great and unsolved problems. Take for instance the problem of black holes – super-dense bodies with gravitational fields so intense that not even light can escape from them. Many people claim to have made observations that provide strong evidence of their existence. Maybe; but many astronomers express great caution on the subject; some will say "of course" and some "of course not". It may be that when the problem of the nature of gravity is solved, then the problem of whether black holes exist or not may completely disappear.

'It's worth emphasizing that some very fundamental assumptions about modern astronomy are not universally accepted. It's generally accepted, for instance, that the observed redshift is a proof that the Universe is expanding. But not everybody believes this. Some astronomers preserve an open mind on this question, and wonder whether the redshift may not have some other explanation – such that the atomic and gravitational constants vary with time. It may be that the existence of the background radiation itself is not incontrovertable evidence for the expansion of the Universe. I take it to be so and so do almost all of my scientific colleagues. But new discoveries may yet prove us wrong.

On the significance of space travel

'It's true that the potential for scientific achievement in space – especially after the launch of the US Space Shuttle early next year – is immense. I will turn to that in a moment, but it's first worth noting that the impetus for the US and Russian space programmes has been largely military. Last year, 1978, the percentage of military operations in space by the Russians was just under 80 per cent and in the previous two years it had been 95 per cent. It is the military authorities who control the money and the important developments depend on the defence budgets of the major powers. It's true that over the next two decades, space technology in both the US and in Russia will make possible the development of very large space platforms, in particular it has been suggested that such platforms might beam down solar power to Earth. It may well be that such devices will enable us in part to overcome the energy crisis that currently faces mankind. But it would be naive to see such expensive schemes purely in terms of the peaceful use of solar energy.

'For one thing, it would be relatively easy to redirect the microwave beams produced by such devices to modify the weather over countries other than one's own. Many experts take the view that to develop solar power satellites would be foolish because it would be far easier to collect the power on earth. They miss the point: whether or not it is economical for power, it is likely to be done because the satellites have military potential.

'Then there is the critical question as to whether the developments for commercial and military purposes in space will destroy the possibility of scientific investigation. Take for example the problem of the solar power satellites. I'm told the cost of development is no greater than that which has gone into North Sea Oil. It is quite possible that by the end of the century there will be 10-gigawatt transmitters in space. But it would take eight of these to service the UK alone. If the rest of the Western world is similarly supplied by power from satellites, then the pursuit of radio-astronomy on many areas of the Earth would become impossible.

'It has been suggested that radio-astronomers would then have a fine opportunity to build large, new radio-telescopes in space. I do not believe that such a development will inevitably follow the building of power satellites. As always, the dividing line between good and evil is a narrow one, and the dangers, as revealed in the history of scientific development, need no emphasis. Why were rockets developed? Well, 98 per cent of all rockets used for launching space vehicles were originally developed for military purposes. So it's no good being starry-eyed about the possibility of scientific investigation in space. Yes, much is possible. But it will not be easy to achieve.

'Nevertheless, I'm not entirely pessimistic. There will, I think, be dramatic scientific advances. A very important step in astronomy will, all being well, occur in 1983, when the Shuttle will take into orbit a large space-telescope. This telescope, free from the limitations that confront earth-based astronomers, may well have the resolving power to solve the question as to whether there are planets revolving around our nearest stars. This will be a major step forward in the solution to another problem which astronomers are currently confronting: the possibility that planets like our own are commonplace in our Galaxy.

'Another question that the space telescope should resolve is: will we see many more objects in the Universe? Will the numbers go on increasing as they have done so far with every increase in light-gathering power? The space telescope will be able to see objects several magnitudes fainter than any visible to earth-telescopes. One wants to know as a matter of the greatest possible interest how many objects there are in the Universe. Will the images begin to overlap on the photographic plates instead of being dotted about in empty space as we now see them?

'In answering this question, we may be able to penetrate to a remote epoch that is now completely blank observationally – the time when galaxies came into existence from the primordial gas. The observation of the microwave radiation reveals information about the Universe less than a million years after the beginning of the expansion. The observation of the galaxies takes us back a few billion years. But of the time in between – perhaps a billion years before the galaxies began to form from the expanding primeval gas – we know almost nothing.

'Such observations may in turn mean we will be able to calculate more exactly the amount of matter in the Universe and thus what its ultimate destiny will be.

'On reflection about what has happened in astronomy over the last few decades – the discovery of radio galaxies, of quasars, of pulsars, of the microwave background – it seems possible that the space telescopes may well reveal objects of which we know nothing. The space telescope will open a new era in astronomy of a significance we cannot yet envisage.

On the possibility of life elsewhere

'Ten years ago, it was my opinion that organic development would have occurred as part of the normal process of the evolution of a planetary system, whether this Solar System or any other. Certainly there are some strong arguments for the general existence of life, such as the existence of complex molecules in interstellar space and the fact that we believe that planets are common features of the Universe. Yet there are also some interesting arguments against. It may be that life could evolve only in earth-like conditions, and that those conditions are extraordinarily rare if not unique.

'4,500 million years ago, Earth and Venus were probably very similar, yet Venus now has a surface temperature that will melt lead and an atmosphere mainly of carbon dioxide. The atmosphere of Earth, unlike those of the outer planets, is radically different from the composition of the solar nebula from which the planets formed. The terrestrial atmosphere is a secondary one, probably formed by volcanic eruption. Some unknown cataclysmic accident swept away the primordial atmosphere of the inner planets in the very early history of the Solar System. We do not know whether such events are common in planetary systems or whether this has been a unique event in the Solar System.

'Again, it strikes me as extraordinary that life emerged so early in the Earth's history. Traces of life have been found in rocks 3,500 million years old, when the Earth had been in existence a mere 20 per cent of its life-span. Now, the chances of the amino-acids and nucleotides forming in the primeval oceans and then coming together randomly to create enzymes, the substances that cause and direct the numerous chemical reactions that occur in living organisms, has been estimated at something like 10^{78} against. So there is a complete gap in our explanation as to why life has developed at all on Earth. If it is indeed true that the chances against the evolution of life are so large, then the present optimism which leads to the search for life elsewhere in the regions of the Milky Way closest to the Sun may be quite unjustified.'

The three main engines of the Space Shuttle – with a combined thrust of over 600 tons – will open a new chapter in mankind's relationship with space. Scientists will be able to work for long periods in space with instruments that will probe the limits of the Universe. Eventually, members of the public will be able to buy tickets into orbit as if for a commercial airline flight.

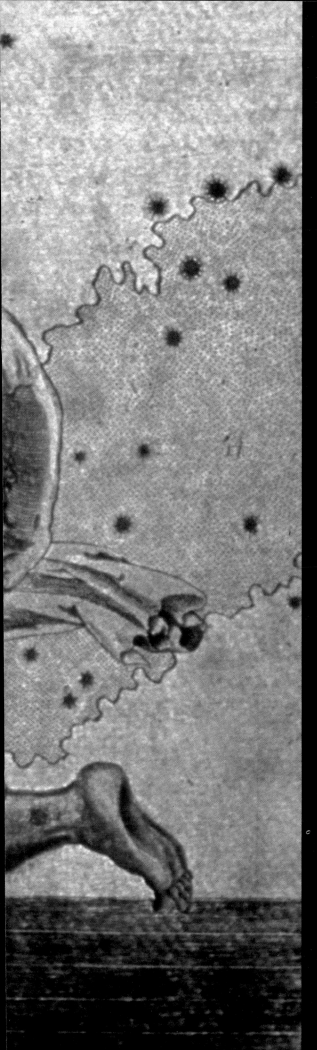

2/THE FRAMEWORK OF THE STARS

NIGEL HENBEST

Stars look like mere pin-pricks of light, set close by in a dome of darkness. For centuries, men sought to understand the Universe within such a cosy framework, sketching pictures in the sky (like the 17th-century version of the constellation Auriga, the Charioteer, at left) to record the stars and provide a reassuring mythology. It seemed common sense that men's destinies and the cycles of the stars were intimately related – astronomy and astrology were different sides of the same coin. Only relatively recently – over the last three centuries – have we come to see that the stars are sun-like bodies set in the vastness of three-dimensional space. And only in this century have we been able to measure the distances between the stars and understand the nature of the Catherine-wheel structure – our Milky Way Galaxy – that binds the stars together.

This 11th-century Arabic version of the figure of Perseus points to the ancient origins of the names and figures associated with the constellations. Perseus was the Greek mythological hero who beheaded Medusa.

In this 14th-century view, two angels crank the fixed stars round the motionless Earth. The idea now seems naive, but (if the motion of the planets is ignored) it makes little difference observationally whether the Earth or the stars are considered to be in motion. The picture also reflects the belief that the wheeling heavens needed repeated inputs of divine energy to keep them in motion.

I t is not surprising that astronomy is the oldest science. While early man's interest in the changeable world around him was highly practical, concerned with the need to survive, his interest in the patterns of the stars and their regular cycles was of a different kind. His imagination could roam the stars unfettered. The sky seemed to roof the world; its exquisite pattern of stars called out for explanation. On this dark canvas, our ancestors depicted huge figures, outlined by scattered stars, much as children of today join up the dots in puzzle books to find the hidden figures. And as the stars bear no numbers to guide the inscribed lines, different cultures have seen different patterns in the same stars.

But even the esoteric field of star-watching had a practical relevance to ancient man. The rising and setting of stars during the night formed a night-clock for the Egyptians, complementing the shadow-stick sundial in use during the day. The appearance of different constellations at different seasons was a simple calendar. And to these early scientists, it was only logical that other appearances in the sky were related to terrestrial events. The Chinese kept a watch for temporarily bright 'guest stars' which surely must signal changes in the Empire; the Babylonians were more concerned with keeping track of the planets, the bright moving 'stars' which wander slowly among the constellations, and should also mirror changes on the Earth. To most early civilizations, there was no hard line between the subjects we would now call the science of astronomy and the occult lore of astrology.

The astrological interpretation of these portents in the sky could only be assessed by their positions relative to the constellation figures. To the Chinese, a guest star 'trespassed' against the constellation in which it appeared, and their astrological records of such happenings are an invaluable source of data on past astronomical events, a virtually complete record stretching back 2,000 years. The Moon and planets travel a well-defined track (see Chapter 4) through a band of constellations, known collectively as the zodiac, whose names have come down to us today to appear in the popular horoscope columns in newspapers and magazines. A glance at one of these will still reveal that important changes in our personal lives should be expected when a planet crosses the division from one zodiacal constellation to the next!

Astronomers now dismiss the claims of astrologers, but the ancient astrological signs have at least given both astronomers and the general public common patterns with which the heavens can be mapped.

The origins of the constellation patterns and names we use today are lost in the mists of the remote past, in the world of the ancient civilizations around the Mediterranean Sea. Our principal constellations were described by the Greek philosopher Aratus around 275 BC, but he was merely passing on an older tradition which probably sprang originally from Sumeria. The Greek astronomers Hipparchus and Ptolemy finalized a list of 48 constellations, which have come down to us in the form of Ptolemy's great work, *The Almagest*. These patterns, well-established throughout Western Europe in medieval times, were outlined by the brighter stars in the sky; between them were uninteresting regions populated by dim stars which were assigned to no particular constellation.

Some of the most striking of these ancient figures can be recognized without too much difficulty. (Ironically enough, the best place to learn the constellation patterns is from a city environment; with only the brightest stars visible, the shapes are much easier to spot.) The figure of Orion, striding through the January skies, is one of the most familiar. Bright stars mark his shoulders and knees, three central stars delineate his belt, while fainter stars make a hanging sword, a raised club and a lion-skin shield. Opposite him in the sky, and only poorly seen from northern latitudes, is the equally striking Scorpio, the scorpion. The bright star Antares marks his heart, and winding below is the body and tail, with a sting at the tip. In front, fainter stars which once represented his claws have now been lopped off as the separate constellation Libra, the scales.

To star-gazers in high northern latitudes, the seven stars of the Plough are an ever-present and familiar sight. They also act as a celestial signpost, for by following the lines of its stars we can readily find other constellations. In particular, the two 'Pointers' leading the Plough

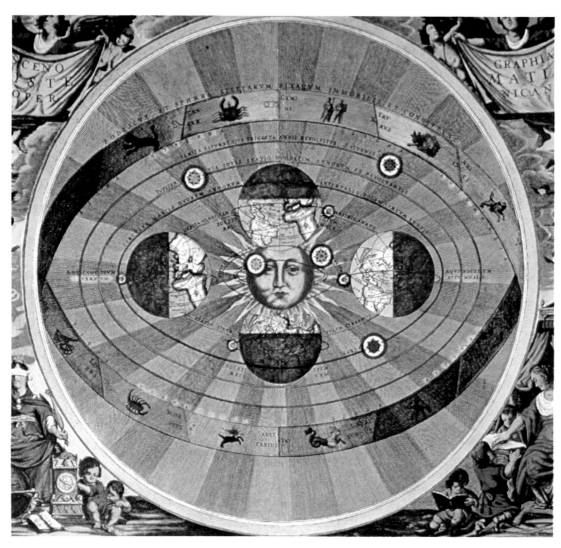

The zodiac, the band along which the Sun, Moon and planets travel as seen from Earth, is shown in this mid-17th century painting. The Solar System is shown correctly with the Sun at its centre. The zodiac was first devised by Mesopotamian astronomers as a system of animal signs as early as 3,000 BC, and the circle of 12 constellations came to be called the *'zodiakos kyklos'* – 'circle of animals' – by the Greeks. Hence its present name, although the signs themselves are no longer all animals.

show the position of the Pole Star, a star which happens to lie almost directly above the Earth's north pole. The southern hemisphere's equivalent of the Plough is the Southern Cross, a superb little group of four bright stars, which can again be used as a pointer to other constellations.

To the amateur star-gazer, learning the constellations is the only way to become familiar with the night sky, and the bright patterns of antiquity are fairly easily recognized. The southern hemisphere astronomer, though, finds that the bright patterns near the south pole of the sky often bear strangely modern-sounding names. This region was invisible from the latitude of the Mediterranean civilizations, and the newly-discovered stars were ordered into 12 new constellations by Johann Bayer of Augsburg in 1603.

Once Bayer had broken with tradition, and boldly introduced his own figures into the sky, others were not slow to follow. Bayer's colleagues Julius Schiller produced a Christian star atlas in 1627, keeping the constellation shapes, but 'improving' the pagan names: Orion became St. Joseph, and the Great Bear the Ship of St. Peter. This innovation did not catch on, however, but other astronomers took the liberty of 'forming' constellations out of the faint stars lying between the classical constellations. Some of these have survived the test of time, while others have fallen into disuse. There is still a Unicorn near Orion, and a Giraffe not far from the Pole Star, and in the southern hemisphere some of the weird

creations of Abbé Nicholas de la Caille have passed into general use. These include a sculptor's workshop, a pendulum clock, an air pump, (perhaps more fittingly) a telescope and microscope, and last – and definitely least – Table Mountain (Mensa), which holds the distinction of being the faintest constellation in the sky. Mercifully, other constellations proposed by Johann Bode were soon dropped, otherwise our skies would also be cluttered with a log-line, an air-balloon, a printing press and an electrical machine!

Modern professional astronomers need to know positions far more accurately than simply by constellation: a position of the famous Crab Nebula given as 'above the southern horn of Taurus (the bull)' is of little use today. The sky, though still being treated as a sphere, has therefore been divided by imaginary lines, corresponding exactly to latitude and longitude on Earth. The latter run from the north to the south point of the celestial sphere (the points above the Earth's north and south poles), the 24 lines being spaced at 15° intervals. As the Earth rotates, these lines of *right ascension* pass overhead at the rate of one per hour: they are accordingly named as 'hours', and the intermediate spaces divided into 60 'minutes'. The equivalent of latitude is the astronomer's *declination*, measured in degrees north (+) or south (−) of the celestial equator towards the poles. So a position in the sky can be specified by two co-ordinates, just as

Mapping the Stars

At first sight, understanding the way stars are pinpointed on a star map seems forbidding. The details are, indeed, complex; but the principles are not. The reality of three-dimensional space is ignored and the stars are seen as points on the inside of a sphere, which is mapped with a grid of lines equivalent to those of latitude and longitude (right). In this way the arbitrary patterns of the constellations (below) can be set in a fixed frame of reference.

The celestial equivalent of latitude – which measures distance north and south of the equator – is *declination*. The equator, which is set at $23\frac{1}{2}°$ to the plane of the orbit, is projected on to the 'celestial sphere'. A star's position north or south is then measured in degrees, between 0° at the equator and 90° at the poles (+ =north; − =south).

The second coordinate – the equivalent of longitude – is more problematical. As with longitude, a north–south baseline (like the Greenwich Meridian) must be fixed. The celestial meridian is defined by the point at which the path of the Sun crosses the celestial equator at the beginning of spring in the northern hemisphere, when day and night are of equal length – the vernal or spring equinox ('equal night'). From this point, a line is drawn north and south to the celestial poles to form the *prime meridian*.

Measurements can then be taken from the prime meridian on a grid drawn from the celestial equator. Since the stars' positions in the sky depend on the Earth's 24-hour period of rotation, the units of this coordinate are usually hours (although degrees are also used; 1 hr = 15°).

In practice, the system works as follows: An observer sees the points in the sky defined by the prime meridian swing above his head from east to west. The other stars follow hour by hour as the Earth turns. Thus the position of any star can be established by the amount of time it lags behind the meridian. This coordinate is known as *right ascension*. ('Right' has nothing to do with direction; it derives from an archaic use of 'right' meaning 'vertical' as in 'right angle'.

But there are added difficulties. The Earth's axis is not quite stable. It wobbles, like an unsteady top, and the point of the axis *precesses* in a circle over 26,000 years. The equinoxes are therefore not at exactly the same time every year, and the celestial coordinates must therefore be given according to a particular year (at present, either 1950 or 2000).

Finally, when the system of coordinates was established 2,000 years ago by the Greeks, the spring equinox lay at a position in the constellation of Aries known as the First Point. Confusingly, the meridian is still known as the First Point of Aries, although precession has since brought it into Pisces.

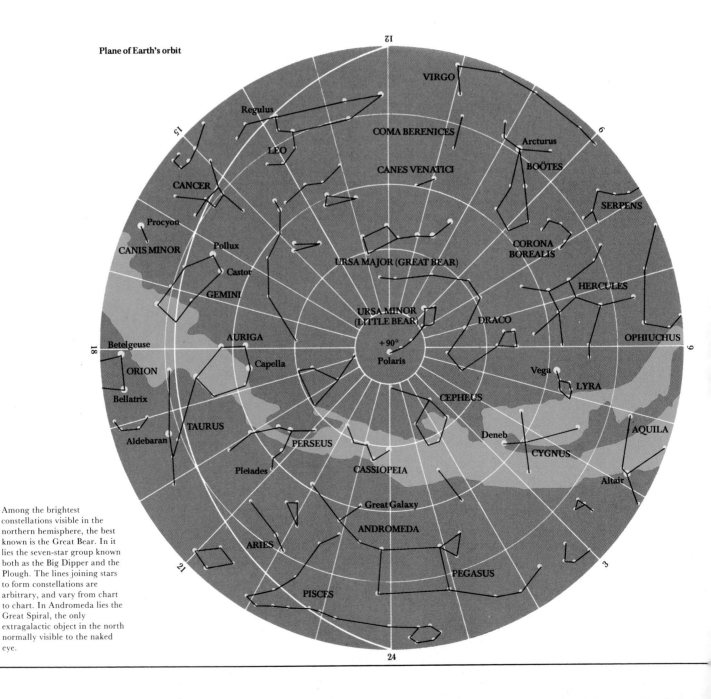

Plane of Earth's orbit

Among the brightest constellations visible in the northern hemisphere, the best known is the Great Bear. In it lies the seven-star group known both as the Big Dipper and the Plough. The lines joining stars to form constellations are arbitrary, and vary from chart to chart. In Andromeda lies the Great Spiral, the only extragalactic object in the north normally visible to the naked eye.

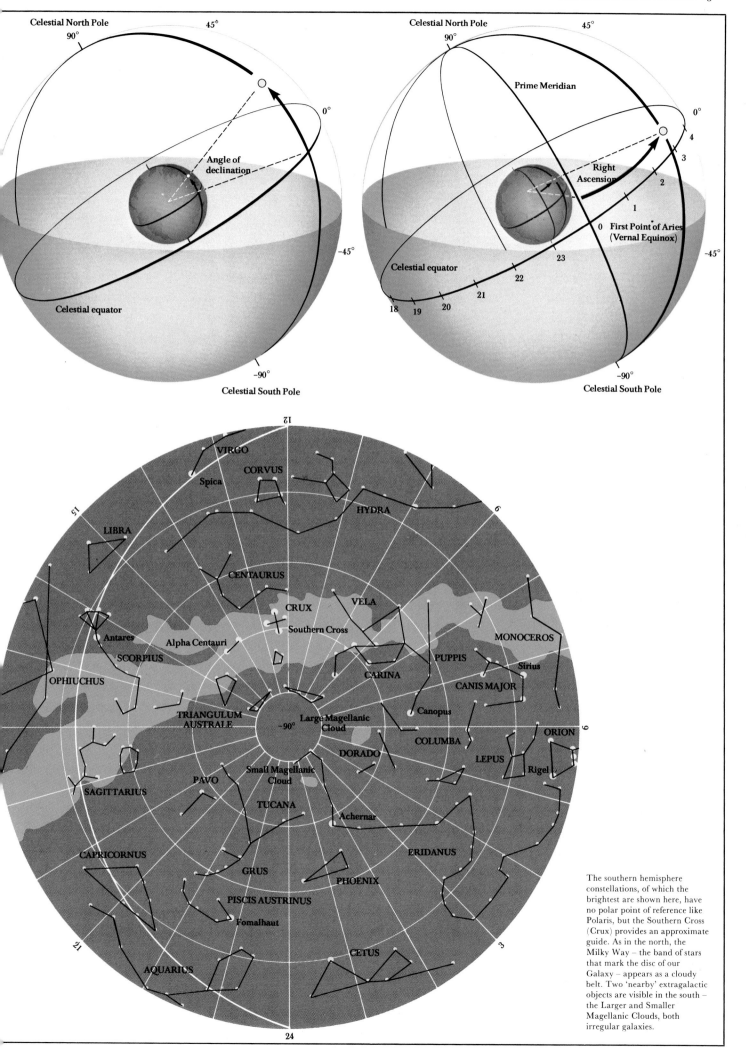

Celestial North Pole
90°
45°
0°
Angle of
declination
Celestial equator
−45°
−90°
Celestial South Pole

Celestial North Pole
90°
45°
Prime Meridian
0°
4
3
2
Right
Ascension
1
0   First Point of Aries
(Vernal Equinox)
Celestial equator
23
22
21
20
19
18
−45°
−90°
Celestial South Pole

12
VIRGO
CORVUS
Spica
HYDRA
15
LIBRA
CENTAURUS
9
VELA
CRUX
Southern Cross
MONOCEROS
Antares
Alpha Centauri
SCORPIUS
CARINA
PUPPIS
Sirius
CANIS MAJOR
OPHIUCHUS
Canopus
ORION
TRIANGULUM
AUSTRALE
−90°   Large Magellanic
Cloud
COLUMBA
6
DORADO
LEPUS
Rigel
PAVO   Small Magellanic
Cloud
SAGITTARIUS
TUCANA
Achernar
CAPRICORNUS
ERIDANUS
GRUS
PHOENIX
21
PISCIS AUSTRINUS
Fomalhaut
3
AQUARIUS
CETUS
24

The southern hemisphere
constellations, of which the
brightest are shown here, have
no polar point of reference like
Polaris, but the Southern Cross
(Crux) provides an approximate
guide. As in the north, the
Milky Way – the band of stars
that mark the disc of our
Galaxy – appears as a cloudy
belt. Two 'nearby' extragalactic
objects are visible in the south –
the Larger and Smaller
Magellanic Clouds, both
irregular galaxies.

This late Medieval sketch of the zodiacal sign Virgo shows her carrying in her right hand Spica – the 'ear of corn' star – which is still used as an alternative name for the constellation's brightest member, Alpha Virginis.

a place on Earth is pin-pointed by its latitude and longitude. The Crab Nebula, above the bull's horn, is better – if less poetically – described as having a right ascension of 5 hours 31 minutes and a declination of $+22°$.

Knowing the co-ordinates of an object, a professional astronomer can swing his telescope to it without any knowledge of the constellations. Indeed, it is rare to find a professional astronomer today with more than a fleeting acquaintance with the constellation patterns and names. In today's astronomy, too, instruments for detecting radio waves, X-rays and other radiations from space must be precisely aligned on the astronomical objects emitting them, and here the co-ordinate system is even more vital. After the position of a celestial X-ray source, for example, has been accurately measured, it can be compared with the positions of stars to identify the culprit 'X-ray star'.

The long tradition of astronomy is not altogether broken, however. In the early days of a new branch of astronomy, positions cannot be measured particularly accurately, and constellation names are often invoked to indicate a general region of sky. An X-ray source which is now believed to surround a black hole, one of the most controversial and exciting objects predicted by the new astronomy, is known by the name Cygnus X-1, the first X-ray source to be found in the constellation of the Swan.

Generally, however, modern astronomy has little use for the constellation patterns, for they have no existence in three-dimensional reality. The stars are not simply studded on the inside of a huge black vault, but are remote suns lying at different distances from us, and the patterns they appear to make are clearly fortuitous. Some of the stars making a constellation will be nearby, feeble stars; others are very distant superluminous beacons. With the departure of astrology as a non-science, astronomy has left behind the study of the constellations to move into the investigation of the real Universe, and the stars themselves are the jumping-off point for the investigation.

The brightest stars have always had individual names: some historians have argued that these stars were recognized before the constellations, and it is quite likely that the constellation Canis Major, the Great Dog, was named after its chief star, Sirius, the Dog Star. Most of our star names are Arabic, tacked on to Greek astronomical knowledge that passed through Arab lands during the Dark Ages, to reappear in Spain in the early Renaissance. So we find that the brightest star in the obviously classical constellation of Hercules bears the outlandish name Rasalgethi – 'Head of the Kneeler' in Arabic.

Johann Bayer, the constellation inventor, introduced the first systematic star names in his catalogue of 1603. In each constellation he distributed the letters of the Greek alphabet, generally in order of brightness. The most prominent star in the constellation Virgo is Spica – meaning 'ear of corn' – and in Bayer's catalogue this star becomes Alpha Virginis. (Notice that despite the mixture of the Greek and Latin languages, classical grammar is strictly followed, and the constellation name appears in the genitive case. Incidently, this system also applies to constellations like the telescope and microscope, instruments unknown in classical times, whose concocted Latin names have invented genitive cases: Telescopium – Telescopii, for example.)

Bayer's Greek letters are still used for the brightest stars, apart from the score with well established names of their own. The nearest bright star, for example, is Alpha Centauri; its alternative name of Rigil Kent, used by air navigators, has never caught on in the astronomical world. But the Greek alphabet runs to only 24 letters – alpha to omega – and other star catalogues have grown up to name the multitude of fainter stars. When the position of the X-ray source Cygnus X-1 was pinned down precisely, for instance, it was found to coincide with a star already catalogued 40 years earlier, and bearing the number HDE 226868.

Measuring the positions of stars is a fundamental branch of astronomy, with its own name – astrometry – and a multitude of new instruments to refine the positions. But to those not engaged in astrometry, a range of other questions seems more pressing: How far away is the star? How luminous is it? How does it work?

So far, we have been thinking largely in the ancient star-gazer's terms, measuring positions 'on the celestial sphere', that great black bowl which seems to surround us at night. But the stars are not points of light: they are searingly hot, blindingly bright suns, much like our own Sun but reduced to gentle glow-worms as their light spreads out on the enormously long journey to us. To understand the Universe, we must fill in the third dimension. Knowing the stars' distances, we can find out how they are arranged in space around us, and we can also calculate their intrinsic properties. Then the full battery of modern physical theories can be brought to bear on the stars, to force from them the secrets of their inner workings, and of their birth and death.

Adding the third dimension

At first sight, measuring the distances of the stars seems to be a stab at the impossible. Even in the most powerful telescopes, few stars appear more than unresolved points. One point of light in the sky is so much like any other that it is difficult to see how an astronomer can confidently assert that one is three, 10 or 100 times further away than another – and give definite distances for them. And when we realize just how large these distances are in everyday terms, the problem seems even more intractable.

The Sun, our local star, is 150 million km. (93 million miles) from the Earth. This distance is huge compared with our everyday experience: in a supersonic airliner a one-way trip to the Sun would take all of ten years! Yet when we venture into the stellar universe, this is a microscopic distance. If we took the Sun away so far that its light dwindled to that of a star like Sirius, we would have moved it 100,000 times further away – a jet journey of a million years. We are now dealing with distances on the grand scale – even a million million kilometres is too small a unit to use conveniently. Instead, astronomers resort to a unit which happens to be about the right size, the distance which light travels in one year. The speed of light is unimaginably high – at 300,000 km. (186,000 miles) per second, a light beam would travel round the Earth seven times in less than one second. Over the course of a year, light travels almost 10 million million kilometres – and this distance, one *light-year* is the fundamental unit we shall use for measuring stellar space.

In terms of light-years, the nearby stars have cosily small distances. Alpha Centauri is 4.3 light-years away; its faint companion Proxima is very slightly nearer and so makes the grade as the nearest star to the Sun. Other well-known stars have distances of a few dozen, or hundreds, of light-years, out to about 900 light-years for Rigel, the bright bluish-white star in Orion.

These distances are daunting on the human scale, but they can be plumbed by astronomers. There are in fact several ways of measuring star distances – some applying only to particular kinds of objects like double stars or variable stars – but one stands out above all others: the method of parallax. The basic principle is fairly familiar. If you hold a finger in front of your eyes, and then open and close your eyes alternately, the finger seems to move to and fro against the

Our nearest stellar neighbours are seen here set in a cube with sides 12 light years in length. Many of them, like the close-by Barnard's Star, are too dim to be visible to the naked eye. Some of the brighter objects – Alpha Centauri, Sirius, 61 Cygni – are in fact double stars whose members are impossible to separate on the scale of this diagram.

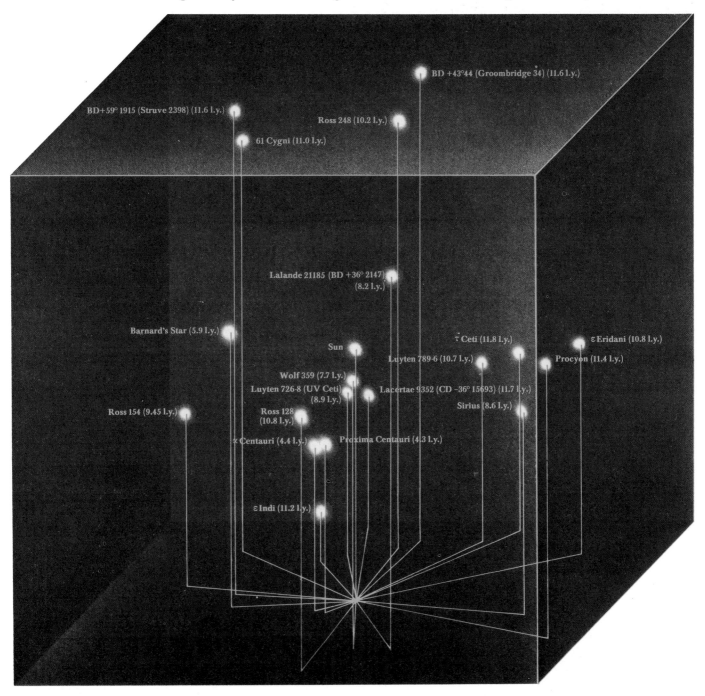

background scenery. This is obviously just a perspective effect, due to your seeing the finger from the different positions of the two eyes. Move the finger further away, and the shift between the views is less. You could in fact calculate the distance to your finger by measuring the angle it seems to shift through (and knowing the separation of your eyes). This is very similar to a surveyor's triangulation method, where he measures angles from each end of a baseline of known length to calculate the position of a distant object.

Turning to the stars, we can replace the finger by a nearby star, and the background by the scattering of very distant stars appearing near it on the celestial sphere. Photographs taken at different observatories should – we might think – show the nearby star at slightly different positions relative to the background stars. In practice, this shift is much too small to measure, because the observatories are too close together in comparison with the star's distance. A longer baseline is required, and the longest available to us is the diameter of the Earth's orbit. And by taking advantage of the Earth's movement in orbit – which establishes a base-line of 300 million km. (192 million miles) – astronomers can economize on observatories, using one instead of two.

A photograph is taken of the star concerned, and a second at the same observatory six months later, when the Earth is at the far point of its orbit. If the star is indeed quite close, careful

Celestial Geometry to Gauge the Stars

Stars that are closer to us seem to move slightly against the background of more distant stars as the Earth swings round the Sun. The same effect can be created by holding up a thumb and opening first one eye, then the other. The point of observation is altered by a couple of inches and the thumb's position changes against the background. Both the change and the angle of the change are known as *parallax*.

The angle a star seems to move through as the Earth swings round the Sun *decreases* in exact proportion to its distance. Parallax angles (in astronomy, actually defined as *half* the apparent shift, i.e. the radius of the Earth's orbit) are measured in seconds of arc. One second (1″) is 1/60 of a minute or 1/3600 of a degree. A star with a parallax of 0.5″ would be twice as far as a star whose parallax is 1″. This system of measurement has given rise to the unit of distance known as the *parsec* (*parallax second*): one parsec is the distance at which a star has a parallax of 1″. In the example, this is the distance of the second star; the first would be at a distance of two parsecs.

One parsec is 31 million million kilometres (19 million million miles). A light-year, the distance light travels in one year, is 9.5 million million kilometres (5.9 million million miles). Both units of distance are used in astronomy, and they can be converted by the equation: 1 parsec = 3.26 light-years.

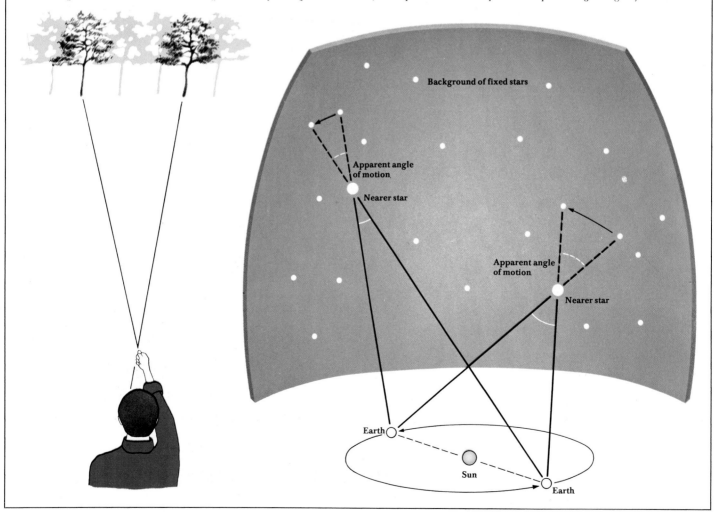

Background of fixed stars

Apparent angle of motion

Nearer star

Apparent angle of motion

Nearer star

Earth

Sun

Earth

Scales of Stellar Brightness

The brightness of heavenly objects is measured on a scale of magnitude. There are two kinds of magnitude:
- —*apparent magnitude*, which is the brightness of stars as they appear to us on Earth; but since the more distant a star, the fainter it seems, apparent magnitude often bears little relationship to its —
- —*absolute magnitude*, or its real brightness. Astronomers usually work in apparent magnitudes because that is the most useful guide to identification.

The Greek astronomer Hipparchus ranked the visible stars into six magnitudes – first magnitude for the brightest stars, and sixth for those only just visible to the eye. The stars, though, cover a whole range of brightness, and modern instruments can easily distinguish differences imperceptible to the eye. So Hipparchus's system has been tightened up, and magnitudes are quoted to the nearest hundredth part. Regulus, for example, has a magnitude of 1.36.

In this brightness scale, a difference of five magnitudes between stars corresponds to a brightness ratio of 100:1. The faintest stars visible to the unaided eye, at magnitude 6.5, are thus just over a hundred times fainter than Regulus.

Note that, following Hipparchus, the *brightest* stars have the *smallest* magnitudes. Sirius is so bright that its magnitude becomes a negative number. – 1.45.

At the other end of the scale, the telescope reveals stars fainter than magnitude 6.5, and these have correspondingly large numbers. Photographs taken with large telescopes can pick up stars as faint as magnitude 24 – about 10,000 million times fainter than Sirius.

measurements can reveal a slight shift in the star's position. Even so, the shift is small, and the measurement a correspondingly delicate one. To draw once again on the analogy of blinking at one's finger, the positional change of even the nearest stars is equivalent to the shift of a finger placed 10 kilometres from the eyes!

After decades of patient observations, parallaxes of thousands of stars are now on record. The distances of the nearest stars are accurately tied down, and the framework of the nearby stars is secure. Among the closest 20 stars we have Alpha and Proxima Centauri, our next-door neighbours in space; Sirius, a not particularly brilliant beacon in the celestial league, but the brightest star in our skies because it is only 8.6 light-years away; and Procyon, the bright star in the constellation Canis Minor, the Little Dog.

Most of our neighbours in space are dim creatures, though. Despite their proximity to us, they are visible only in a telescope. In real terms, then, they are shining extremely feebly, often with a thousandth or less of the Sun's light output. Since our region of space is undoubtedly quite typical, these faint stars must be the commonest kind in space. Although we think of the Sun as a typical star, it is in fact brighter than the average denizens of space.

When we push out our frontiers to the further stars, parallax measurements become less and less reliable. The further a star is, the smaller is its shift between the six-monthly photographs, and so the more difficult it is to measure its distance. If a star is more than about a hundred light-years away, its distance can be ascertained only within fairly wide limits. Yet a volume of space stretching out only a hundred light-years is only a fraction of stellar space. Fortunately, astronomers can supplement the parallax method by others, not so accurate for nearby stars, but with a greater range in space.

Most of these rely on the fact that stars are not stationary. All stars are moving, and with speeds which by ordinary standards are extraordinarily high – around 100,000 km. (60,000 miles) per hour which is several times faster than any spaceprobes launched from Earth. But because they lie at such huge distances, the stars' apparent movements across the sky – *proper motion* to astronomers – are very small. A nearby star like Sirius moves only 1/20th the Moon's apparent diameter in a human lifetime, and the further stars seem to be even slower. The celestial gold-medal sprinter, incidently, is the second-nearest star, a faint telescopic object called Barnard's Star, which moves through the Moon's apparent diameter ($\frac{1}{2}°$) in only 180 years. The constellation patterns are in fact gradually changing because of the stars' proper motions: the skies of 1,000,000 BC or AD 1,000,000 would be totally unfamiliar to us.

Although the stars' motions are small, modern techniques can pick them up readily and allow astronomers to follow the stars' tracks. Indeed, the yearly proper motion of a star is usually greater than its parallax, complicating the measurement of the latter. In practical parallax measurements, the star has to be followed during the course of the year for several successive years before the straight track due to its motion through space can be disentangled from the apparent 'wobble' in position caused by our changing viewpoint on the moving Earth. Without knowing a star's distance, its proper motion does not tell us how fast it is moving sideways through space – it could be a nearby, slow star, or a very distant one which has to travel phenomenally fast to cover the same angle in the sky in the same time. To measure its velocity we need to know its distance; or conversely, as we shall see, we can measure a star's distance from its proper motion if we have some independent way of knowing its actual speed across the sky.

The proper motion of a star also tells us nothing of how it is moving towards or away from us. By analysing the *spectrum* of a star's light, however, we can calculate precisely its speed along our line-of-sight, for the wavelengths of light are affected by the star's motion. By measuring this Doppler effect, the star's *radial velocity* can be worked out directly in kilometres per second. The star's real motion in space is evidently a combination of this radial velocity towards or away from us, and its velocity across the sky. A star travelling directly away from us will have no apparent velocity *across* the sky; one travelling at right angles to our line-of-sight (that is directly across the sky) will have zero radial velocity but a high proper motion. A star moving at 45° to the line of sight will have an equal amount of velocity directed along the line-of-sight (as radial velocity) as it has across the sky.

From these relationships between angle of motion, radial velocity, and proper motion, astronomers can measure star distances, albeit in a rather more roundabout way than the parallax method. But such techniques can be extended to stars whose parallax shift is too small to measure accurately.

Usually the actual direction of a star's motion through space is unknown, and astronomers must take an average over several stars which are intrinsically similar. It is however easy to measure the direction a *cluster* of stars is moving in. The stars in a cluster appear to converge towards a distant point, whose position depends on the cluster's direction of motion in space, just as the parallel lanes of a motorway appear to converge at the horizon. With a cluster of stars, then, astronomers can measure its distance without recourse to parallax at all. Distances to star clusters, measured by this *moving cluster method*, are very important in today's astronomy,

for they reach well beyond the range of parallax, and add all the members of a particular cluster to our list of stars with known distances. Some of these clusters also contain types of stars, like Cepheid variables, which are so rare that none are near enough for direct parallax distances.

From such a compilation of stars with distances accurately determined by one method or another, astronomers have been able to piece together a tremendously detailed picture of how stars work, how they are born and how they eventually die. The next chapter details this fascinating detective story of truly cosmic proportions; what is important here is that any two stars which are identical in particular details of their light (when examined spectroscopically) have exactly the same total output of light and other radiations (luminosity). Although stars cover a wide range in luminosity, from stupendous beacons a million times brighter than the Sun to glow-worms with a millionth the solar luminosity, as soon as

The Message of Light

Ordinary light consists of a range of colours, as shown when it is split up as a rainbow. The spread of light according to colours is called a *spectrum*. When light passes through anything that bends the rays (whether a glass prism or a raindrop), the light separates according to the wavelengths it contains. 'Colour' is just our eyes' response to light of different wavelengths. Ordinary white light contains all the visible wavelengths, and the spectrum is a continuous bright band, stretching from violet to red.

In the early 19th century, the German physicist Joseph von Fraunhofer discovered that any chemical suspended in a gas absorbs its own particular wavelength of light. The absorbtion appears as a dark line – a sort of 'shadow' of the missing radiation – in the

spectrum. Fraunhofer mapped 576 such lines, and discovered that they were all associated with their own radiation frequency, and thus always showed up in the same position in a spectrum.

In a star, the outer layers contain many chemical elements and these absorb particular wavelengths of the star's light. When the light is spread into a spectrum, Fraunhofer's dark spectral lines – of which some 25,000 are now known – appear. Spectral lines are invaluable to astronomers. For a start, they 'fingerprint' each star and show what elements the star's atmosphere contains; in addition, a detailed study reveals its temperature, whether it is a giant or a dwarf star (and hence its luminosity), and whether it has a magnetic field.

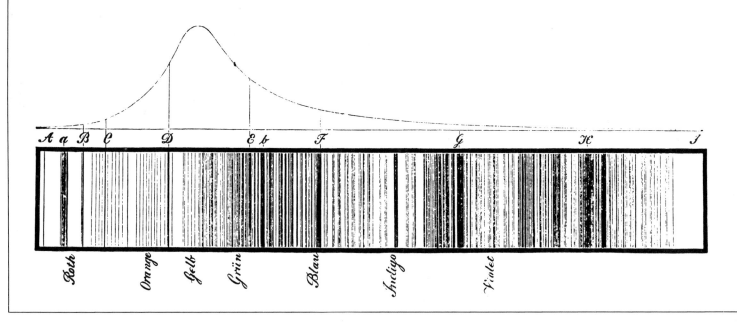

we have a star's spectrum we can match it with the spectrum of a nearby star whose characteristics are well known, and hence we deduce its luminosity. This star's apparent brightness in our sky depends both on its luminosity and how far away it is: the greater the distance, the more its light spreads out and weakens on the journey. The star's distance can thus be calculated by comparing its apparent brightness with its calculated luminosity.

So from our framework of local stars, whose distances are deduced from the parallax shift, and from nearby clusters of stars, with distances plumbed more indirectly, astronomers have interpreted the 'fingerprints' of star spectra, and from this scheme the distance to *any* star can be determined. All that is needed is its spectrum and a measurement of its apparent brightness. Such is the strength of modern astronomy that the third dimension of our star-filled night is firmly welded into place. Furthermore, the speeds of the stars as they travel their separate journeys is also open to the astronomers analysis. We are in a position now to describe the framework within which the stars exist and see how it fits into the entire scheme of the Universe.

The Grand Design of the Stars

Our eyes show us some 3,000 stars on a clear night – a total of 6,000 in the whole sky. Yet despite the impression of crowding we get on a truly brilliant star-lit night, these are only the nearest and brightest of a vast throng of stars. A telescope reveals multitudes of fainter stars, apparently increasing without limit. These distant stars are not scattered at random. They concentrate towards a broad band running right around the sky, and in the band itself so many faint stars are concentrated that, although each is individually invisible to the eye, the combined light makes a pearly glow as striking as the constellation patterns themselves. It, too, has been woven into

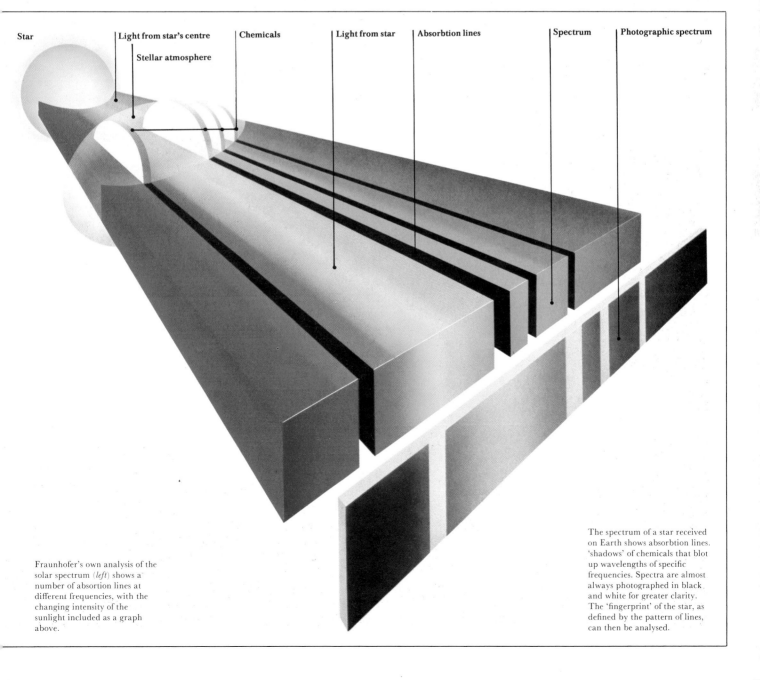

Star | Light from star's centre | Chemicals | Light from star | Absorbtion lines | Spectrum | Photographic spectrum

Stellar atmosphere

Fraunhofer's own analysis of the solar spectrum (*left*) shows a number of absortion lines at different frequencies, with the changing intensity of the sunlight included as a graph above.

The spectrum of a star received on Earth shows absorbtion lines. 'shadows' of chemicals that blot up wavelengths of specific frequencies. Spectra are almost always photographed in black and white for greater clarity. The 'fingerprint' of the star, as defined by the pattern of lines, can then be analysed.

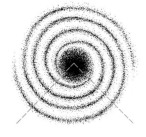

legend. To the Greeks, it was milk split from the breast of Hera when she suckled the infant Herakles (Hercules); hence its modern name, the Milky Way.

The crowding of the stars towards the Milky Way is simply a perspective effect. The stars are not actually more closely packed there, but the stars of our system simply extend farther in that direction. The Milky Way's glowing band is telling us about the overall structure of the star system of which the Sun is one member, a star system named the Galaxy, from the Greek word for milk. Even before any star distances were known, 18th-century scientists such as Thomas Wright of Durham and the great astronomer Sir

William Herschel had realized the message of the Milky Way. Herschel counted the number of stars visible in his telescope in different directions in the sky, and deduced that if the stars were uniformly spread, they must be confined to a grindstone or lens-shaped system, with the Sun near the centre. When we look anywhere along the circle formed by the grindstone shape, we see light from many distant stars, which blend to produce the Milky Way; but looking up or down we see only the nearer stars with a dark sky behind, for there are no distant stars in these directions to give a background glow. Herschel's reasoning was entirely correct, but for other reasons his theory was dropped by 19th-century

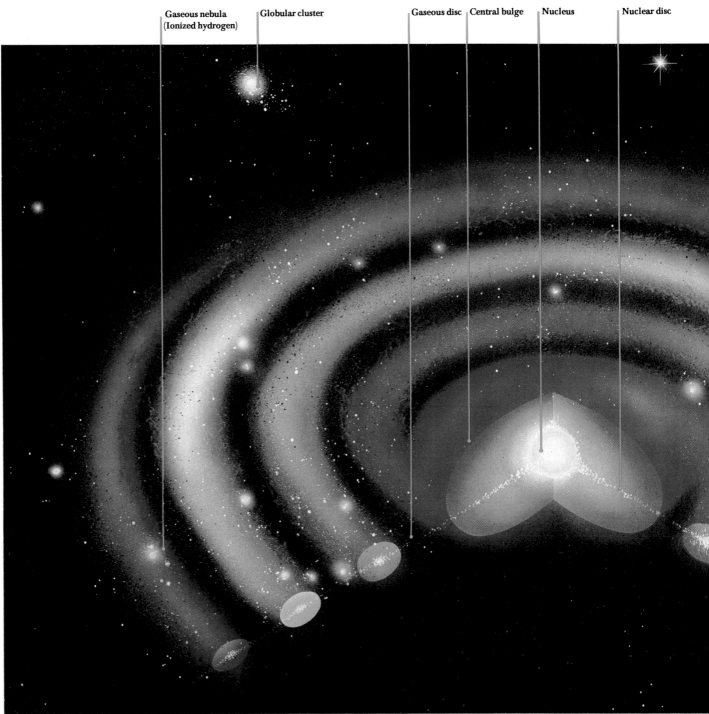

Gaseous nebula
(Ionized hydrogen) Globular cluster Gaseous disc Central bulge Nucleus Nuclear disc

astronomers. Its bones were resurrected early this century, and clothed in new observations to give us the true picture of our Milky Way Galaxy of stars.

It is indeed a lens-shaped star system. But the Sun is not central; it lies two-thirds of the way to one edge. Herschel and his successors were misled by a phenomenon unrecognized until the 20th century: the space between the stars is dusty. Small dust grains (mixed in with invisible gas) dim the light coming from distant stars, so that we cannot see even the centre of our Galaxy, let alone the 50 per cent of stars which lie on the far side.

The gravitational effect of all these stars,

whether or not our telescopes can detect them, is what shapes the Galaxy. In all, our Milky Way system consists of about 100 billion stars. Our Sun is just one member of a huge family of stars of all types, ranging from red giants hundreds of times larger to white dwarfs a hundred times smaller. Every star is pulling on every other by its gravity; and every star has its motion through space.

Yet all is not chaos. It we could stand back and look at the Milky Way system as a whole, we could see that it is a remarkably orderly system. Although our system is roughly lens-shaped, the modern view is not that it just tapers off in thickness as we go out to the edge. The outer parts of the Galaxy, where the Sun lies, actually constitute a thin disc whose thickness is roughly constant. Its proportions are similar to a gramophone record, but on an enormous scale: from edge to edge it measures 100,000 light years. This is a scale too tremendous to comprehend. Even the nearest star is four light years away, and if we imagine a scale model where Alpha Centauri is literally our next-door neighbour, the Milky Way Galaxy is a super-city entirely covering an area the size of Greece. And to be accurate, our city would also have to extend upwards several hundred metres to represent the thickness of the disc.

Towards its centre, the Milky Way system bulges up and down, like the yolk in the middle of a fried egg, its thickness increasing from 2,000 light years in the disc to about 10,000 light years at the centre. Before looking at the central regions, though, let us start by investigating the Sun's neighbourhood in the outer disc.

The framework of star distances, so painstakingly built-up, puts our region of the disc into perspective. At first sight it seems that the nearby stars are not going anywhere in particular: each is moving with its own random velocity, without any overall pattern. This is pure illusion. It is natural to think of the Sun as being at rest, but since our planet Earth will be carried along with it however the Sun is moving, this may not be so. And indeed indirect measurements show that the Sun, and most of the nearby stars are speeding through space at a rate of 250 km. (156 miles) per second.

All the stars in the disc of the Galaxy are revolving in orbits around the galactic centre, and the motion of the Sun and nearby stars is just part of this general merry-go-round. The orbits are kept in rein by the gravitational pull of the Galaxy as a whole – in other words, all the stars towards the centre contribute a gravitational tug which helps to hold the Sun in its path. Unlike the Solar System, whose planets are controlled by one massive central body (the Sun), the stars moving around the Galaxy feel the smoothed-out pull of billions of other stars. Despite the Sun's enormous speed, the Milky Way system is so huge that one complete orbit takes 250 million years to complete.

The random motions of the other nearby stars

Sun   **Spiral arm**   **Globular cluster**

A cut-away, three-quarter view of our Galaxy reveals its spiral structure, caused by the different speeds at which the various parts rotate. Our Sun takes 200–250 million years to complete one revolution. In the interstellar gas, which lies along the plane of the disc and between the arms, very hot stars create a scattering of hot-spots of ionized hydrogen. Around the outside of the Galaxy hang ball-like bodies of stars known as globular clusters.

can now be seen as a result merely of their slightly different orbits. A racing car driver sees the other cars slowly approach and recede from him as they speed on their slightly different paths around the track. Taking each car relative to the others, their common high speed does not matter: the deviations from parallel paths shows up as comparatively slow, random motions. And so it is with stars. Our neighbours are not pursuing exactly circular paths around the Galaxy's centre. They therefore have apparent motions which reveal themselves to astronomers as *proper motions* across the sky and *radial velocities* towards and away from us, but these speeds are much smaller than the stars' common velocity in orbit.

The Sun itself is not in a circular orbit. Its path deviates towards a point in the sky known as the 'solar apex', which lies near the bright star Vega. Sir William Herschel first measured this motion of the Sun, by averaging out the proper motions of nearby stars, but only in this century did astronomers realize that it is not a very important part of the Sun's motion. We know now it is just our stellar 'car's' sideways drift relative to the others on the course; far more significant is the fact that all the 'cars' are racing round the track at high speed in roughly the same direction. All the local stars, including the Sun, are heading in the direction of the constellation Cygnus.

The gravitational pull of the Galaxy's stars on one another gives the stars basically quite circular, orderly orbits. But there is one rather strange, and beautiful, result of this combination of gravitational attractions. The stars of the disc cannot stay uniformly spread out, but become clumped together in two spiral arms, which appear to wind out from the central nucleus. Deep within our Galaxy's disc as we are, it is difficult to see this pattern, just as the outline of a wood is not obvious when we are in amongst the trees. But there are millions of other galaxies, each – like our own Galaxy – consisting of billions of stars. Galaxies stretch out into the depths of space beyond the Milky Way, and in these distant galaxies we almost invariably find that the stars constituting the disc have clumped in a beautiful winding tracery of double spiral arms. Photographs overemphasize the appearance of the arms at the expense of the rest of the disc, and disc-galaxies like ours have been given the general name of *spiral galaxies*.

Although all the stars of our Galaxy's disc appear stacked behind one another in hopeless confusion, astronomers have been able to pick out local stretches of spiral arms relatively near to the Sun. The framework of star distances reaches out some 10,000 light years for the brighter stars, and spiral arms happen to be the abodes of the brightest stars in a galaxy. Astrono-

The Galactic Halo

Knowing the distance from the Sun to the galactic centre is one of the most vital facts we want to know if we are to have an accurate model of the galaxy. But the distance cannot be measured directly. Although astronomers can assess the distances of many stars (by working out their absolute magnitudes and comparing this to their apparent magnitudes), they cannot see to the centre of our Galaxy, which is obscured from even the largest telescopes by interstellar dust.

The solution to the problem lies in the 'globular clusters', stars bound by their own gravity. These clusters – like M13 (right) – lie above and below the Galaxy's disc, in a halo, and they are not obscured by dust; and as they contain around a million stars each, even those on the far side on the Galaxy can be detected with a large telescope. We know by observing other galaxies that clusters are spread roughly uniformly around the galactic centre. By ascertaining the middle of the distribution of globular clusters, astronomers can find the position and distance of the galactic centre (above right).

American astronomer Harlow Shapley first fixed the centre of our Galaxy around 1920, using this method. The nearest globular clusters lie close enough to us for individual stars to be detected in them. Their distance can thus be measured directly. Shapley calculated the distances to the further ones by comparing their apparent

sizes with those of their nearby brothers, and worked out that the centre of the globular cluster distribution – and hence the Galaxy's centre – lies 50,000 light years away, in the direction of the constellation Sagittarius. Later research has refined Shapley's figure, reducing the distance to 30,000 light years. Our Sun lies about two-thirds of the way out from the centre towards the galactic edge.

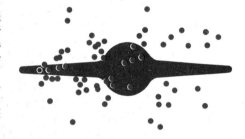

mers have located a spiral arm some 7,000 light years farther out than the Sun, running through the constellations Perseus and Cassiopeia, while towards the galactic centre another runs by at about the same distance, appearing in the constellations Sagittarius and Carina. These seem to be turns of the two major arms of our Galaxy. The bright, relatively nearby – on this scale – stars of the Orion region were once thought to be a local arm in which the Sun is placed, but astronomers now think it is merely a spur projecting from the inner edge of the ragged Perseus arm.

The disc – and especially the spiral arm regions within it – are where the action is in a galaxy. Stars are being born here, and stars are dying spectacular deaths as supernovae. These stories concern not just stars, though, but also the gas and dust lying between the stars, and the full story of the disc must take into account these insubstantial but vital components. Before leaving the basic framework determined by the stars, however, we should turn to the central parts of the Galaxy.

The nucleus is the geriatric ward of the Milky Way. No stars are being born here at the present time, nor have any since the original formation of the Galaxy. These stars are all around 15,000 million years old, three times the Sun's age, and they make up a fairly rounded bulge at the centre of the disc. The Galaxy's 'yolk' of old stars is some 10,000 light years thick and 20,000 light years in diameter, and its shape is a reminder that the Galaxy originally condensed from a colossal almost spherical gas cloud. The first stars to form from this gas did not travel in orderly circular orbits, but took elongated paths arranged higgledy-piggledy around the Galaxy's centre. Still pursuing these paths determined in their youth, the stars of the nucleus obstinately retain the original shape of the Galaxy, while the remaining gas smoothed out its turbulent motions and settled down as a smoothly rotating disc.

The original gas cloud was rather larger than the Milky Way system's present size. As it collapsed under its own gravity, the stars of the nucleus formed at the centre, where the gas was densest. But – for reasons which astronomers still do not really understand – condensations of gas further out collapsed individually to make *globular clusters*, each a closely packed ball of about a million stars. The 125 known globular clusters are spread out around our present Milky Way Galaxy in a huge spherical volume called the *halo*, the ghost of the original gas cloud. The two brightest globular clusters are actually visible to the unaided eye, appearing as 'fuzzy stars', and they were indeed originally catalogued in the same way as stars, by the names Omega Centauri and 47 Tucanae.

The halo region also contains a host of individual old faint stars. Their light output is so feeble that halo stars outside the giant globular clusters are difficult to detect. But these stars, like their contemporaries in the nucleus, are pursuing elongated, randomly orientated orbits around the galactic centre, and in doing so they must at some time pass through the disc. Some halo stars thus come within spitting distance of the Sun, and astronomers can study them if they can spot them amongst the multitude of disc stars. In fact, their orbits give them away. Since they are not partaking of the disc stars' circular dash around the Galaxy, these stars are being rapidly left behind. From our vantage point, they seem to be shooting backwards. Since these stars were identified before astronomers realized the Sun's high speed around the Galaxy, the slow-moving halo stars were dubbed 'high-velocity star' – and with the innate conservatism of astronomical names, this misnomer is still in use.

The old regions of the Galaxy – the nucleus and the halo – seem staid and unexciting. But in recent years, some astronomers have suggested that we have misjudged our Galaxy's halo. There are some indirect reasons for thinking it may contain a lot more mass than can be accounted for by the dim stars within it. If this suggestion – and it is only a suggestion – is correct, then the halo could contain more matter than the rest of the Galaxy. And this matter cannot be in the form of stars: it would have to be bound up in some dark type of object, perhaps a multitude of free planets, or even black holes.

The old region at the centre of the nucleus also has surprises in store. Radio astronomers have found evidence for an explosion, or series of explosions, there; and an intense central radio source. In some ways, the very centre of our Galaxy resembles – on a very much smaller scale – the explosions of the distant and enigmatic quasars. Later chapters will deal with the surprising violence that astronomers are finding in the hearts of many galaxies, so let us now return to our local neighbourhood, the Galaxy's disc, where star-life and death continues in a regular cycle.

Between the Stars
The yawning gulfs between the stars are by no means 'empty space': nowhere in nature is there a perfect vacuum, a region devoid of all matter. A scientist's 'vacuum' in the laboratory is not completely empty of gas, even though this residual gas is very tenuous in comparison with the crowding of molecules in air. Interstellar gas is a much better 'vacuum' than this, but it is still filled with extremely tenuous gas, mainly hydrogen, spread out so thinly that there is on average only one atom to every five cubic centimetres of space – half a dozen atoms in a small matchbox volume. Our Galaxy is so huge, though, that the total amount of gas between the stars in its disc would be enough to make 10 billion stars – in another words, 10 per cent of the matter of the Galaxy is not bound up in stars but is spread out as gas between them.

In the early history of our Galaxy, after the nucleus and halo stars had formed, the remaining gas settled down into a rotating disc. Most of this gas formed into the disc stars, which still pursue the circular paths dictated by the earlier gas disc, but some of the gas has always remained. As we shall see in the next chapter, dying stars eject gas back into space; this mixes in with the original primordial gas, and eventually ends up in new stars. The disc is a dynamic place, with stars always being born out of gas, and in dying replenishing the surrounding gas reservoir.

The gas from a dying star, however, always contains slightly less of the hydrogen and helium which were the sole constituents of the original gas disc, and in their place contributes 'heavier' elements like oxygen, carbon and iron. The interstellar gas is thus gradually changing in composition, becoming 'enriched' in heavy elements. Some of these elements do not appear as gas, though, but as small solid grains of dust which obscure the light of distant stars.

To the traditional astronomer, observing the light from the stars, the dust is mainly a nuisance. Although there is a hundred times more gas than dust, weight for weight, the gas is transparent while dust is very efficient at blocking off light. The Milky Way in the region of the constellations Cygnus and Aquila, for example, seems split in two lengthways, but this is nothing to do with the distribution of the stars themselves. The culprit is the narrow layer of dust lying along the midplane of the disc, absorbing light from the distant stars near the central line of the Milky Way.

From southern latitudes, an even more prominent black cloud – the Coal Sack – stands out near the Southern Cross. Such dark clouds – and photographs reveal hundreds of them silhouetted against the Milky Way – show that the dust and gas in the disc are not uniformly spread. The distribution has been clumped into denser clouds, like the Coal Sack, by a variety of forces. The pressure of light from intensely bright young stars forces gas away from them, as does the explosion of an old star as a supernova, and in both cases a shell of denser gas piles up around the star. Changes in the surrounding gravitational field can also compress gas into a cloud, particularly when the interstellar gas is moving through a spiral arm of the Galaxy.

At the centre of a gas and dust cloud forced to contract in this way, the gas may become so compressed that its own gravity pulls it unremittingly together, and the gas condenses into a cluster of stars. Radiation from these young stars will eventually blow the residual gas away, but first it makes the gas glow. Ultraviolet radiation is mainly responsible. Just as teeth, or a freshly laundered shirt will shine in the 'black light' ultraviolet lighting of a party or discotheque, so the gas in the clouds glows in the radiation from the central stars. The most famous example of such a bright *nebula* (a word meaning cloud) is the great Orion Nebula, 15 light years across, and lit up by four young stars in a central group called the Trapezium.

Nebulae and the bright stars lighting them

occur throughout the Galaxy's disc, although they are most common in the spiral arms. They are some of the most beautiful objects in the skies. A small telescope shows the wisps in the Orion Nebula, while long exposure photographs reveal hundreds of beautiful nebulae, in a fascinating variety of shapes. Their outlines are determined not only by the edge of the gas cloud itself, but by the extent to which the radiation from the central stars can reach. If dark clouds on the near side are out of the radiation, they will block light from the nebula behind, producing fantastic black silhouettes, often in the form of long conical 'elephant trunks'. The most famous dark patch in a nebula is the Horsehead Nebula in Orion (not far from the Orion Nebula), an uncannily realistic equine head silhouetted against a diffuse background glow. Dust patterns stretching over whole nebulae have produced glowing clouds in such forms as the trisected Trifid Nebula, the cartographically perfect North America Nebula, and the swirling Lagoon Nebula.

The Horsehead Nebula in Orion is a dark cloud that obscures lighter regions of hotter gas and countless stars. The stars in the lower half of the picture are in front of the cloud. At lower left in the panorama is another, hotter gas cloud reflecting the light of a hidden star.

The message of light from space tells us not very much about the transparent interstellar gas itself, nor about how stars form deep within the obscuring cocoons of a nebula's dark dust clouds. Today's astronomers are however not limited to investigating merely the light from space. In Chapter 9 we shall find how other radiations from space, ranging from radio and infrared to ultraviolet, X-rays and gamma rays can be picked up and analysed. Since modern physics can tell us what is emitting the radiations, astronomers have effectively opened several more eyes on the Universe.

For the investigation of interstellar space, radio waves and infrared have the enormous advantage that they can travel unimpeded through the dust. Thus we can pick up radio waves from natural transmitters right the other side of the Galaxy. Natural radio broadcasters take several forms, but the most interesting for astronomers probing the Galaxy is hydrogen gas. A hydrogen atom broadcasts quite spontaneously at an exact wavelength of 21.1 cm. Since hydrogen is the commonest gas in interstellar space, a radio telescope tuned to 21 cm. receives a cacophony from hydrogen atoms throughout the disc of the Galaxy.

Astronomers can however interpret the message of the 21 cm. radiation. Like the spectral lines in a star's spectrum, the 21 cm. radiation is slightly displaced in wavelength if its source is moving towards or away from us, because of the Doppler effect. Radio astronomers use this effect to measure how fast the gas is moving throughout the Galaxy's disc. The gas is orbiting under the gravitational influence of the stars, and by investigating the orbital velocity of the hydrogen in different parts of the Galaxy, astronomers can probe the Galaxy's gravitational field at distant points.

Once we know the orbital speed of the disc at different distances from the galactic centre, we can turn the argument round and determine the distances of individual hydrogen clouds from

The Pleiades 'open cluster' – a young group some 60 million years old – is a dense mass of stars still in the process of formation. The gas that surrounds them like a fog will eventually be dispersed. The group is commonly known as the Seven Sisters, because that is the number of stars usually visible to the naked eye. But long-exposure photography has revealed that there are several hundred stars in the group, which will gradually break up as its members age.

their measured radial velocities. In this way, radio astronomers can provide distances to radio sources anywhere in the Galaxy, well beyond the optical astronomers framework.

Moreover, interstellar gas is more concentrated in spiral arms, so the arms emit stronger radio signals. By measuring the distances to the more powerful hydrogen clouds, radio astronomers have mapped the entire spiral structure of our own Galaxy, showing it to be very similar to other spiral galaxies.

These radio observations have also shown that the interstellar gas between the obvious nebulae is certainly not uniform – unlike the gas of the Earth's atmosphere which is all at roughly the same density and temperature (at sea level). Travelling through interstellar space, a voyager of the future could find himself passing through regions a hundred times denser than average, and at a temperature of $-200°C$, and in a few light years emerge into a stretch of space filled with gas a hundred times thinner, at $3,000°C$, (though the very low density of gas everywhere in space means that his spacecraft would not actually be frozen and heated).

Threading through all the gas in the disc is an enormously extended magnetic field, 100,000 times weaker than the Earth's field. Astronomers have mapped the Galaxy's magnetism by its effect on radio waves from distant sources; and by its effect on the dust grains in space.

The final components of the disc are the insubstantial 'cosmic ray' particles. These are electrically charged fragments of atoms speeding along at almost the velocity of light. Nine in ten are protons, the nuclei of hydrogen atoms, and the rest mainly helium nuclei, while one in a hundred is an oppositely charged electron. Astronomers believe that these atoms have been shattered and the fragments flung up to enormous speeds by the colossal explosions of supernovae, either in the explosion itself or else by the ultra-energetic pulsars left afterwards.

At their enormous speeds, cosmic rays should quickly shoot out of the Galaxy, but the magnetic field bottles them in, so that they keep retracing their paths in the disc. As a result, some penetrate the Solar System and can be studied from the Earth or from space probes travelling between the planets. Astronomers can also investigate cosmic rays far out in the disc, for as these electrically charged particles gyrate around the lines of magnetic force they generate radio waves. Unlike the 21 cm. broadcast of hydrogen, these *synchrotron* radio waves come out at all wavelengths, like noise all the way along the receiver

single atoms. Only sheltered within an ultra-violet-absorbing umbrella of dust can the molecules survive intact.) Unfortunately, hydrogen molecules are not radio broadcasters. Astronomers were long in ignorance as to exactly how much hydrogen there is in the centres of clouds, let alone its temperature and density. The answer came in the 1960s by an indirect route, when radio astronomers picked up radiations of precisely defined wavelengths, which were *not* 21 cm. These radio waves are emitted by a variety of molecules. First identified was hydroxyl, a bound oxygen-hydrogen pair, which is essentially water with one hydrogen atom missing. Other discoveries soon followed, and the total number of molecules found in space now stands at around fifty. The largest molecule presently known contains a string of 11 atoms, and familiar compounds so far identified include carbon monoxide, formaldehyde (the medical preservative) and ethyl alcohol. Although molecules are very rare in space when compared to hydrogen, itself very tenuous by Earth standards, the size of astronomical objects means that the total amount of these compounds is enormous in everyday terms: an interstellar dark cloud contains enough alcohol to fill the Earth, were it hollow!

The vast majority of these molecules are found only in dark clouds, where the dust protects them from being broken down into atoms by the ultraviolet radiation in space. They probably form on the surface of the dust grains, where atoms can come together and combine into molecules.

In the cloud centres, the gas is packed a million times more tightly than it is in the general interstellar medium, and these are just the right conditions for stars to form. As we shall see in the next chapter, even the first stages of a star's life produce a great deal of light and heat, which is trapped by the surrounding dust, producing infrared ('heat') radiation which can penetrate the dust and escape from the cloud.

Later on, these stars will disperse the dust around them, and light up the surrounding gas as a nebula. When the nebula disperses, the stars will be left as a cluster, like the well-known Pleiades (Seven Sisters) cluster in Taurus. The Galaxy's disc contains an estimated 10,000 such 'open' clusters, all formed relatively recently on the cosmic timescale. The Pleiades, for example, are 'only' 60 million years old, born at the time the dinosaurs perished on Earth.

Unlike the old compact globular clusters of the halo, the disc's open clusters are only loosely bound together by their own gravitation, and will eventually disperse as individual stars. The Sun, and its family of planets, must have begun in such a cluster some 4,600 million years ago, though the Sun's siblings have long since dispersed to other parts of the disc. The cluster stars spreading out to populate the disc will eventually die and return some of their gases to space to continue the cycle of star life and death, which we shall follow in the next chapter.

band of a home radio in contrast to a broadcast at a specific wavelength. The newest tool in the study of cosmic rays are the gamma-ray satellites. When the superspeed cosmic rays smash into stationary hydrogen atoms in space, they emit the very short wavelength gamma-rays. By mapping the directions from which gamma-rays originate, astronomers are learning about the distribution of cosmic rays in the farther reaches of the Galaxy's disc.

And all this does not take into account the dense gas clouds, found especially in the spiral arms. Deep within these clouds, hidden from the optical astronomers by wreaths of dust, stars are forming. Modern astronomy is now laying bare the process of star birth itself – invisible in ordinary light – and revealing the stages between the collapse of gas as a dark cloud and the spectacular unveiling of the young, fully formed stars at the centre of a nebula.

The clouds of particular interest to astronomers, for in them are formed surprisingly complex molecules. In the centres of the clouds, hydrogen atoms combine in pairs to make hydrogen molecules. (These hydrogen siamese-twins are the usual form of hydrogen we encounter on Earth, but in ordinary interstellar space the energetic ultraviolet radiation splits them into

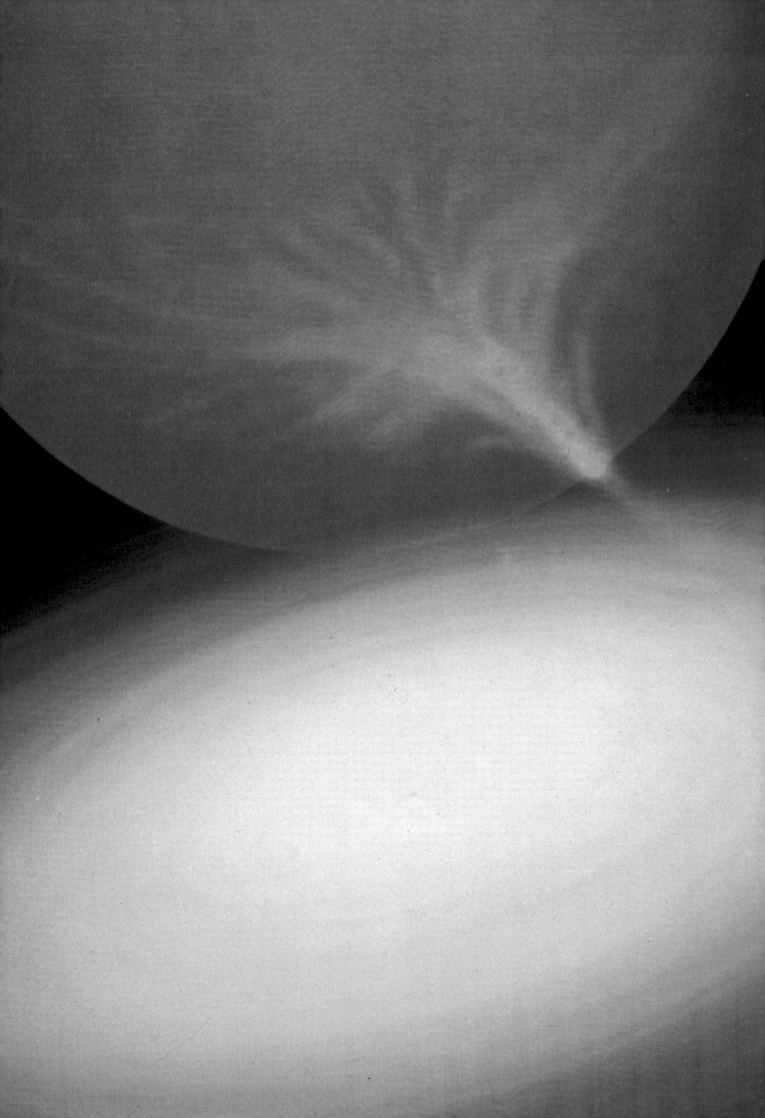

3/THE LIFE AND DEATH OF STARS

GEOFFREY BATH

Stars are the crucibles of the Universe. Each pin-prick of light in the night sky is a furnace, fuelled by the gas that tenuously pervades the voids of interstellar space. Burning with the energy of countless hydrogen bombs, it is the stars that from their fuel form almost all the heavier elements familiar on Earth. These elemental furnaces have many different forms. Some are small, and burn out to a cold cinder. Some end their lives in stupendous explosions, blasting their gases back into space. Some are so massive that they collapse under the force of their own gravity into super-dense black holes, swallowing matter swept from any neighbouring star (left) and binding to them even their own light.

Stars are the basic cells of the astronomical world. They are the furnaces in which the future of the Galaxy and other galaxies is being forged, for stars create almost all those elements that are more complex than hydrogen and helium. This building up of heavier elements is slowly changing the chemical composition of the Universe as gas is processed through stars and returned to the interstellar medium. Not only is the chemical state of matter changed in this way, but the very process of creating heavier elements provides the energy which gives life to stars and makes them luminous objects.

Stars originate in interstellar space itself. To the naked eye, the space between the stars appears to be an empty void. In one sense this is true. Most of interstellar space is a vacuum more rarefied than any that can be produced in a laboratory on earth. But space is not totally empty. It contains atoms, dust and electro-magnetic radiation.

The typical distance between individual atoms in interstellar space is about a centimetre: on average there is only one atom in a cube the size of a sugar lump. However the size of the Galaxy is so vast that even this extremely low density gas contains an enormous mass of matter.

It is important to realize how enormous this mass is. The gas lies in the plane of the galactic disc in a thin sheet about 300 light years thick, less than one two hundredth of the radius of the Galaxy. The volume of the gas is equivalent to 10^{67} sugar lumps, that is 10^{67} cubic centimetres, each cubic centimetre containing one atom on average. The total number of atoms in the gas is therefore 10^{67}. The gas is composed mainly of hydrogen, each atom of which has a mass of about 10^{-24} g. (one million-million-million-millionth of a gram). Hence the total mass of gas in the Galaxy is about $10^{-24} \times 10^{67}$, that is 10^{43} g., or about 10,000 million (10^{10}) times the mass of our own Sun! Since the Galaxy contains about 10^{11} stars, each typically with a mass something like that of the Sun's, we see that about 10 per cent of the material in the Galaxy is in the form of gas – hydrogen and helium, with a small admixture (about two per cent by mass) of heavier elements (principally carbon, nitrogen and oxygen).

In addition to this gas, the interstellar medium contains minute grains of solid dust particles. These dust grains are extremely small, less than one hundred-thousandth of a centimetre across. The presence of this fine dust can be deduced from the effect it has on the light of stars shining through it. The dusty regions of space have two effects on observations. Firstly they cause the stars behind them to appear fainter than they would if that light travelled unhindered through the Galaxy. The reason for this is that starlight is scattered and absorbed by the dust and is thus reduced in intensity. Hence stars at larger and larger distances are not only fainter because of their distances, but also fainter because they shine through increasingly thick layers of dust.

The scattering of starlight has a second, subsidiary effect. Just as the sun appears redder when seen through the smoke of an evening bonfire, so the light of stars is reddened by passage through the dusty regions of space. The reason for this is that, although all light is scattered when it shines through clouds of dusty material, the red light from a star is scattered less than the blue light, and so the observer sees more of the red light and the star appears redder. Conversely, if a particular dense cloud of dust lies near a star, the blue light which it preferentially scatters will cause the cloud to glow as a faint blue nebula.

The interstellar medium between the stars also contains very high energy particles – rapidly moving electrons, protons and atomic nuclei – called cosmic rays. These cosmic rays are charged particles, and spiral round the weak magnetic field lines in the Galaxy. This spiralling motion generates radiation at radio wavelengths thus producing a diffuse source of radio emission throughout the Galaxy, with strongest emission in the plane of the Milky Way.

The Birth of Stars

The gas and dust between the stars is not spread uniformly throughout the galactic disc. Instead it is clumped in a largely chaotic way, with concentrations along the spiral arms in the form of clouds and nebulae. It is this deviation from a uniform distribution that triggers the birth of stars. For, once a deviation from a uniform spread of gas is large enough, the force of gravity takes command. The subsequent evolution depends on the balance between gravity and internal pressure generated by heat: the hotter the cloud, the greater the outward push of pressure; the more massive the cloud, the greater the inward pull of gravity. Small, low mass and hot clouds tend to expand like a balloon being blown up, with pressure dominating gravity, whereas large, dense and cool clouds collapse like a balloon deflating.

At the temperatures and densities typical of the gas clouds in the Galaxy the critical mass which is required for collapse to occur, known as the Jeans mass, after the British physicist Sir James Jeans, is several thousand solar masses. If collapse were to occur without further changes in the cloud, stars would be formed which were several thousand times more massive than the Sun. However, once started, the process of collapse, which takes many millions of years, becomes unstable. The cloud breaks up into sub-condensations. The increasing density of the initial cloud as it contracts leads to an increasing gravitational force. Since it is transparent in these early stages it can leak energy out and thus remain cool. The outward push of pressure remains the same but the gravitational pull gets stronger. The critical Jeans mass decreases and the cloud fragments into smaller and smaller sub-clouds.

This break-up into smaller fragments cannot continue indefinitely. Eventually the density of

The dense mass of stars crammed into the disc of our Galaxy testifies to the immense bulk of matter from which such stars condense – and to which they will finally return most of their gases.

each sub-cloud rises to a point where dust grains start to trap heat. The temperature, and therefore the pressure, rises, and pressure starts to win the battle against further fragmentation. The sub-cloud has then reached its final stable mass and becomes a protostar.

From this point onwards each protostar evolves as a separate coherent object. The inner regions contract to form a hot nucleus and the outer regions settle onto this central core over the next million years. By this stage each protostar has become opaque, and heat can no longer leak out easily. As the protostar continues to contract so its temperature increases steadily. Starting as a cool distended object with a surface temperature of 1,000° radiating strongly in the infra-red region of the electromagnetic spectrum, it contracts and increases in temperature to become gradually brighter at visible wavelengths.

What stops protostars from contracting indefinitely? The barrier which eventually halts this contraction is thermonuclear fusion. The same source of energy that is liberated in the hydrogen bomb acts to stabilize and halt the collapse of protostars. At a temperature of about 10 million degrees centigrade in the centre of the star, nuclear fusion of hydrogen into helium starts to occur. The energy released in these fusion reactions keeps the central regions hot, and continually replenishes the energy which is slowly leaking out through the star. In this way a balance is achieved between gravity pulling inward and the pressure pushing out, a balance which can continue to be maintained so long as thermonuclear fuel is available. The stage when nuclear 'burning' starts to liberate thermonuclear energy is the stage when a star is truly born – a cool, transparent cloud has fragmented into a cluster of a hundred shining stars.

Stars in Middle Age

The newly-formed clusters of young stars are visible in the Galaxy as associations and open clusters. Because they are born with random motion, the stars in a cluster will slowly disperse, spreading themselves throughout the Galaxy. As we shall see, more massive stars, with masses 20 or 30 times that of the Sun, cannot survive longer than a few million years before consuming the bulk of their nuclear fuel. In this time, with the stars moving off on their separate trajectories, the typical cluster will be broken up. Thus the massive O and B spectral class stars tend to be found mostly in open clusters. By the time the cluster has dispersed these stars have already reached the end of their life.

Clusters containing young, massive O and B stars are found more frequently in the spiral arms of the Galaxy. For this reason it is believed that the spiral arms act as triggers for star formation. Although the Galaxy looks as if it should be revolving like a giant Catherine-wheel, its motion is not as simple as the image suggests. The gas, dust and stars in the spiral arms do not move round the Galaxy at the same rate as the spiral arms themselves. Rather, the arms are thought to be waves, moving around the Galaxy rather like water waves moving across the sea. But in contrast to water waves, the spiral arms move around the Galaxy more slowly than the gas and stars. The passage of these arms disturbs the interstellar medium and generates the inhomogeneities – the regions of denser gas – that induce the collapse of interstellar clouds. In this way young stellar clusters are formed largely in the arms of the galactic disc.

All the stars in one of these young clusters are born at the same time, from a gas cloud of uniform chemical composition. The only major difference between them is their mass. A whole range of stellar masses are formed, varying from the lowest masses of less than a tenth of the mass of the Sun, up to the highest of about 50 solar masses. However, the process of fragmentation leads to many more low mass stars than high mass stars being formed.

The intrinsic brightness of these stars in a single galactic cluster also varies from one to another, but in a way which is found to be closely correlated with their mass – the more massive stars are brighter. Furthermore the brightest stars are also hottest, with a surface temperature of 40,000°C. They are bluish O-type stars. The fainter stars are quite cool, going down to temperatures of 3,000°C. Their spectral appearance is again different: they are faint 'red dwarfs'.

The brightness, or luminosity, of a star is normally measured in units known as 'absolute magnitudes' (i.e. the star's actual brightness, rather than its 'apparent magnitude', which varies according to the star's distance from us). If the absolute magnitudes of all the stars in a young open cluster are plotted against their surface temperatures, the relation between brightness and temperature is shown up in a striking fashion. All the stars lie in a narrow band extending from the hot, bright stars in the top left-hand corner down to the cool, faint stars in the bottom right-hand corner. Stars in this band are said to lie on the 'main sequence', and such a diagram is called a Hertzsprung-Russell diagram after the two astronomers who independently emphasized its importance as a technique for studying stellar properties. The Hertzsprung-Russell diagram plays a major role in stellar evolution theory as a code containing all the main clues to the way stars evolve.

The Hertzsprung-Russell diagram relates the inherent brightness, or absolute luminosity of stars to their colour. Bright stars are shown at the top, dim ones at the bottom. Hot blue stars are on the left, cool red ones on the right. (The main spectral types are classified by letters often recalled by the mnemonic: O, Be A Fine Girl; Kiss Me.) Most stars show a close link between colour and luminosity, and fall on a line known as the 'main sequence'. Other types are abnormal. Some, although cool, are very bright and red as a result of their giant size. Others are dull and small, but with a high surface temperature. These are ancient, dying white dwarfs. The line shows the probable evolution of the Sun, first to a red giant 10–100 times its present size and finally to a white dwarf one-hundredth its present size, all its energy spent.

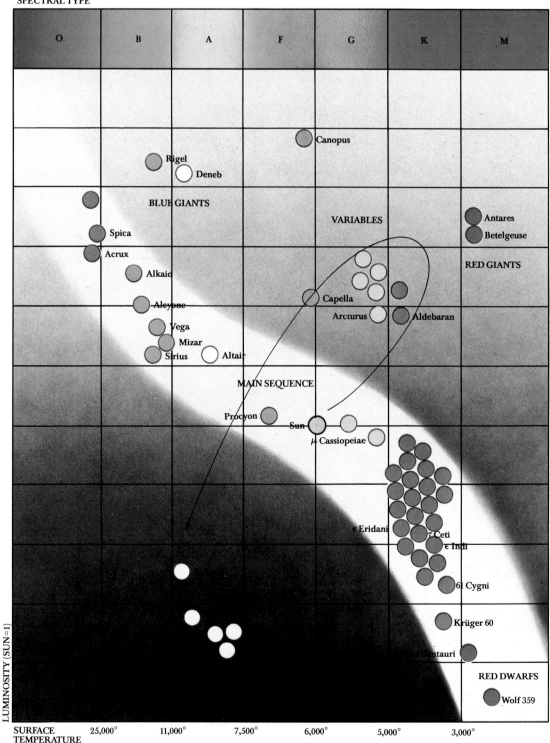

SPECTRAL TYPE

O   B   A   F   G   K   M

Canopus

Rigel

Deneb

BLUE GIANTS

VARIABLES

Antares

Betelgeuse

Spica

RED GIANTS

Acrux

Alkaid

Alcyone

Capella

Aldebaran

Vega

Arcturus

Mizar

Sirius

Altair

MAIN SEQUENCE

Procyon

Sun

μ Cassiopeiae

ε Eridani

τ Ceti

ε Indi

61 Cygni

Krüger 60

Centauri

RED DWARFS

Wolf 359

LUMINOSITY (SUN=1)

SURFACE TEMPERATURE   25,000°   11,000°   7,500°   6,000°   5,000°   3,000°

Like the stars in an open cluster, and indeed like the majority of stars in our Galaxy, the Sun is a main-sequence star, and sits right in the middle of the main-sequence band. It has an average surface temperature, and an average brightness. But the Sun has not always been the way we see it now. Five thousand million years ago the Sun was born as a result of some slight disturbance of the interstellar medium – a neighbouring supernova explosion or the passage of a spiral arm perhaps acting as the trigger.

The main sequence stars have reached the adult stage of the stellar life cycle. Once nuclear burning reactions commence, a star settles down into a stable, quiescent state, steadily converting its hydrogen into helium. Its brightness and surface temperature depend mainly on its mass, and will remain unchanged for a very long time. (For example, our Sun has been burning much as we now see it for some 5,000 million years, a little longer than the age of the Earth; it will go on burning for about another 5,500 million years without much outward change.) There the star will sit during this phase of its life, in the same position on the main sequence band of the Hertzsprung-Russell diagram. Only when the hydrogen fuel begins to be burnt out in the centre will any fundamental change take place.

In order for a star to remain on the main sequence three conditions must be satisfied. Firstly, there must be an energy source. Secondly, this source must supply energy at just the right rate to balance the energy which is leaking out from the hot interior, and which eventually escapes as the radiation we observe at the stellar surface. And thirdly, the pressure force pushing outwards must balance the gravitational force pulling inwards. The internal structure of a star is determined by the need to satisfy these three requirements.

The nuclear reactions that provide the energy work in the following way. Hydrogen nuclei fuse to form a helium nucleus, liberating energy in the form of high energy photons (or gamma rays) and neutrinos.

The neutrinos are massless particles and are basically unaffected by gravity. Furthermore neutrinos are most elusive particles, for they interact only very weakly with the other elementary particles. Once produced in the centres of stars they escape freely, disappearing into space at the speed of light.

Photons, on the other hand, though also massless and travelling at the speed of light, interact strongly with the atoms and electrons of the gas in which they are formed. Unlike the atoms and electrons, the massless photons are virtually unaffected by gravity. They are not bound to the star in the same way as the gaseous material, but, like neutrinos, try to fly out into space. However the outward flow of photons is hindered by the gas itself. A photon streaming out from the centre trips up in its flight whenever it hits an atom, which may then absorb it. One of the electrons encircling the atomic nucleus

then bursts out, flying off with some of the photon's energy. The higher the energy of the photon, the faster the atom and electron fly apart. In this way the energetic photon emitted by the nuclear furnace gives up its energy to the gas, keeping the centre of the star hot.

However, the electron itself will soon get captured, jumping back into an orbit around some other atomic nucleus and in turn releasing a new photon. Thus the gas of atoms and electrons engage in a complicated dance, sometimes absorbing photons as they fly apart, sometimes emitting them as they jump together. It is not the same photon which is involved in each collision and re-emission by an atom. Each photon is like a runner in a relay race, passing energy out like a baton after every interaction with an atom. The gas itself is held firmly in place by gravity, the atoms and electrons dashing about in such a way that there are always the same number on average at any one point in the star. However the photons do slowly drift outwards to the surface. Here they can escape freely as the light which causes a star to 'shine'. In this way the energy being lost from the surface of a star in the form of the radiation which we see, is being continually replenished by the nuclear furnace in the centre. What we see as points of light in the depths of the night sky are photons that have bumped and bounced their way out through the gaseous matter of the stars.

These interactions slow down the progress of the photons. If the photons emitted in the burning region of the sun were to escape freely they would travel out at the velocity of light, escaping from the surface after a couple of seconds. Instead it takes this radiation something like 50 million years to batter its way out. The energy we see escaping from the Sun today was first released when four hydrogen nuclei fused to become a helium nucleus some 50 million years ago, at a time when Earth's primates were emerging.

We have seen one way in which energy leaks out through a star – by the blundering walk of photons. Energy can also be carried out by another mechanism: by the motion of the gas itself. The hotter material at lower depths in a star may suffer from what is called 'convective instability'. Hot blobs of gas may find themselves surrounded by cooler, denser material. Because the hot blobs are lighter, they will rise up like bubbles of air rising in water. Only when they reach a place where their density is the same as that of their surroundings do they slow down, come to rest and release their heat. Similarly, cool blobs will be travelling downwards, in such numbers that on average the star neither expands nor contracts. In this turbulent region heat is physically carried by the gas from the inside to the outside of the convectively unstable region.

Stars differ in their structure depending on their mass. Theoretical studies show that the central nuclear burning region of a massive, main-sequence star is extremely compact, producing a small, high temperature centre. The

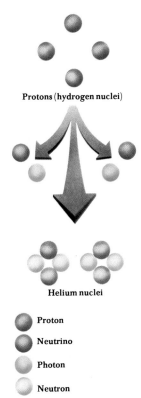

Protons (hydrogen nuclei)

Helium nuclei

● Proton

● Neutrino

○ Photon

○ Neutron

In the nuclear furnaces at the heart of stars, the nuclei of hydrogen atoms (protons) are fused in a series of complex reactions (here much simplified) to produce helium nuclei. In the reactions, energy in the form of neutrinos and light (photons) is released.

temperature drops rapidly away from the centre. The rapid drop in temperature makes the centres of massive stars convectively unstable. But their turbulent cores are surrounded by stable layers of gas. Photons slowly leak out through this quiet outer region to the surface.

Lower mass main-sequence stars, like the Sun, are different in every way. Lower mass stars have less extreme temperature changes outside their burning regions and are quite stable in their centres. But the outside is in violent motion. Low mass main-sequence stars are relatively cool, with surface temperatures less than 10,000°C. Below this temperature radiation finds it hard to escape because the stellar gas is opaque, rather like a very thick fog. With the photons dammed up in this fog, convection takes over as the most effective means of transporting energy outwards. So the outside of these stars is in rapid, turbulent, convective motion, with plumes of gas rising and falling as they carry heat out from the interior.

Our conclusions about the outside of low mass main-sequence stars are confirmed when the Sun is examined by telescope. The Sun is the only star whose surface we can examine closely. All the other stars remain as unresolved points of light through even the largest optical telescopes. The Sun, on the other hand, is so close that we can see details like sunspots even with the naked eye (though the sun *should never be looked at directly* as this seriously damages the retina of the eye). The Sun is seen to be covered by a fine cellular pattern of light and dark regions, of bright spots surrounded by darker dividing walls. The whole mass of cells, referred to by astronomers as the solar granulation, is seething, like a bubbling basin of rice pudding. These cells are the topmost layers of the convective zone that envelops the inner regions of the Sun.

What about the insides of stars? Is there any way in which we could directly confirm our conclusions about what is going on now in the nuclear furnace of the Sun? Since it takes 50 million years for radiation to escape, and the photons which escape are totally different from the photons which were first released in the central thermonuclear reaction, this would appear impossible. But there is another source of information: those elusive neutrinos. The neutrinos fly out unaffected by the surrounding gas and

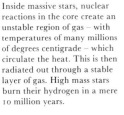

Inside massive stars, nuclear reactions in the core create an unstable region of gas – with temperatures of many millions of degrees centigrade – which circulate the heat. This is then radiated out through a stable layer of gas. High mass stars burn their hydrogen in a mere 10 million years.

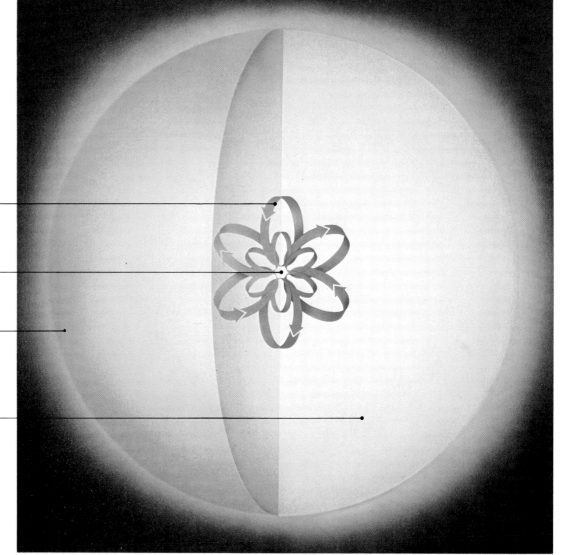

Convective gases

Core

Corona

Radiation

disappear into space. If they could be detected, the neutrinos would tell their story about the nuclear burning fires where they were created.

Detailed analysis of the nuclear fusion reactions occurring in the Sun predict a vast flood of neutrinos. Something like 100 million million are passing through this page every second. An American astronomer, R. Davis Jnr. has been searching for these neutrinos in a most extraordinary place – at the bottom of the Homestake gold mine in South Dakota, USA. The neutrino 'telescope' is a huge tank, containing several hundred thousand litres of dry cleaning fluid, perchloroethylene. This is a compound of carbon and chlorine; some of the chlorine is in the form of the isotope (an atom of different structure but the same chemical properties) chlorine 37. This isotope reacts weakly with neutrinos and is converted into the gas argon. The tank of fluid is kept a mile below ground to shield it from bombardment by other sources such as cosmic rays, which produce unwanted signals in the detector. By counting the tiny number of argon atoms produced in the tank, Davis is able to determine whether the Sun is emitting the

number of neutrinos that theory predicts.

It is not! At the moment the sun is producing something like one-third of the neutrinos it should be. This is very worrying for theoretical astronomers, who thought they understood the structure of the Sun. Some of the details of the nuclear reaction calculations used to construct theoretical models of the Sun, and of stars in general, seem to be wrong, or else more complex processes are occurring than we presently understand. Though the general basis of stellar structure is well understood, and the properties we have described so far are well established, the solar neutrino experiment is a disturbing reminder that nothing in science can be taken for granted.

Old Age
All stars are mortal. Indeed nothing in the Universe is unchanging except absolutely cold, dead, stellar remnants. Stars sitting on the main sequence are doomed to die because their reserves of energy cannot last forever. As they eat up the hydrogen in their centres they slowly exhaust the fuel which allows them to shine.

How fast a star runs through its adult life on

A mile down in the Homestake Goldmine, South Dakota, shielded from all normal cosmic radiation, this huge tank, containing 400,000 litres of dry-cleaning fluid is used to trap elusive neutrinos, produced in the interior of the Sun.

A low-mass star, like our Sun, burns its hydrogen over 10,000 million years. The central regions, at relatively low temperatures, are stable, and radiation leaks away from the core until blocked by the outer gases, which circulate like boiling oil to release the star's heat at the surface.

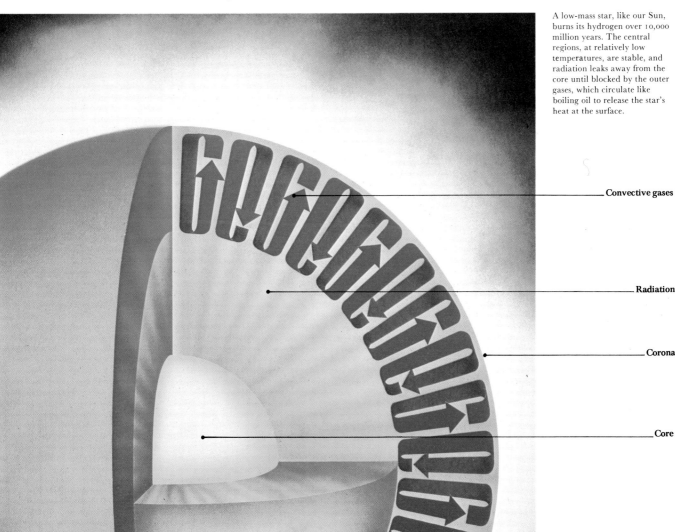

Convective gases

Radiation

Corona

Core

the main sequence depends on its mass. Though more massive stars have a greater supply of fuel, they are also very much brighter. They are profligate in their consumption of energy, and eat up their fuel stores rapidly. For example, a main-sequence star with a mass ten times that of the Sun is roughly a thousand times brighter. Consequently it is consuming its total supply of hydrogen a hundred times faster than the Sun, and it can only live for a hundredth of the Sun's life span. Contrary to expectation, it would not be wise to bet on the longevity of a massive star; choose a faint, cool, low mass star if you seek a long, stable, quiet life.

Detailed calculations show that a star like the Sun will exist as a main-sequence star for

The Seething Sun

Right : A false colour photograph of a solar flare licking thousands of miles out into space shows the hot-spot (white) associated with such events.

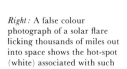

Below : A solar flare billows up in 1969, releasing a stream of particles that caused a severe magnetic storm on Earth when they arrived a few hours after the outburst.

At the Sun's surface, gases carrying away heat from the interior seeth and bubble at about 6,000°C. But the normal turbulence of convection is often disturbed by magnetic storms that tear the surface, or photosphere, into sunspots, cooler regions sometimes thousands of miles across.

Associated with these are sweeping flares and prominences that flame anything up to half a million miles into space, curving back along lines of magnetic force. Some quiescent prominences are semi-permanent features. These violent events contribute to the streams of charged particles – the 'solar wind' – that drive across the Solar System.

On Earth, they cause magnetic storms and (possibly) changes in the weather. For reasons unknown, the surface disturbance, as measured by sunspot activity, peaks about every $11\frac{1}{2}$ years (see graph).

about 10,000 million years. The Sun itself has reached the half-way stage in its main-sequence life – in about 5,500 million years it will have burnt the last of its central hydrogen and will die. In contrast the bright, massive stars live for less than 10 million years and are dead and gone before the lower mass stars have really got started. As we mentioned previously, this differ-ence accounts for the concentration of bright, massive O and B stars in young, open clusters of stars; they cannot escape far from the cluster of their birth before they die.

So much for the majority of stars born in the main body of the Galaxy. But there are stars found elsewhere, in the Galactic suburbs, as it were. These stars are grouped in clusters that are

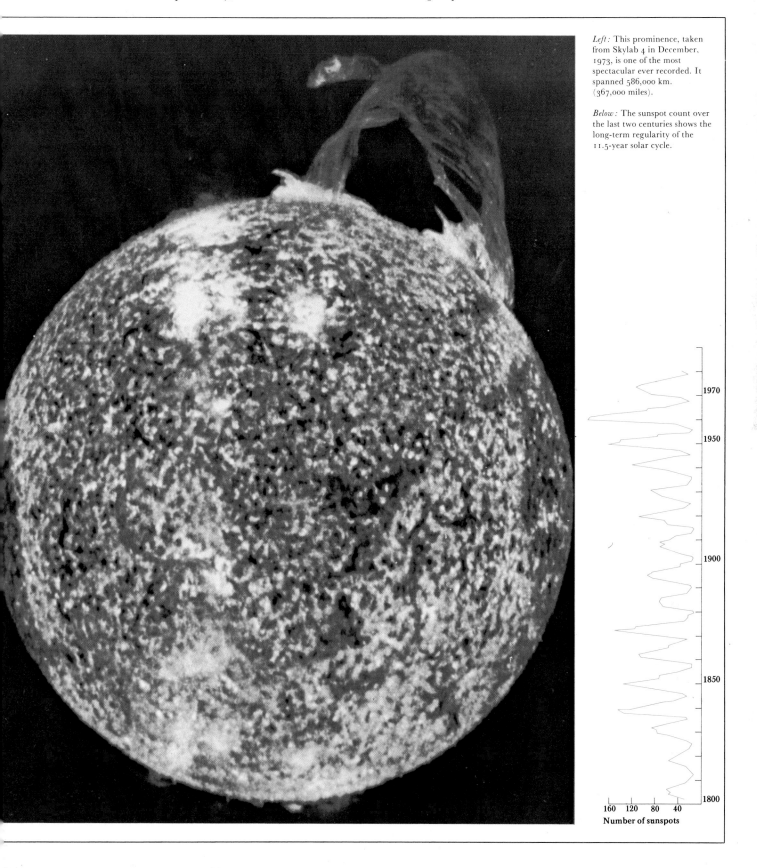

Left: This prominence, taken from Skylab 4 in December, 1973, is one of the most spectacular ever recorded. It spanned 586,000 km. (367,000 miles).

Below: The sunspot count over the last two centuries shows the long-term regularity of the 11.5-year solar cycle.

Number of sunspots

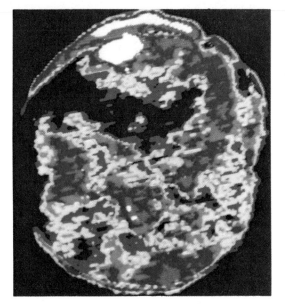
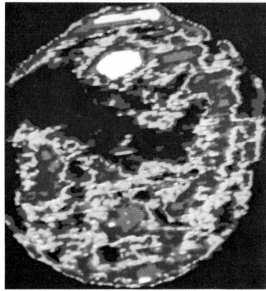

Two false-colour images of the Sun, taken from Skylab 3 on August 19 and 20, 1973, reveal holes in the Sun's corona caused by irregularities in the surface, and also show the Sun's daily rotation.

not associated with the spiral arms of the Galaxy, or even with the galactic disc. These clusters are spread in a sphere above and below the disc in a swarm nearly two hundred strong. This region surrounding the galactic disc is called the galactic halo. Each cluster contains thousands or even millions of stars packed in a spherical 'globule' with a bright core containing most of the stars.

Unlike open clusters, the stars in these so-called 'globular clusters' stay together as a group for their whole evolutionary history. Globular clusters contain so many stars – many more than open clusters – that they cannot break up. Each star is tied, or bound, by the gravitational pull of all the other stars in the cluster.

Globular clusters are not only much more permanent groups of stars than the open clusters, they are also much older. They were formed in the early stages of the evolution of the Galaxy. The Hertzsprung-Russell diagram of a globular cluster does not consist of main-sequence stars only. Indeed, a typical Hertzsprung-Russell diagram for a globular cluster has no stars at the top of the main-sequence band at all. Instead the brighter stars 'turn off' towards the right and climb in a band up the cool side of the diagram. Instead of a diagonal strip across the diagram, the stars form a reversed S shape.

These cool, bright stars that have turned off the main sequence are very large and are called 'red giants'. A typical red giant is about a hundred times larger than the Sun. To compare a red giant with the Sun is like comparing Concorde with a paper dart.

The red giant state is the first step taken by a star on the path towards its death. When all the hydrogen is exhausted in the stellar core of a main-sequence star, its weight is no longer supported by energy generation in the interior.

At first, hydrogen burns in a shell, rather like an eggshell, surrounding the core of helium left behind by the main-sequence burning stage. The shell of burning hydrogen dumps its 'ashes' onto this core, building it up progressively with supplies of more and more helium. But the core, with no energy sources of its own, cannot grow indefinitely. When it reaches a critical mass, the centre collapses and the outer layers of the star

expand, stretching out into space and growing in size one hundredfold. The central regions continue trying to disappear down to a central point, collapsing towards higher densities and increasing in temperature, until the centre becomes hot enough for helium to 'ignite' and undergo fusion reactions which yield more complex elements, principally carbon and oxygen. Just as the collapse of the initial protostar was halted by hydrogen fusing to helium in the centre, so the collapse of the core is halted by helium starting to fuse into heavier elements. Thus helium, formed as the ashes of hydrogen burning during the main-sequence stage, becomes the fuel for the next chapter in a star's life, the red giant stage.

As the central regions are taken over by helium burning, the outer regions expand in the way we have described. The star has mushroomed into a huge, inflated gas-bag with a tiny, dense nuclear burning region in the centre.

In this way, the cool red giant stars found in globular clusters are produced. The red giants have exhausted their central hydrogen fuel and are on the road to senility and death.

What about the even more massive stars? Globular clusters are groups of stars which are so old that the most massive stars within them long ago ceased to be even red giants. The most massive stars have died completely, and any remnants form a stellar graveyard of super-dense dead stars – peculiar objects with peculiar names: white dwarfs, neutron stars, and perhaps even black holes.

Senility and Death

In massive stars the ignition of helium is only the first in a whole series of stages in which more complex nuclei are successively burnt. As each nuclear fuel is burnt out and exhausted in the centre, collapse of the inner core sets in again. Collapse continues until the temperature is high enough for the ashes of the previous burning stage to ignite. Helium burns to carbon and oxygen; carbon and oxygen burn to silicon; silicon burns to iron.

What halts this sequence of burnings to even more complex nuclei? At the point when iron is being forged in the centre of a massive star, with surrounding shells of silicon, oxygen, carbon, helium and hydrogen, the end of the road is near. When the iron core collapses there is no hope of a new reaction stepping in to save the star, for it is not possible to fuse iron into more complex elements (such as uranium) *and* release energy at the same time. Indeed, elements much more complex than iron have a tendency to undergo fission, splitting apart spontaneously into less complex elements.

The explanation of what happens to a star at this point is believed to account for the origin of elements like uranium also. The collapsing iron core is thought by astronomers to explode, with explosive nuclear reactions releasing vast amounts of energy very suddenly. This sudden outpouring of energy blows off the outermost layers of a star, and causes the brightness to increase enormously. The vast supernova explosions observed in our own and neighbouring galaxies, in which a single star becomes as bright as a whole galaxy in a couple of days, are due to a star bursting apart in this way at the end of its life. Such supernovae are the last dramatic gasp of a star in its death throes. (The name is derived from another type of stellar explosion – a 'nova', or new star – but the processes involved are different, and the two names should not be associated with each other.)

It is during supernova explosions that the creation of the more complicated elements like uranium is thought to occur. These, together with the other elements built up from hydrogen over the life of the star, are flung out into space in a vast expanding cloud of gas. The space between the stars is replenished with gas – but not the original hydrogen and helium which collapsed to form the star. Instead it is full of oxygen, nitrogen, copper, manganese, bromine, titanium, gold, silver and all the other elements which make up our world on earth. These elements went into the mixture from which the Sun and the Solar System were later formed.

Massive stars are thus the crucibles in which the bulk of the elements with which we are familiar are created. Without these massive stars the Universe would simply be a mixture of hydrogen and helium, created during the early stages of the Universe before stars or galaxies had formed at all. It is sobering to realize that almost all the elements which make up our familiar world of water, air, earth and living tissue were formed in the deep interior of distant stars. You and I, and this book you are reading, and the ink it is printed with, once went through the raging furnace in the centre of a star.

Probably the most remarkable object in our Galaxy is the remnant of such a supernova explosion. On 4 July AD 1054 Chinese astronomers noted the sudden appearance of a 'guest star', a brilliant apparition clearly visible during daylight for about a month. An enormous amount of gas was expelled in that explosion, equal to the mass of the Sun. We can now see this ejected gas as a tangle of bright pinkish filaments expanding outwards into surrounding space, called the Crab Nebula.

Associated with the filaments of expanding gas are strong magnetic fields. High energy particles spiralling around these magnetic fields produce radio, ultra-violet, X-ray and even gamma-ray radiation within the nebula.

Near the centre of the Crab Nebula lies a most unusual star. In 1967 it was discovered that what appeared to be a relatively normal bright star was behaving in a most extraordinary fashion. It flashed on and off 30 times a second like some galactic warning beacon at the centre of the remnant clouds of the supernova.

What is this pulsing stellar remnant left

Birth of a Supernova

The Crab Nebula, the gaseous remnants of a supernova explosion, was noted by Eastern astronomers when the light from the stupendous blast first reached the Earth on July 4, 1054. The original star, at a distance of 6,300 light years, would have been inconspicuous from Earth. But the brightness of the blast was 200 million times that of the Sun, and it was visible by day for 23 days. A Chinese account records that it was as bright as Venus; 'it had pointed rays on all sides and its colour was reddish-white.'

The object was rediscovered in the late 18th century, and received its name in the 1840's from a crab-like sketch of it by the English astronomer Lord Rosse. Since then, it has assumed a unique place in astronomy as the first known supernova remnant, the first X-ray source and the first known site of a pulsar – the tiny throbbing cinder of the cataclysm.

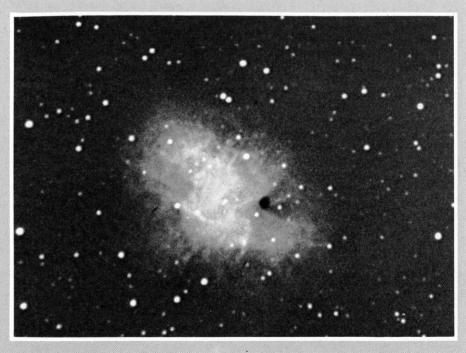

behind by the catastrophic explosion of AD 1054? Astronomers believe this star is a neutron star, spinning 30 times a second, and flashing (like a lighthouse) every time a bright beam of light emitted from near the surface sweeps across our line of sight.

Stellar Remnants

Neutron stars of which the so-called 'pulsars' are special examples are one of three types of stellar remnant formed when a star dies.

The most common stellar remnants are 'white dwarfs': compact stars that are slowly cooling off in space. White dwarfs are made of strange stuff. Normal gas tends to heat up when it is squeezed. It is this response that leads to the sequence of nuclear burning reactions during stellar evolution that we have described. However, when gas is compressed to extremely high densities – such densities that a cigarette packet full of matter would weigh several tons – the electrons start to exert a special kind of pressure. They become what physicists call 'degenerate' and, as long as the star is below a critical mass, refuse to be squashed further. The electrons exert this pressure no matter what their temperature. A star composed of this material can cool without collapsing since the electron pressure can hold the star up against gravity.

Most stars end in this state, as white dwarfs the size of the earth but 100,000 times more massive. The white dwarfs cool off in space, and become steadily fainter and fainter as they slowly freeze. Eventually they form completely dead stellar stumps: worlds which are absolutely cold, and emit no radiation whatsoever.

The pressure exerted by degenerate electrons is not without limit, however. If the star is too massive, the force of gravity will overwhelm the electrons. The maximum mass of a white dwarf (first calculated by S. Chandrasekhar) is a little less than one and a half times the Sun's mass. Stars more massive than this must either lose mass before they reach the end of their life and become white dwarfs, or else they will collapse to form an even more compact star when they die. Such an object is a 'neutron star'.

If the electrons and protons are crushed together even more than they are in a white dwarf, they bind together forming neutrons. Neutrons exert their own 'neutron degeneracy' pressure which is even more powerful than that of electrons. It is able to support a star of up to three solar masses. A neutron star, being crushed even more strongly by the force of gravity, is even more compact than a white dwarf. A neutron star with the mass of the Sun would be about the size of Manhattan.

What happens if a star is squeezed even further, as must happen if the collapsing stellar remnant is more massive than three solar masses? The possibility arises, within the laws of physics as we presently understand them, that collapse continues indefinitely. All the matter showers into a central point, squashed by gravity to

infinite density, and forms a 'black hole'. No star has ever been observed in this state, for before this condition is reached any observer will lose track of the star. It will disappear from view. As gravitational collapse beyond the neutron star state develops, densities are reached which are so enormous that light itself cannot escape from the surface. With the increasingly strong gravitational fields associated with the increasing density, the velocity needed for escape increases. Eventually a stage is reached at which this escape velocity is greater than the velocity of light. Nothing can escape, then, or ever after. The collapsing star disappears behind an impenetrable horizon – the 'event horizon' – and is hidden forever from our eyes.

How then could we 'see' a black hole if no radiation can escape from it? Only by observing the gravitational effect the matter buried inside the black hole has on material around it. As we shall see, the most likely candidate for a black hole has been found by this technique. It can be seen only indirectly, through the effect it has on a companion star in a binary star system.

So far we have been following the death pangs of a massive star. Massive stars end in a catastrophic supernova explosion, leaving behind a remnant neutron star, or a black hole, or they may even disrupt entirely. In that case only the filamentary supernova remnant expanding into the interstellar medium remains to tell the tale of a star's death.

The less massive stars die less dramatically, but form just as beautiful structures in space. Stars like the Sun never even reach the stage of burning carbon in their interior. At some point the whole outer envelope lifts off and drifts out into space. The outer region of the star becomes less and less dense as it expands outwards, and eventually the hot, central core of the star is revealed surrounded by a tenuous halo of gas like some vast, spherical, cosmic smoke ring. Such structures are observed as planetary nebulae. The central star in a planetary nebula is con-

There is a 100-million-fold difference in size between the largest and smallest types of star. This diagram compares types that are all of one solar mass. In fact, some red giants are even bigger. If Betelgeuse, for instance, were placed in our Solar System, its surface would lie beyond the orbit of Mars. White dwarfs and neutron stars have well defined limits, but black holes (the very existence of which is still controversial) could be of any size.

Star type (1 solar mass) and diameter:
Red giant: 140 million km./ 87 million miles.
Sun: 1,392,000 km./ 870,000 miles.
White dwarf: 13,000 km./ 8,000 miles.
Neutron star: 16 km./ 10 miles.
Black hole: 2.5 km./ 1.5 miles.

tracting to become a white dwarf, and will finally cool to become a lifeless stellar stump.

Pulsating Stars

In addition to the red giants, the white dwarfs, the neutron stars, and the planetary nebulae, there exist a number of special types of star whose strange properties are caused by peculiarities in their structure. Among the most fascinating are the pulsating stars, stars whose surface layers rise up and sink down just as if the stars were breathing. Various classes of pulsating stars exist, including the RR Lyrae stars, the Mira variables, and the Cepheids. The Cepheids are particularly important because the scale of the Universe has been determined to a large extent using their pulsations as a yardstick.

As a Cepheid pulsates its brightness, surface temperature and radius all vary. In 1908 Henrietta Leavitt discovered that amongst the Cepheids in the Small Magellanic Cloud (a small galaxy, neighbour to our own) the brightest had the longest period of pulsation. She showed that the correlation between the period and the intrinsic brightness was so good that, by measuring the period of a Cepheid, its intrinsic, or absolute, brightness could be deduced. By observing the apparent brightness of a Cepheid as seen from Earth, and comparing this with the intrinsic brightness deduced from its period, the distance of a Cepheid may be derived. This method is exactly the same as the method one might use to judge the distance of a car from the brightness of its headlights. When applied to Cepheids the technique has proved to be one of the most important methods for determining the size and structure of our own Galaxy and the distance of nearby galaxies.

The pulsating Cepheids oscillate with typical periods of about five days, whereas the RR Lyrae stars oscillate with periods of less than a day, and the Mira variables with periods as long as several years. Each type of oscillating star sits in a particular region of the Hertzsprung-Russell diagram and each is at a particular stage of post main-sequence evolution.

Why should a star oscillate like this, bouncing in and out like the lid on a pan of boiling water? The pan lid lifts up because of the pressure of the steam trapped underneath. The steam then escapes and the lid drops down. Again the pressure builds up and the lid rises, only to drop once more when the supporting steam escapes. The lid will bounce up and down just so long as there is a source of steam, that is, as long as water is boiling in the pan.

In a similar way the outer layers of pulsating stars act as a trap on the radiation inside, damming up the photons in their flight to the surface. The barrier presented by these outer layers causes the temperature, and the pressure, to build up and the star expands. In this expanded state the radiation escapes more easily – the lid has lifted off. With more radiation escaping, the brightness of the star increases, and the outer layers cool. The surface layer, no longer hot enough to support itself, collapses and radiation is once more dammed up. Radiation, trying to escape, forces the star to oscillate at its natural frequency of vibration. The brightness, surface temperature and radius of the star all vary during these pulsations.

Cannibalism Amongst Stars

Stars not only pulsate, they also consume one another. In the past few years cannibalism amongst stars has become recognized as quite common.

Stars can only consume one another in a binary system (a double star system in which two stars orbit around their common centre of gravity, in exactly the same way as the earth and moon orbit about one another). If the two stars in a binary system are close enough, they may run into a serious difficulty as they evolve. A star evolving off the main sequence expands and eventually becomes a red giant 100 times larger than its youthful main-sequence state. As the two stars in a binary system age, the more massive star, or primary as it is called, will burn out its supplies of central hydrogen fuel first. It will start to expand. However, if its less massive companion, or secondary, is close enough, it will start to interfere with the expansion of the more massive star. It will disturb the primary star's envelope because of the clutch of its gravitational attraction. A star is held together by the gravitational attraction of the gas *inside* pulling inwards on the gas *outside*. If the companion star is too close, the gravitational attraction of the companion will overwhelm the inward pull that holds the primary star together. The primary takes on a pear-shaped form, the point of the pear pointing towards the secondary star.

When the primary expands further the secondary's feast begins. For gas at the point of the pear is ripped off and falls in a continuous stream onto the secondary. The secondary will continue to consume its fellow star so long as it tries to evolve. The so-called Algol binary star systems

The pulse of a typical Cepheid variable star – named after the prototype in the constellation Cepheus – is related to brightness and size. Apparent brightness can therefore be used as a guide to distance. The surface, like the lid on a boiling kettle, lifts as energy builds up internally over the course of several days. Maximum brightness coincides with the bluest colour, but size, which varies by 20 per cent over the cycle, increases beyond that point as energy is dissipated. The surface then collapses again to renew the cycle.

Colour and relative size of pulsating star

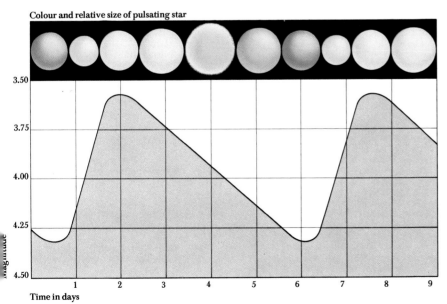

Time in days

Cygnus X-1, a powerful X-ray source, is a binary – a blue supergiant 20 times the mass of the Sun and a black hole with four times the mass of the Sun which orbits its companion every few days just a million or so miles away. The enormous gravitational field of the invisible black hole (the 'hole' would be no more than two or three miles across) tears gas away from the companion. The gas is sucked into the hole as if into a whirlpool, and particles in the whirling accretion disc are heated by friction. In the process, X-rays are released.

makes up the Earth's atmosphere, is dumped over in a sudden burst onto the disc, where it slowly spirals into the white dwarf. With each burst the disc brightens, and the binary goes through a small-scale nova eruption.

The second class of novae, called 'classical novae', is much more spectacular. In this case the white dwarf erupts, flinging a vast cloud of gas away from the whole binary system. In its brightest state the binary system cannot even be seen. It is completely enveloped in a cloud of expanding gas, flooding out at a speed close to a million miles an hour. Only as the nova fades does the cloud of gas disperse, gradually thinning out and revealing the central binary system at its centre.

These classical nova explosions are thought to be true explosions, gigantic nuclear bombs burning off on the white dwarf surface. The gas collected by the white dwarf from the companion star is hydrogen, a potential nuclear fuel. When sufficient hydrogen has collected on the surface of the white dwarf, nuclear reactions start to burn in the accreted hydrogen envelope. These fusion reactions are not controlled in the way they are in the centre of the star. They may grow rapidly, liberating energy in an explosive detonation, and eject the outer envelope of the white dwarf into space.

Finally, perhaps the most extraordinary cannibalistic stars are the recently discovered X-ray sources. These are rather like the novae, but in this case the accreting star is a neutron star rather than a white dwarf. The neutron star is so small that the infalling gas is moving extremely fast by the time it reaches the stellar surface. The gas is so hot that the radiation escaping from the disc is emitted predominantly in the X-ray region of the electromagnetic spectrum. When we observe these systems with X-ray telescopes we are seeing gas at temperatures approaching those attained in the centres of stars.

Many of these X-ray sources have fascinating properties, but probably the most intriguing of them all is called Cyg X-1. Cyg X-1 was one of the earliest X-ray sources to be discovered, in the constellation of Cygnus. In this X-ray binary the accreting star is calculated as being greater than three times the Sun's mass, which is too massive for a neutron star. Of all the objects known Cyg X-1 is thus the most likely candidate to contain a black hole, the first to be detected. Cyg X-1 may contain the most extreme possible state of a stellar remnant.

If so, many millions of years ago the black hole in Cyg X-1 was a normal star, shining through space as a result of nuclear fusion reactions in its centre. Before that it was a cloud of gas, spread out like a veil between the other stars. Now all that is left is a puzzling, massive, gravitational centre, feasting itself on its neighbouring companion, and emitting X-rays from the hot accreting gas as it falls towards, and eventually disappears behind, the black hole's impenetrable and unknowable event horizon.

are in this state.

Other cannibalistic binary stars are even more spectacular – they explode as novae. It has been discovered over the past 20 years that all novae are binary stars in which gas is being exchanged. In the novae the parasitic star is a white dwarf which is slowly consuming its main-sequence companion. The gas does not fall straight on the white dwarf once it is ripped off the main-sequence star. Instead it swings around the back of the white dwarf, and slowly spirals inwards onto the surface. The spiralling gas forms a disc within which the gas circulates like water spiralling around the plughole of a bath. The gas slowly falls inwards in the disc, circling faster and faster and getting hotter and hotter as it approaches the white dwarf. Finally, reaching the surface, the infalling gas is arrested. Like sparks flying from a grinding wheel, the sudden arrest of this gas liberates energy and radiation. When we look at these nova binaries we see optical radiation which is predominantly produced by this accretion process. The disc of gas is brighter than either of the two stars.

Two classes of novae exist. One class, called 'dwarf novae', erupts repeatedly in a series of small explosive outbursts. Every 30 days or so the disc grows brighter, and then slowly fades. It is thought that these irregular eruptions are due to bursts of gas falling onto the white dwarf in a series of repetitive splashes. Instead of gas passing smoothly between the two stars, the main-sequence star suffers a sequence of hiccups. A quantity of gas, about the same as that which

STARCLOUDS: DEATH AND REBIRTH

The space between the stars is emptier than any vacuum on Earth, but it is still rich in clouds of gas and dust. These clouds, or nebulae, are both remnants of ancient explosions and seed-beds in which new stars are born.

Some nebulae are the outer blankets of red giants of about the mass of the Sun. As heat builds up in their nuclear furnaces, such stars become unstable and throw off a smoke-ring of gas. These planetary nebulae, of which some 1,000 are known, disperse over 10,000 years, leaving their parent stars to decline into compact senility.

In other cases, two stars – a white dwarf and a large companion – interact; the white dwarf's intense gravity tears tidal-waves of gas from the companion and ejects them into space in what is known as a nova ('new star').

Two or three times a century in our Galaxy, massive stars explode in supernovae with a phenomenal release of energy, brightening to the luminosity of a billion Suns. The gas they eject – travelling at up to 10,000 km (6,000 miles) per second – plays a crucial role, for it contains many of the heavier elements familiar on Earth. In the process, the supernova stars either blast themselves to pieces or are left as tiny, dense neutron stars or black holes.

From the gas clouds of interstellar space, new stars will eventually emerge. The clouds collapse under their own gravitational influence to continue the cycle of life and death.

In this planetary nebula, the Ring, the central star glows strongly, its thermal equilibrium restored by the ejection of the gas shell.

Right: The splendour of the Carina Nebula make it one of the wonders of the southern sky. At its centre is a variable star, Eta Carinae, which apparently ejected the cloud of gas in what may have been a peculiar form of supernova explosion.

Overleaf: This diaphonous filigree of gas is part of the Veil Nebula, or Cygnus Loop. It represents the last scattered remnants of a supernova explosion some 50,000 years ago. In 25,000 years time, it will have vanished.

Below: The Trifid Nebula in Sagittarius, lit to a hot blue and cooler red by nearby stars, is thought to contain collapsing masses of gas and dust that are on their way to becoming new stars.

Bottom: In the Orion Nebula, the brightest in the sky, new stars are in the process of formation from the red loops and filaments of hydrogen.

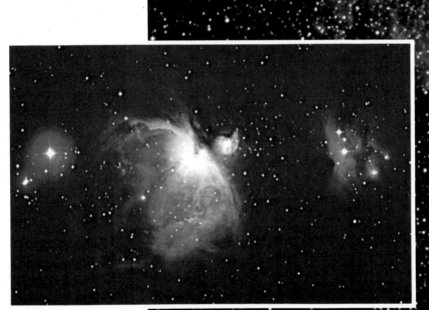

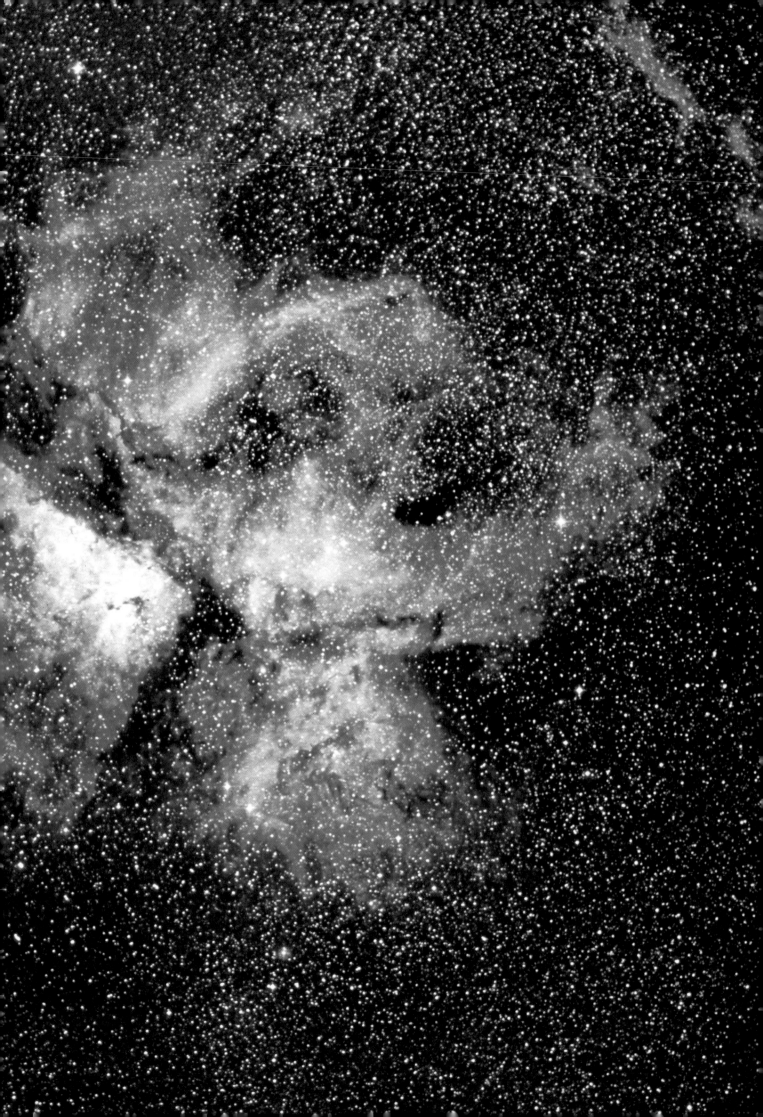

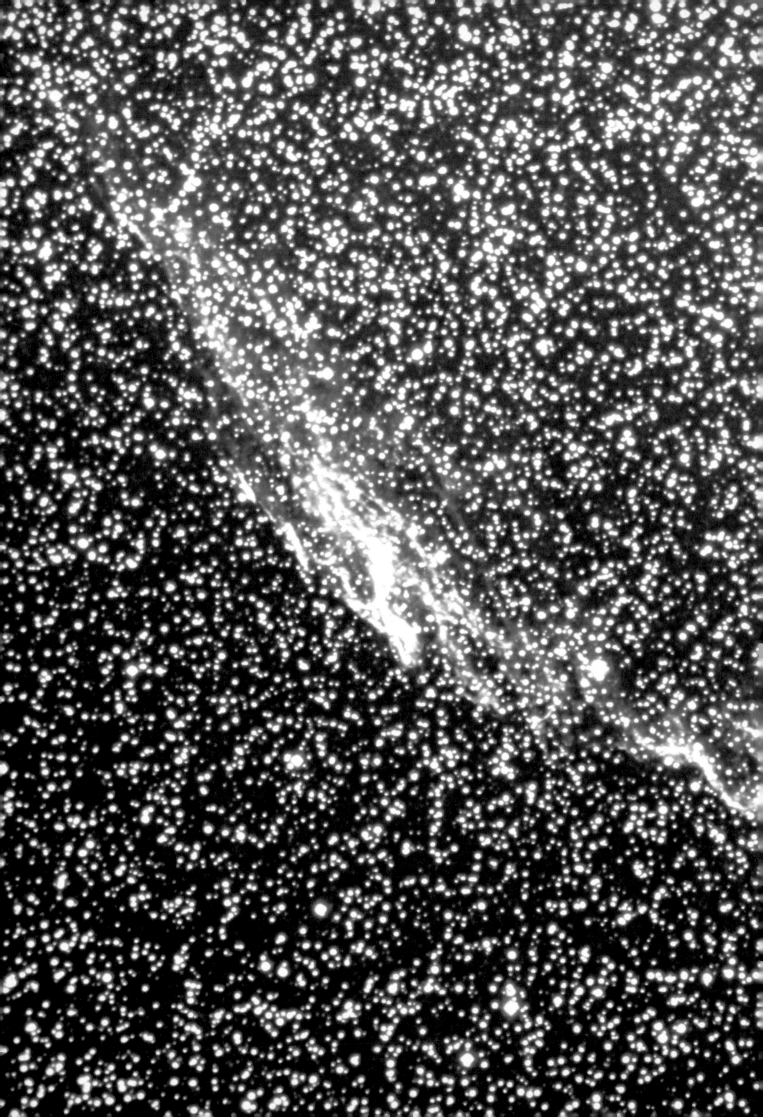

4/THE SOLAR FAMILY

PETER FRANCIS

Our Solar System is a collection of planets, comets and assorted chunks of rock spinning about the Sun, all on much the same plane. It seems likely that most of these elements are debris left-over from the formation of the Sun 5,500 million years ago – larger versions of the chunks of rock and dust that form the rings of Saturn (left). Despite the similarity of their origin, however, there are startling differences between the various bodies in the Solar System, and in particular between the individual planets.

It is something of a paradox that 20th-century mortals have less direct experience of the members of the Solar System than did our distant ancestors. The 20th-century man in the street lives in large cities, where the night sky is swamped by the glare of street lights, and the air is thick with pollutants. His ancestors had the advantages of undefiled skies, without glare or fumes, and could see the heavens with a clarity that is now almost unattainable. Apart from this, 20th-century man is also a long way divorced intellectually from the Solar System: it is no longer important for him to know what a point of light in the night sky is, or how it moves, or where it will be in a month's time. In the past, such factors were of critical importance, and were used astrologically both in constructing calendars and in the planning of future actions.

The visible Solar System

What, then could one of those ancient observers have seen of the Solar System? What can *we* see, if we choose to make the effort and spend some hours on crisp, clear nights scanning the skies?

First, and most obvious, there is the Moon. One could hardly miss that, and its presence is taken for granted by most people. The ancients, however, were seriously puzzled by the Moon, because of its apparent changes of shape from a thin crescent to a full, glowing disc, and different tribes and civilizations devised a myriad of myths and legends to account for these, and for the curious mottled markings on its surface. Men have now been to the Moon, so much of its mystery has been lost, but there are still a great many misconceptions about it amongst people who should know better: there are some who still think that the Moon only 'comes out' at night, an error an ancient observer would never have made.

Next in importance is the great vault of the heavens, the starry firmament. The stars themselves do not concern us here (they are dealt with in Chapters 2 and 3), except in so far as five of them were singled out as anomalous by even the earliest observers of which we have record. These were the five 'wandering' stars, or planets. (The very word *planet* comes from the

The planets and their orbits drawn to scale reveal the pattern of the Solar System. The inner or terrestrial planets are all close to the Sun and small, the lighter gases probably driven off as they formed. The outer planets are all gaseous giants, except for Pluto. Comparable in size to our Moon, Pluto's path is eccentric – for some 20 years in each orbit it is closer to the Earth than Neptune (e.g. January 1979–March 1999). The Sun is ten times the diameter of Jupiter (*below*).

Mercury Venus Earth Mars Jupiter

Mercury	Venus	Earth	Mars	Jupiter
58/36	108/67	150/93	228/142	778/486
0.24	0.61	1.0	1.88	11.86
59 days	243 days	23.9 hrs.	24.6 hrs.	9.8 hrs.
4,880/3,050	12,040/7,600	12,750/7,900	6,787/4,242	143,000/89,000
0.05	0.81	1.0	0.11	317.8

original Greek word for *wanderer*).

Of the five planets, Venus is much the most conspicuous. It often hangs as a brilliant jewel in the evening or morning sky, shining so brightly that it may cast its own shadow, and is regularly mis-identified as a UFO. Venus is also conspicuous because it wanders across the sky rapidly relative to the starry background and to the Sun. On one occasion, it may blaze brightly high in the sky long after the Sun has set; over the course of months it will appear to move nearer and nearer the Sun, and will set sooner and sooner after it. Eventually, it will become all but invisible, but will reappear shortly after as a *morning* star, rising before the Sun. This strange behaviour confused many early observers, not surprisingly, and many of them thought there were *two* separate bodies involved – the astronomers of ancient Greece retained two names for Venus for a long time, calling it Hesperus as an evening star, and Phosphorus as a morning star.

Next most obvious of the planets is Jupiter, shining like a particularly brilliant yellow star. Its movements, however, are relatively regular –

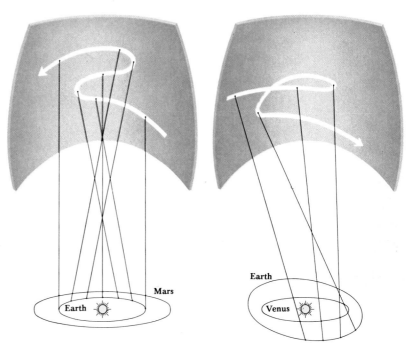

Motion of Mars   **Motion of Venus**

These diagrams show why a planet's motion seems peculiar when seen from Earth. As the Earth overtakes – or is overtaken by – another planet, an observer will see the planet reverse its direction against the background of fixed stars. It is not surprising the Greeks called our solar companions 'planetes', or wanderers.

Saturn	**Uranus**	**Neptune**	**Pluto**	
1,427/892	2,870/1,800	4,496/2,800	5,900/3,700	Millions km./miles from Sun
29.45	84.01	164.79	247.7	Year (Earth years)
10.2 hrs.	10.8 hrs.	15 hrs.	6.4 days	Rotation period (Earth units)
120,000/75,000	52,000/32,000	50,000/31,250	6,000/3,750	Diameter (km./miles)
95.1	14.5	17.2	0.11?	Mass (Earth = 1)

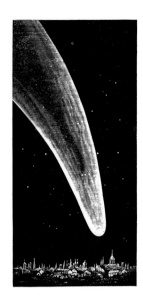

Comets – like this one seen over France in 1811 – have always been matters of note, and usually of astrological significance, partly because they were so little understood. Until the 16th century, they were considered to be atmospheric phenomena.

The Moon in its double orbit – round the Earth and round the Sun – describes a curved path through space and shows a series of light-and-shadow phases to an Earth-based observer.

it drifts slowly against the starry background, shifting almost imperceptibly from night to night, but steadily. Mars is much more erratic. Sometimes it blazes out much more strongly than Jupiter, with the fiery red colour that has led Mars always to be associated with blood and war. It also moves highly erratically, sometimes moving slowly and steadily in one direction, then slowing down, appearing to come to a complete stop before moving backwards for a short way, and then resuming its original steady drift against the starry background.

Saturn is much less bright than either Mars or Jupiter, and looks like an ordinary star. Its planetary nature, however, is soon revealed by its movement, which is slow and regular. Much the most difficult planet to observe is Mercury. It is always very close to the Sun, and is only visible shortly before sunrise, and shortly after sunset. Usually, it is lost against the glare of the Sun, and this has caused some astronomers, perhaps plagued with more than usually bad weather, never to have observed Mercury in their lifetime. With ordinary luck, however, Mercury can often by picked out as a tiny pink lamp gleaming against the glow of the morning or evening sky. Like Venus, its appearance as either a morning or evening star confused the ancients into thinking that two bodies were present. As for Venus, the Greeks had two names for Mercury: Hermes and Apollo.

To the first observers, then, the Solar System consisted of the Moon and five points of light, moving across the heavens. Notice that we have said nothing about sizes or distances or masses or densities, or even about why the planets appear to move as they do: they are just point of light. But what else can be seen? What else is there in the Solar System?

For the naked eye observer, there are only two other visible kinds of objects: meteors and comets. On any clear, crisp night a patient watcher will see several brief flashes of light, or shooting stars, sweeping across the sky, vanishing even as they are perceived. More rarely, a great fireball may be seen, lighting up the whole sky, and more rarely still ear-splitting detonations may be heard, and solid fragments (meteorites) may reach the ground. It was not until the 19th century, however, that any link was established between these exceptionally rare 'stones from the sky' and shooting stars, which are commonplace.

Rarest of all objects in the night sky are comets, which appear as diffuse points of light trailing great banners or tails across the sky. The greatest comets are perhaps the grandest of all natural spectacles, and throughout history they have been associated with profound events affecting man. It may even have been a comet that the Three Wise Men saw when they 'followed the Star' to the birthplace of Christ. More prosaically, the word 'comet' comes from a Greek root meaning something like 'long-haired star', an apt description, but not one that gives much of a clue to the nature of comets.

The visible Solar System, then, consists of planets, meteors and comets, all of them points of light moving against the great canopy of the night sky. But what is the relationship between these very disparate bodies?

We know now that the planets all move around the Sun, in regular, nearly circular orbits. It may even seem obvious that they do. But it has not always been so. For most of recorded history, men have assumed that the Earth is the centre of the Solar System. About 300 BC, one of the greatest of Greek astronomers, Aristarchus of Samos, realized that the Sun, not the Earth, lay at the centre of the Solar System, and that all the planets revolved around the Sun. Sadly for science, his arguments were not accepted, and for nearly the next 2,000 years men continued to believe what seemed to them to be obvious and

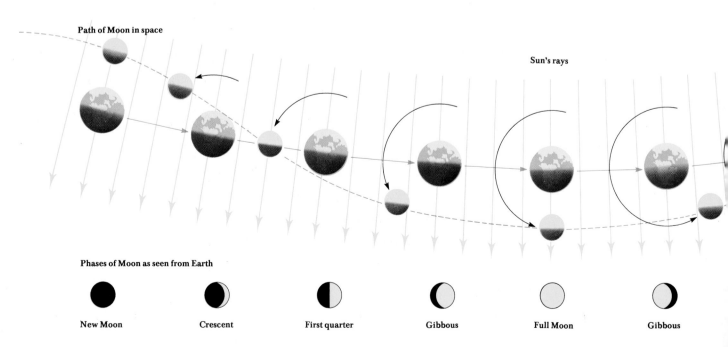

Path of Moon in space

Sun's rays

Phases of Moon as seen from Earth

New Moon   Crescent   First quarter   Gibbous   Full Moon   Gibbous

unquestionable: that the Earth is the centre of all things.

It was not until the 16th century that a Polish monk, Nicholas Copernicus, revived Aristarchus' ideas, and began to argue for a Sun-centred Solar System. Even Copernicus, who brought astronomy out of the Dark Ages, still laboured under some of the misconceptions implanted by the revered classical Greek astronomers and mathematicians: like them, he believed that all motion in the Solar System had to be in perfect circles at uniform speed. Copernicus did not have much observational data to go on, and therefore could theorize freely. He explained the oddities of planetary motions, such as Mars's brief about-turns, by invoking a series of elaborately elegant epicycles, or mini-orbits superimposed on the planets' main orbits.

The true nature of planetary motions was ultimately perceived by Johannes Kepler (1751–1630) who did something that no previous astronomer had done: he tried to fit theories to observations, rather than *vice versa*, and thus created a great watershed in science. He realized that the planets move in elliptical, not circular, orbits, and laid down three great laws of planetary motion which gave rise to Newton's fundamental work on motion and gravitation and thus laid down the foundations of modern physics.

So much for the history. The five 'wanderers' all move in regular elliptical orbits around the Sun. But what about the meteorites and comets? And what about the three other planets, Uranus, Neptune and Pluto, not known to the ancients or to Copernicus or Kepler? These too, all move in elliptical orbits. Everything in the Solar System moves on elliptical orbits. The planets themselves have orbits which are very nearly circular, but the meteorites and comets are much more highly elliptical, sweeping across the Solar System towards the Sun, and then streak-ing away again. Many comets, in fact, have such highly elliptical orbits that they disappear from view on the outer fringes of the Solar System for long periods before returning, and some seem to disappear altogether.

There is one component of the Solar System that we have not yet discussed: the asteroids. None of these is visible to the naked eye, but there are tens of thousands of them. Most are clustered in a well defined belt midway between Jupiter and Saturn, but a few move on highly elliptical orbits which bring them sweeping towards the Sun, across the orbits of Mars and Earth. (There is a remote possibility that one day one of these asteroids could collide with the Earth.)

The complete Solar System, then, consists of nine planets and innumerable meteorites, comets and asteroids gyrating round the Sun in regular elliptical orbits. Each of these members of the Solar System has been studied minutely since the invention of the telescope, but our knowledge of them has advanced far more in the last decade than it did in the previous three millennia, thanks to the advent of spacecraft. Before going on to examine how the different members of the Solar System relate to one another, and how the Solar System as a whole originated, therefore, we shall examine what is known at present of each of the main members.

The Moon

The Moon is our nearest neighbour in space, orbiting the Earth at a mere 380,000 km. (240,000 miles) distance. With a diameter of only 3,476 km. (2,172 miles), it is quite a small body by the standards of the Solar System, but its closeness makes it appear to be almost exactly the same size as the Sun, which is 150 *million* km. (93 million miles) distant. But while the Moon is an insignificant scrap of cosmic dust compared with the Sun, it is exceptionally large as a planetary satellite. Both Jupiter and Saturn have bigger moons than ours, but these are very much smaller in proportion to the size of their parent planet than is our own Moon.

It is, of course, relatively easy to observe the Moon from Earth – the smallest pair of binoculars will reveal some of the magnificent craters that pock-mark its surface. But it is surprising how little was learned about the Moon prior to the Apollo landings. In part, this was because even the largest telescopes could not resolve features much less than about 500 metres across, since turbulence in the Earth's atmosphere causes telescopic images to appear to boil and swirl, blotting out completely all fine detail. In part, it was also because astronomers were faced with the philosophical problems of trying to interpret surface features with which they had no direct contact, and of which there are no comparable examples on Earth. To be sure, highly detailed maps of surface features were made, but there was little progress in interpreting *how* these features were formed. Most importantly, the origin of the lunar craters remained in doubt until the

The almost perfect prints of the sole of an astronaut's boot in the Moon's surface show that the lunar soil is very fine-grained and cohesive, like damp beach sand.

quarter   Crescent   New Moon

The moon, dotted with craters formed by meteorites, shows the distinctive dark *maria* (seas) – in fact, the most massive craters of all. The Moon keeps one face turned towards the Earth. Soon after its formation, the Earth's gravity created tidal forces in its rocks which slowed its rotation by friction until it revolved on its axis once for every orbit round the Earth. *Maria* are largely absent from the far side and are thought to be related to the Moon's orientation.

Apollo programme was well under way. Some scientists thought that they were exclusively of volcanic origin, others that they were produced by meteorite impacts.

The Apollo missions led to an enormous increase in our understanding of the Moon, and also of the whole history of the Solar System. If one had to select the single most significant discovery to arise from Apollo, it was probably that of the *age* of the Moon. Highly sophisticated studies of the decay products of radioactive elements in rock samples returned from the Moon have shown that the *youngest* rocks on the Moon are no less than 3,200 *million* years old. This was a staggering discovery. The youngest rocks are volcanic lavas which cover the smooth, grey coloured areas of the Moon – the dark patches or 'seas'. The lighter coloured parts of the Moon are often more ancient, some parts of them being 4,600 million years old.

To keep these ages in perspective, one has to recall that the *oldest* known rocks on Earth are only 3,800 million years old, and that almost all the surface rocks are much younger than this. Advanced life forms did not evolve until 600 million years ago, by which time the Moon had been geologically 'dead' for 2,600 million years. The Apollo missions did record some exceedingly weak 'moonquakes' taking place deep within the Moon, but otherwise it is a totally sterile, inactive mass of rock, with none of the Earth's internal warmth and volcanic activity.

The only events still taking place on the Moon are very occasional meteorite impacts, which gouge out fresh craters on the already heavily cratered surface. It is now known that *all* the craters on the Moon were produced by impacts, some of them involving bodies many kilometres across, and travelling at speeds of tens of kilometres per second. Some of these impacts were so huge that their effects penetrated deep into

the Moon's interior, and caused partial melting of the rocks beneath. The molten rocks flooded out to the surface along cracks and fissures and completely filled the vast circular basins created by the impacts. Thus the great lunar 'seas' were created – not 'seas' as we understand them on Earth, and as early astronomers believed, but ancient 'seas' of molten rock, which froze solid aeons before the Earth's continents and oceans took up their present recognizeable configurations.

What is the Moon made of? For many decades prior to the Apollo missions, scientists knew the Moon was rocky; moreover it was possible to calculate the Moon's mass from the effects it has on the Earth's ocean tides; and from its mass, its density, which turns out to be somewhat less than that of the Earth. While the volcanic lavas filling the lunar 'seas' are very similar to the basalt lavas which are spewed out by terrestrial volcanoes such as Mt. Etna, and while the Moon has a mantle closely similar to the Earth's consisting of silicate minerals such as olivine and pyroxene, it does *not* have a dense, metallic core like the Earth's. The Earth has a density about 1.67 that of the Moon. The striking difference is due largely to the Earth's high-density iron and nickel core.

It has been argued that the Moon formed by fission from the Earth; that it literally split off from the Earth at an early date in the history of the Solar System. It has also been argued that the Moon was 'captured' by the Earth, and that the Earth and Moon formed separately, but in the same region of the Solar System. The differences in their deep structures and in their bulk chemistry are important in trying to determine which of these theories is correct, but so far no firm agreement has been reached.

Mercury and Mariner 10

Because it is so close to the Sun, never reaching further than 70 million km. (43.5 million miles) from it, and is so difficult to observe telescopically, Mercury was one of the least known planets prior to the Space Age. An indication of just how little was known is the fact that for decades its rotation period was believed to be 88 days, based on observations of fuzzy markings dimly perceived on its surface. In 1965, radar beams were first successfully bounced off the surface of the planet, and these echoes were used to establish the genuine rotation period: 59 days.

One reason for supposing the rotation period to have been 88 days was that this is also the period taken by Mercury to go once round the Sun. In other words, it was thought that the Mercurian year was the same as the Mercurian day, and that Mercury always keeps one face turned towards the Sun, just as the Moon keeps one face always turned towards the Earth.

The correct rotation period, 59 days, reveals an even more interesting relationship. Fifty-nine is almost exactly two-thirds of 88, which means that Mercury rotates on its axis *three* times, while

it goes round the Sun *twice*. Thus Mercury does not have one searingly hot hemisphere, and one cold one, as was long supposed, but, as the result of the shape of its orbit and its unique rotation period, it has two 'hot' poles and two 'warm' ones, all of them located on the planet's equator! Temperatures at the hot poles range from over 400°C at Mercurian noon, to a chilly −170°C during the night. This extraordinary range of temperatures must give rise to unusual effects on the surface rocks, but these remain to be explored.

Radar studies provided the only reliable data on Mercury until 1973, when the Mariner 10 mission was launched. This was a magnificently planned and executed mission, which transformed Mercury from a poorly known little planet, swamped by the Sun's glare, into a well-known world. The spacecraft's trajectory was carefully calculated so that it initially swung very close to Venus, and used Venus' gravitational field to sling it into a fresh trajectory, which put it into orbit around the Sun, and ensured that it encountered Mercury on no less than three separate occasions over a period of a year. On

the third it practically grazed the surface at a distance of 327 km. (200 miles). On each of the three encounters, the spacecraft's cameras obtained and transmitted back to Earth a stream of first class pictures, which provided the first ever close up view of the planet.

The pictures revealed a planet strikingly similar to our own Moon, only larger. (Its diameter is 4,880 km. (3,050 miles).) Like the Moon, much of its surface is scarred with great craters and impact basins, the largest of them, known as the Caloris Basin, is no less than 1,300 km. (810 miles) in diameter. The impact which produced this structure was so gigantic that powerful shock waves penetrated every part of the planet's interior, and, on the point at the surface immediately *opposite* the impact site, caused a jumble of ridges and hills to be thrown up, which the Mariner 10 scientists mapped as 'weird' terrain. Weird it certainly is.

Because Mercury's heavily cratered surface is so similar to that of the Moon, it seems certain that Mercury must be as old as the oldest parts of the Moon's surface, namely, more than 4,000 million years old. Whereas, however, the Moon

Mercury, about which information was scanty until Mariner 10 photographed it in 1974–5, is now mapped in detail. In this view – a combination of two panoramas – the planet reveals its startling similarity to the Moon. On Mercury, however, the larger features do not form such obvious *maria*, and there are huge scarps 3.2 km. (2 miles) high and several hundred miles long, perhaps the result of contractions as once liquid rock cooled.

Even in close-up Venus remains an enigma. The clouds recorded by Mariner 10 in 1974 on its joint Mercury–Venus mission separate into streaks as they swirl, but never break to reveal the surface.

experienced large scale volcanic activity until about 3,200 million years ago, there are no great 'seas' of basalt lava on Mercury, and it seems that it may have experienced even less geological history than the Moon. It may have been quite extinct since about 4,000 million years ago.

Mercury is the smallest of the planets, with the possible exception of Pluto, but it is only slightly less dense than the Earth. This indicates that Mercury must have a large, metallic core like the Earth's, and theoretical studies suggest that this core may be very nearly as large as the entire Moon. The rest of Mercury is probably composed of stony, silicate materials closely similar to the Earth's mantle.

Further confirmation of Mercury's internal structure came from an observation whose significance is not immediately obvious. Mariner 10 discovered that Mercury has quite a strong magnetic field around it, whereas the Moon, by contrast, has only the weakest of fields; and the Earth has an extremely powerful field, which is why we are able to use magnetic compasses as direction finders. The origin of the Earth's field is not fully understood, but it does seem clear that it has something to do with movements taking place within its liquid metallic core (see Chapter 5). Mercury's much weaker field may be produced in a similar manner. The Moon, lacking a liquid metallic core, therefore cannot have a magnetic field.

Venus: the Veiled Planet

Venus orbits at a mean distance of 108 million km. (67 million miles) from the Sun; it can come within 40 million km. (25 million miles) of the Earth, far closer than any other planet. It is also the same size as Earth, and its surface is entirely covered with clouds which reflect the

Sun's light. Little wonder, then, that Venus shines so powerfully down on us. As it is so bright and so close, it may seem that Venus should be the easiest planet of all to observe. Certainly one can *see* it easily enough, but all that the most powerful telescope can reveal is the bland, blank surface of the all-enveloping cloud layers. The surface is literally shrouded in mystery.

Since it prevented direct observation of the surface of Venus, the cloud blanket provoked endless speculation about what *might* be hidden beneath. An extraordinary range of ideas evolved, suggesting at one extreme that Venus was dripping wet, covered with swamps and luxuriant vegetation, and at the other that it was a barren desert, perpetually swept by dust storms. The uncertainty about surface conditions led, predictably, to a large crop of science fiction stories about green Venusians, flying saucers, and invasions of the Earth.

So completely do clouds conceal the surface that, as with Mercury, the rotation period of Venus remained entirely unknown until the advent of radar studies. When these were completed in 1962, they revealed a major surprise. Venus not only rotates extremely slowly on its axis, taking 243 days to go around once, but it also rotates *backwards* compared to every other planet in the Solar System! This unique quirk is still not at all understood, but undoubtedly has profound implications for the origin of Venus and its geological evolution.

Prior to the advent of space missions, there was little that could be learned about Venus except from some studies of its atmosphere. By the time the first direct spacecraft investigations took place, these had already begun to indicate that Venus was a rather hostile planet, and that in particular its atmosphere was rich in carbon dioxide, and poor in water and oxygen.

Spacecraft investigations have produced an enormous volume of new data on Venus, and have revealed that surface conditions are much more frightful there than was imagined by even the most extreme of the science fiction writers. No less than 12 Russian spacecraft have been sent to Venus (not all of them successful), two of which landed on the surface and obtained the first pictures. The Americans have also sent many spacecraft to Venus; in 1978 alone two separate missions were launched, one of which released four separate probes to penetrate and investigate the atmosphere.

The results from this armada of spacecraft add up to show that Venus, far from being the peaceful, romantic place conceived of by poets is much more like medieval views of Hell: it is extremely hot, with surface temperatures of nearly 600°C, such that objects shaded from direct light would glow dull red with their own heat. The atmospheric pressure at the surface is crushing – more than 90 times the Earth's. The atmosphere is mostly carbon dioxide, suffocating to terrestrial forms of life. Worst of all, the clouds that appear so bland and innocuous are made up of droplets of sulphuric acid, while some of the lower layers of cloud may contain droplets of liquid sulphur. It is hard to imagine a more hostile environment. There can be no question of life existing on Venus, and it seems that if ever Man leaves his own planet to explore others, Venus may be about the last place he would wish to visit.

Appalling though its surface conditions may be, these facts raise some profoundly important questions for science, and most particularly, why should Venus be so totally different from Earth, its nearest neighbour in space, and one which it closely resembles in size and mass? This is a question which will require much further research to solve. One thing is already clear, however: the extremely high surface temperature is a consequence of the thick carbon dioxide atmosphere. When light and heat from the Sun fall on Venus, most of it is reflected back into space by the clouds, accounting for Venus' spectacular brightness. Some of the Sun's rays penetrate, however, and illuminate and heat the surface. The warming of the surface means that some energy is *re-radiated* at longer wavelengths. And this radiation is blocked or absorbed by the carbon dioxide; it cannot escape to space. Thus Venus' atmosphere acts like a greenhouse, preventing the escape of heat, and as a consequence, the surface is maintained perpetually at its scorching temperature. (Fortunately for us, the Earth has only a small quantity of carbon dioxide in its atmosphere, but here too a greenhouse effect operates; if the amount of carbon dioxide were to be increased by continued burning of fossil fuels, the Earth's surface temperature would be raised by several degrees.)

Apart from their different atmospheres, Earth and Venus are different in other important aspects. Although the Russian landers obtained surface pictures, which revealed a bleak, stony landscape, the only information about the geography of Venus comes from radar maps of the surface, painstakingly built up by Earth-based studies.

These reveal a rather confusing state of affairs. The surface seems to contain some very large impact craters, suggesting that it has been undisturbed for billions of years, like those of Mercury or the Moon, but there may also be some very large volcanoes and rift valleys. If these do indeed exist, it suggests that Venus' internal processes may be similar to Earth's, and that it has a hot, mobile interior, capable of giving rise to all the complexities of plate tectonics and continental drift.

Because it is so similar in size and mass to the Earth, Venus ought also to possess a dense, liquid metallic core. The planet seems, however, to lack a magnetic field completely, arguing against the existence of a mobile core, although some scientists attribute this in turn to Venus' very slow rotation period. Thus Venus remains a thoroughly mysterious planet whose secrets, once unravelled, will explain much that still remains unknown about the origin of the Earth. We can be sure that there will be many more spacecraft missions to this strange, hostile neighbour of ours.

Mars: the Red Planet

Mars has always been the most notorious of the planets. In the days before Copernicus and Kepler the about-turns with which it interrupted its steady drift from west to east were regarded as most peculiar; but even when its orbital motion had been understood and telescopic observations made, it remained an enigmatic planet.

At the heart of the mystery was the fact that Mars, like the Earth, has two well defined polar caps. It experiences a regular procession of four seasons, and there are dusky markings on the surface which appear to show regular variations. Together these facts encouraged belief that there was life on Mars. Specifically, the variations in the dusky markings were interpreted in terms of seasonal changes in vegetation.

Such beliefs received apparently solid support from the observations in the 19th century of the astronomer Giovanni Schiaparelli who mapped the presence of numerous strange linear features on the surface, which he called *canali*, and which he thought were associated with liquid water flowing in channels. He did not, however, suggest that the canals were in any way artificial.

His observations became widely known, however, and in particular they inspired a wealthy American, Percival Lowell, to establish his own observatory at Flagstaff, Arizona, and to spend much of the rest of his life in observing Mars. He became firmly convinced that the *canali* were indeed canals; that they were artificial, and that they were constructed by an intelligent Martian civilization struggling to survive in the face of the progressive desiccation of the planet.

Lowell died in 1916. Long after his death, there was a widespread willingness to believe that life existed on Mars, if not in intelligent form,

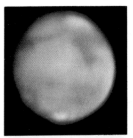

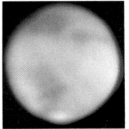

Through the telescope, Mars is an angry, patchy red. The patches – including the icy poles – vary slightly with the passage of the seasons. But it is impossible to resolve details less than 500 km. across with Earth-based telescopes.

To an approaching spacecraft, Mars begins to resolve itself into a planet with a rugged, Moon-like surface of a variety of temperatures, as this false colour view shows.

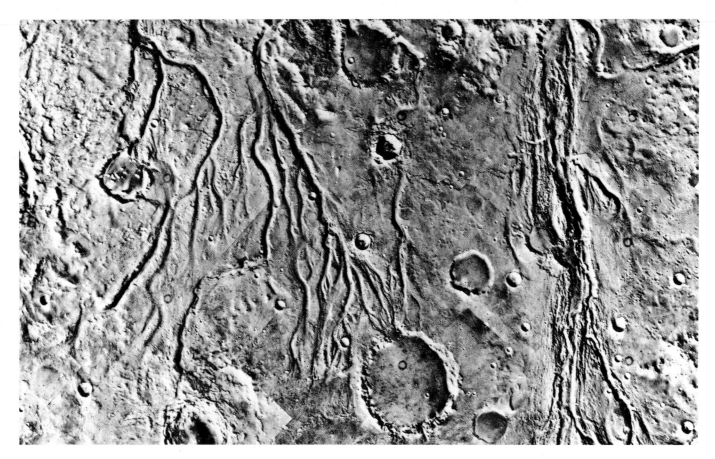

In close-up photographs, Mars proved a geological treasure-trove.

Above: Floodwaters once swept over this region. The river beds – dry for aeons – cross more ancient craters.

Far right: This puzzling material, suitably known simply as 'white rock', lies at the bottom of a crater. Some 14 km. (8.5 miles) across, the rock appears to have flowed or been splattered, but its origin and composition are unknown.

Right: At sunrise in a high plateau region, white clouds lie in canyons. As the heat increases, they probably clear.

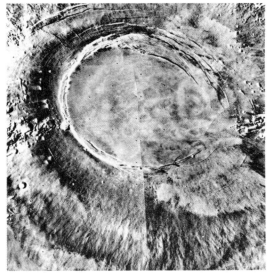

Right: The huge caldera of Arsia Mons is 120 km. (75 miles) across. One side of its walls has collapsed at some time in the past, releasing a flood of lava.

Far right: Possibly the Solar System's largest volcano, Olympus Mons is over 500 km. (310 miles) across at its base and 29 km. (18 miles) high (see following page for a reconstruction).

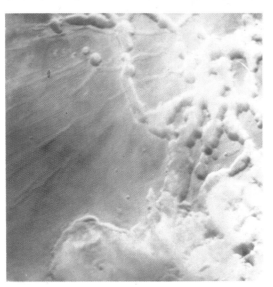

As this reconstruction shows, a Martian river-bed and valley wall is very similar to dried-up desert rivers on Earth.

then at least of a lowly, vegetable kind. As telescopic observations improved, and more scientific techniques became available, it became clear that Mars was, at the very least, a difficult place for life-forms to survive. Early estimates of the atmospheric pressure suggested an atmosphere quite as dense as the Earth's. With the passage of time and more observations, these estimates had to be steadily whittled down, until spacecraft data finally indicated a surface atmospheric pressure of only six thousandths of that of the Earth.

The first successful spacecraft mission to Mars was Mariner 4 in 1964. The 22 pictures it obtained showed a drab, cratered surface, strongly similar to the Moon. This was a great disappointment since many scientists had hoped that Mars would turn out to be an active, dynamic planet, not a senescent or dead one. Mariners 6 and 7 in 1969 obtained many more pictures, which did little to change this rather dull impression.

A startling change took place in 1971, when the Mariner 9 spacecraft was put into orbit around the planet, and photographed almost its entire surface. Overnight, the depressing, static picture of Mars obtained from the rather limited coverage of the earlier missions was swept away. Mariner 9 showed that Mars was indeed an exciting planet. Not only did it reveal the presence of giant volcanoes, the largest in the Solar System, and a great rift valley slashing across half a hemisphere of the planet, but also broad river channels winding their way across the

In reconstruction, Olympus Mons's cone shape – leading up to the 65-km. (40-mile) wide caldera – forms a broad shield, somewhat similar to those of the Hawaiian volcanoes.

surface. The channels shown in the pictures were quite dry, and other observations showed that it is physically impossible for liquid water to exist on the surface of Mars today – it would simply evaporate away immediately. The immediate questions posed were thus all about *age*. When could the rivers have flowed? Could life have evolved when the rivers were active? When were the great volcanoes constructed? Could they still be active now?

None of these questions can yet be answered with certainty, and they will probably remain unanswered until samples can be returned to Earth for dating. The very fact, however, that large parts of Mars are as heavily cratered as the Moon means that these areas must be extremely ancient, probably more than 4,000 million years old. At least some of the channels are comparably ancient, and many of the volcanoes themselves show impact craters, which indicate that, while not as ancient as the oldest parts of the surface, they are still very old compared with terrestrial volcanoes.

Notwithstanding the uncertainties about the ages of the river channels, their discovery gave fresh impetus to the quest for life on Mars. Now, however, it was not a question of looking for large animals, but for tiny organisms, and even for fossil evidence of the previous existence of life. This was the main objective of the Viking missions of 1976. Two identical missions took place, both consisting of two spacecraft, an orbiter and a lander. The orbiters provided a whole series of new pictures of the surface, far surpassing in quality those of Mariner 9, and both landers successfully touched down on the surface and sent back hundreds of pictures and a mass of detail about surface conditions.

The crucial instrument on board the Viking landers were those designed to search for life (for details, see Chapter 6). A sampling arm on each lander was extended from the spacecraft to scoop up samples of surface soil. These were then subjected to rigorous analytical processes in a miniaturized laboratory within the spacecraft. Four separate sets of tests were carried out. In three out of the four sets, the results obtained were far less clear cut than the mission scientists had hoped – some thought that life had been detected, others that the peculiar chemistry of the Martian soil may have been responsible for the results observed.

One instrument, however, called the Gas Chromatograph Mass Spectrometer, yielded very clear results: it found no signs whatever of any organic compounds on Mars. This was a very serious blow for those who wanted to find life on Mars, but it still does not rule out the possibility completely – there may be a few tiny localities on the surface where conditions are sufficiently good for life to have evolved. These localities must be extremely restricted, though, and the life forms very primitive.

Mars, therefore, remains an exceptionally interesting and important planet. It has clearly

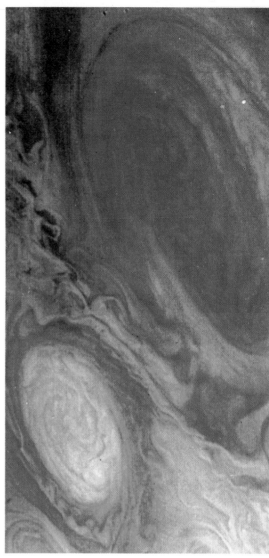

In March 1979 Voyager I revealed startling details of Jupiter, a gaseous giant twice the mass of all the other planets combined, and of its four major inner satellites. *Top:* From 47 million km. (29 million miles) Jupiter's banded atmosphere and two satellites, Ganymede and Europa, are clearly visible. *Far right:* The bands are probably caused by gas columns rising from below and 'smeared' by the planet's rapid rotation. One peculiarity, shown here, is the Red Spot, probably the vortex of a huge cyclone. Though the spot changes in colour and intensity, it has been a permanent feature for at least 300 years. *Above left:* Europa is a bright, icy moon a little smaller than our own and Callisto (*above right*) is dark and strangely mottled.

had a much more complex history than either the Moon or Mercury, and is much more Earth-like than either. There are major differences though. Mars is much smaller than the Earth, with a diameter of only 6,787 km. (4,242 miles) and has a density close to that of the Moon. This indicates that Mars lacks a dense metallic core, and must be made up largely of rocky, silicate materials with perhaps some metal sulphides in its core.

Jupiter: the Greatest Planet

A great empty gap in the Solar System yawns between the orbits of Mars, 228 million km. (142 million miles) from the Sun, and Jupiter at 778 million km. (486 million miles). This important gap is filled only by the asteroid belt, of which more later. The planets on either side of the gap could not be more different. Mars is small, solid, rocky and fairly Earth-like. Jupiter is *enormous*, made up mostly of liquid and solid gases, and is not even remotely Earth-like. Saturn, Uranus, Neptune and Pluto, although much smaller than Jupiter are similar in that they too are low-density planets consisting mostly of frozen gases. Effectively, then, there are two kinds of planet in the Solar System; the inner four which are dense and rocky, and which are often called the terrestrial planets, and the outer five, which are low-density and are composed almost totally of condensed gases.

It is important to grasp just how vast Jupiter is. Its diameter is 143,000 km. (89,000 miles), about one tenth that of the Sun, and ten times greater than that of the Earth. In terms of mass and volume, it is greater than all the rest of the planets put together. Despite its great mass, its density is only a quarter that of the Earth, and this is a clear guide to the nature of its interior. Beneath an atmospheric layer about 1,000 km. (625 miles) thick there is probably a layer of liquid hydrogen, and below that a core of liquid *metallic* hydrogen. At the very centre of the core there may be a small kernel of rocky, silicate material, but even this is not certain. It does seem likely, however, that there is nothing that one could describe as a solid 'surface' to the planet.

The only parts of the planet that are accessible to observation are the extreme outer layers of the atmosphere, which consist mostly of ammonia and hydrogen with smaller amounts of methane and water. Even a modest telescope will reveal a series of dark bands or stripes running across the yellowish disc of the planet; these are interpreted in terms of circulating currents of gases sweeping along parallel with the equator, but in different directions. It is certain that enormously powerful winds blow in the upper atmosphere of Jupiter.

In part, this is due to the extremely rapid rotation of the planet, which revolves on its axis is only nine hours and 50 minutes. This rapid rotation is also responsible for another characteristic feature of Jupiter: it is visibly flattened, so that it looks a little like a squashed orange. The centrifugal force induced by the rapid rotation forces material outwards in the equatorial regions, causing this to bulge relative to the polar regions.

Another of Jupiter's most famous features is also an indirect result of the rapid rotation. The powerful counter-currents of gases set up in the upper atmosphere give rise to long lasting spiralling systems, a little like the cyclones which are well known in tropical regions of the Earth. Jupiter's Great Red Spot, originally interpreted as a great volcano rearing above the surface, is almost certainly one of these circulating atmospheric disturbances, although the reason for its red colour is not fully understood. The Great Red Spot has been continuously visible since it was first discovered by Giovanni Domenico Cassini in 1665; no other single feature has been visible for as long.

At the time of publication, Jupiter has been visited by only four spacecraft, Pioneers 10 and 11, and Voyagers 1 and 2.

Although the Pioneers succeeded in obtaining some superb pictures of the details of Jupiter's atmosphere, and the Voyagers obtained even more – even recording the existence of a thin, diaphonous ring system – much of the interest in the missions lies in the data on Jupiter's moons.

There are no less than 13 satellites, and more may well be discovered by the visiting spacecraft. The interest in them lies in the fact that many of them are large bodies in their own right, and some are more solid and rockier than Jupiter itself. The four largest satellites, Io, Europa, Callisto and Ganymede, are known as the Galilean satellites, since they were discovered by the great Galileo himself, as soon as he turned his first telescope towards the planet. Of these four, Ganymede and Callisto are the biggest, with diameters of over 5,000 km. (3,125 miles). They are therefore bigger than the planet Mercury. There is a considerable range in density amongst the four, ranging from Io (about .65 Earth's), to Callisto (about 0.3 Earth's). Thus it is thought that Io may be made up predominantly of rocky material closely similar to our own Moon, while Callisto and the other low-density satellites are probably composed of ice and rocky materials.

The Galilean satellites, then, are large and interesting bodies in their own right. They have an additional attraction, however, in that since it is impossible to land directly on Jupiter, one of the satellites would make an excellent place to set up an observatory to study Jupiter. It will probably be many decades before this can be accomplished, but the technical difficulties are not very great. It will be a fascinating stage in the exploration of the Solar System.

The Outermost Planets

Saturn is closely similar to Jupiter in many respects. It is rather smaller, with a diameter of 120,000 km. (75,000 miles), is even less dense, and shows an even greater degree of polar

Like its huge neighbour, Saturn is largely gas, and also shows banding. It is best known for its rings, which are now known not to be unique. (Both Uranus and Jupiter have some.) The rings are thin bands – they almost vanish when seen edge on – 275,000 km. (170,000 miles) across in all. Consisting of swarms of rocky particles (see reconstruction on pp. 58–9), the rings are divided into three main sections and two minor ones by their different orbital speeds.

Pluto, hardly bigger than our own Moon, is almost invisible against the starry background. It was the first planet to be found by photography. Suspecting the presence of a ninth planet from perturbations in the motion of Uranus, astronomers at the Lowell Observatory, Arizona, began a systematic search in 1929. A year later, Clyde Tombaugh spotted a minute difference between two plates. A point of light had moved (arrowed), and Pluto had been identified.

flattening. It also has a large retinue of 10 satellites. The great glory of Saturn, though, is its ring system. There can be few of us who have not gazed in wonder at photographs of the wonderfully elegant disc that encircles the planet. Yet there is literally almost nothing to the rings: when seen edge on, they are so thin that they are almost invisible, and cannot be more than a few kilometres thick. Although they look solid, they are very definitely not. In fact, from time to time, stars can be seen through some of the thinner parts of the rings.

It is clear that the rings contain only a tiny fraction of the total mass of Saturn, and that they must consist of myriads of tiny particles orbiting around the planet. It seems most likely that the particles are small chunks of ice, perhaps with small quantities of rocky material, but how and when they came to form such an elegant system of rings remains to be clarified.

Since the nature of Saturn's rings was first described in 1659 by Christian Huygens, it was thought that the magnificent rings were unique in the Solar System. In March 1977, however, the totally unexpected discovery was made that Uranus too has a set of rings. This discovery emerged from a carefully planned set of simultaneous observations from a telescope carried in a high flying aircraft, and a ground observatory at Perth in Western Australia. Uranus' rings are by no means as spectacular as those of Saturn, and can only be detected by refined techniques, but they do confirm what some scientists had been saying for some while: that rings are not an exceptionally unlikely phenomenon, and indeed are predictable results of certain kinds of gravitational interactions between a large planet and objects near it.

Uranus is quite unique in another sense. Whereas all the other planets rotate on their axes in a more or less conventional sense, so that their axes are tilted at only a small angle to their orbital plane. Uranus' axis lies almost in the plane of its orbit, so that it trundles round the Sun rather like a barrel. The Earth's axis is inclined at $23\frac{1}{2}$ degrees from the vertical; this tilt is sufficient to induce our familiar procession of seasons, without which the Earth would be a strange place, not at all as we know it. With an axial tilt of over 90°, Uranus' seasons are bizarre in the extreme. The Arctic regions of the Earth experience months of constant darkness, but large parts of Uranus experience 20 or more *years* of darkness!

Uranus' diameter is 52,000 km. (32,000 miles), much smaller than Saturn, but still a giant planet by comparison with the Earth. Its density is closely similar to that of Jupiter, and it too is thought to be composed of hydrogen and frozen gases, probably with significant amounts of water ice. Because of its great distance from the Sun (2,870 million km. (1,800 million miles)), Uranus is difficult to observe with the naked eye, and was not discovered until 1781, when Sir William Herschel detected its disc telescopically. (Incidentally, he at first dismissed his new planet as a comet). Although telescopes have improved enormously since Herschel's day, it is still difficult to resolve any details on the surface of Uranus. No spacecraft has yet visited it, nor are they likely to for many years – the voyage itself would last for nearly a decade.

Neptune, with a diameter of 50,000 km. (31,250 miles) is closely similar to Uranus, except that it does not seem to possess a set of rings, and its axis is conventionally inclined to the plane of its orbit. Neptune was not discovered until 1846, when two brilliant young mathematicians, one French and one English, independently predicted its existence and position from observations of Uranus' orbit. Uranus did not follow the orbit originally calculated for it: it was clear that it was experiencing gravitational perturbations from an outside body, and it was these perturbations that led the mathematicians eventually to Neptune. Little is known about Neptune, but it is probable that it is similar to Uranus in internal structure. It is slightly more dense, which suggests that it may have a larger kernel of rocky material.

Pluto, last of the planets, was discovered in 1930 after a long and diligent search. This search was initiated by none other than Percival Lowell, who worked as hard on the search for the new planet as he had done on the canals of Mars. Lowell misled himself (and Science) over the Martian canals, but his search for Pluto was careful and meticulous, and a fine piece of scientific work. Sadly, the discovery itself was made at the Lowell observatory 14 years after his death, but it was made by his colleagues, in a continuation of the project that he had initiated. Appropriately, the first two letters of the planet's name, PL, commemorate the man responsible for its discovery.

Pluto lies on the fringes of the Solar System, 5,900 million km. (3,687.5 million miles) from the Sun (on average; it actually has an unusually elliptical path that at one point brings its orbit *within* Neptune's – it will in fact be closer to us than Neptune between 1979 and 1999). Pluto is so distant that it appears as a mere speck of light in even the largest telescopes, and little can be learned about it. Its size, mass and density remain uncertain, although some recent work suggests that it is very small, with a diameter of only 6,000 km. (3,750 miles) and a density of only 0.3 that of Earth's. If these observations are correct, then Pluto can be little more than a frigid chunk of ice, drifting perpetually around the extreme edge of the Solar System.

Meteorites

Although the arrival of a meteorite may be a spectacular event, accompanied by a blazing fireball, the objects which reach the ground have little superficial appeal – most are rather drab chunks of stone, indistinguishable to the untrained eye from any other chunk of stone. To students of the Solar System, however, meteorites have an interest and importance which is out of all proportion to their abundance and dull appearance, because with the exception of the samples returned from the Moon, they provide the only pieces of the Solar System (besides the Earth itself) that can be studied directly.

Although shooting stars are common, it is only rarely that solid fragments survive their headlong passage through the Earth's atmosphere. Only a handful of fist-sized chunks are retrieved each year, and the total number of known meteorites is only a few thousand. Very occasionally, large meteorites arrive out of the blue – a 2,000 kg. (4,400 lb.) meteorite fell near Pueblito de Allende in Mexico in 1969, and provided a bonanza for meteorite students.

Clearly, a large meteorite could cause considerable damage if it fell in a built-up area, but the fact that there are no recorded deaths due to meteorite impacts anywhere in the world indicates just how rare they are. Many extremely large meteorites which fell in prehistoric times are known. Of these the largest is the Hoba meteorite, which fell in South Africa, and had a mass of at least 60 tonnes. Still larger meteorites are known not from their own remains, but from the huge craters that they blasted out on impact. Best preserved of these is Meteor Crater in Arizona, believed to have been excavated some 20,000 years ago.

Of all the known meteorites, those known as 'stones' are much the most common. They are composed dominantly of silicate minerals like

those in the Earth's mantle. Much better known, because they are so much more distinctive are the 'irons'; these constitute only a small proportion of meteorites observed to fall, but a much larger proportion of them are found. They are composed of iron and nickel· alloys, sometimes associated with silicate minerals. It's thought that the Earth's dense core is probably composed of very similar materials; indeed it was the existence of iron nickel meteorites which gave support to this concept of the core before sophisticated geophysical techniques became available.

Drabbest of all meteorites, but of greatest importance, are the so called 'carbonaceous chondrites'. These dark coloured lumps of stone contain, as their name implies, substantial quantities of carbon. They are important because it is thought that they have a composition very close to that of the Sun, minus the Sun's light gases.

All meteorites are very ancient, around 4,600 million years old, the same as the oldest rocks recovered from the Moon, and it is confidently supposed that the age of meteorites is also the age of the entire Solar System. Thus meteorites are very ancient, primitive components of the Solar System. But whereas the stony and iron meteorites

At first sight, it seems the Earth has somehow escaped battering that has cratered all the other inner planets. On the contrary, Earth has received its fair share of impacts in the remote past. The effects, however, have been largely eroded away. The Solar System is now almost depleted of large-scale wandering interplanetary debris, but some rocks have fallen in recent times, as shown by the meteorite that struck Mexico in 1903 (*top*), and the existence of the huge Meteor Crater in Arizona, the result of an impact some 20,000 years ago.

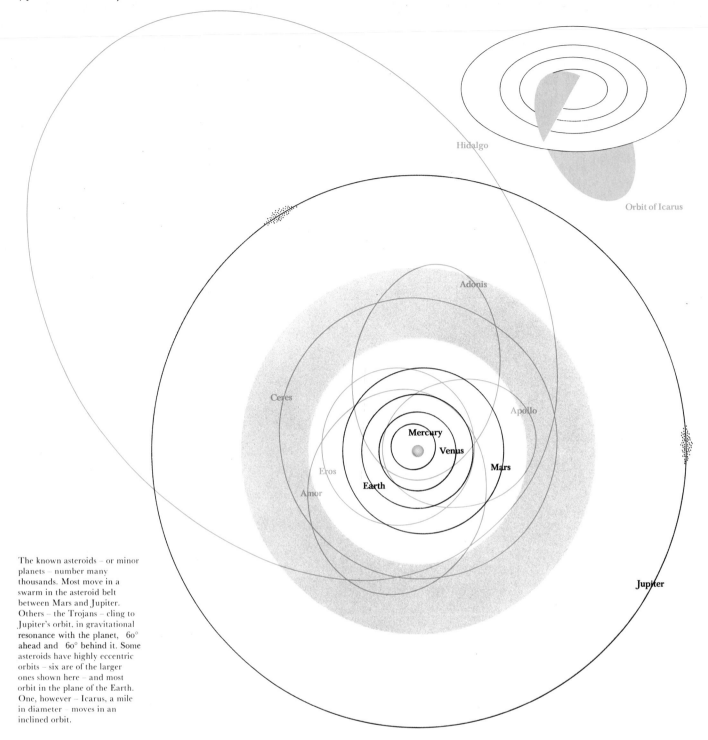

Hidalgo

Orbit of Icarus

Adonis

Ceres

Mercury

Venus

Apollo

Eros

Mars

Amor

Earth

Jupiter

The known asteroids – or minor planets – number many thousands. Most move in a swarm in the asteroid belt between Mars and Jupiter. Others – the Trojans – cling to Jupiter's orbit, in gravitational resonance with the planet, 60° ahead and 60° behind it. Some asteroids have highly eccentric orbits – six are of the larger ones shown here – and most orbit in the plane of the Earth. One, however – Icarus, a mile in diameter – moves in an inclined orbit.

do show evidence of some processes affecting them after the initial formation of the Solar System, the carbonaceous chondrites do not: they are the most primitive material sampled, containing clues to the formation of the Solar System some 5,000 million years ago.

Asteroids

The asteroids are amongst the most poorly known members of the Solar System, which is scarcely surprising in view of the fact that they are perceptible only as tiny points of light in the most powerful telescope. The largest asteroid, Ceres, was also the first to be discovered, on the first day of the 19th century. Ceres is about 1,000 km. (625 miles) in diameter. Since its discovery, smaller asteroids have been found at an ever increasing rate. It is now thought that there may be as many as 70,000 of these objects, most of them less than 100 km. (62.5 miles) in diameter.

The larger ones are more or less conventional, spherical bodies, like mini planets, but most of the rest are irregular chunks of rock.

Finding out what asteroids are made of is clearly a daunting task, given their small size and great distance from the Earth. Over the last few years, however, it has become clear that there are strong similarities between the major classes of meteorites and asteroids. Thus asteroids of stony, iron and carbonaceous chondrite types have all been recognized. This was a great step forward and helps to explain the origin of at least some of the meteorites that reach Earth. Collisions taking place in the crowded traffic lanes of the asteroid belt may cause some fragments resulting from the impact to be flung off in orbits that intersect the Earth's. There are even a few asteroids known to have similar orbits. It may be that the only difference between meteorites and asteroids is one of scale.

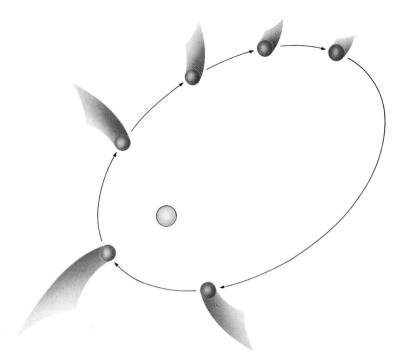

Comets

Although comets which are visible to the naked eye are not common, it is thought that out beyond the orbit of Pluto myriads of comets orbit the Sun in complete anonymity, never coming near enough to be visible. It has even been postulated that there is a great cloud of them, known as the Oort cloud, surrounding the visible parts of the Solar System.

The appearance of comets is well known, and has been portrayed scores of times in different media since the earliest times. One of the best known is the unmistakable shape of Halley's comet on the Bayeux tapestry, made to commemorate the conquest of Britain in 1066.

Superb comets like Halley's consist of a tiny head, and an enormous tail, which may trail over a large part of the night sky. The tail has one very interesting property: it always points directly away from the Sun. This is an important guide to the nature of the tail itself: it is a thoroughly insubstantial cloud of gas stripped off from the head of the comet and streaming out behind it. So insubstantial is the tail that its gas content is similar to that of a hard vacuum on Earth, and on more than one occasion the Earth is believed to have passed right through the tail of a comet, with no untoward effects.

The head of a comet is rather more solid, but still does not amount to much. Many authors on astronomy talk of comets as merely 'dirty snowballs in space'. The heads appear to be loose, fluffy aggregates of ice particles, together with some gravelly silicate material, and other frozen gases like methane. Their composition is, therefore, extremely primitive. Comets resemble carbonaceous chondrite meteorites, except that they have a far higher content of volatile material. Both comets and carbonaceous chondrites may be merely scraps of material left over from the formation of the Solar System.

The origin of the Solar System

In examining the origins of the Solar System we take on one of the largest and most difficult problems in Science. Not only are we dealing with an event that took place over 4,600 million years ago, but are also limited to deductions drawn from telescopic observations, data from a score or so spacecraft, a small quantity of materials from the Moon, and a random selection of meteorites that happen to have reached the Earth.

Any hypothesis for the origin of the Solar System has to be able to account for a number of different factors, and this helps to constrain the possible models:

- the first factor concerns the distribution of mass within the Solar System. The innermost planets are small, dense and rocky, the outer ones large, of low density, and made largely of volatile materials;
- second, and not immediately obvious, is the rate at which different members of the Solar System rotate on their axes. Jupiter,

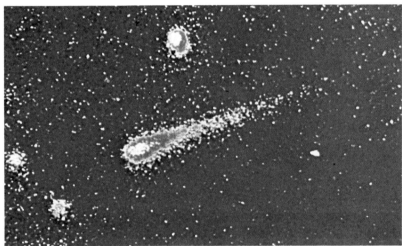

for example, spins much faster than the Sun, which is the opposite of what one might expect from first principles;
- third, is the bulk composition and internal structure of the planets;
- fourth is the actual evidence of early events in the history of the Solar System contained on the surface of the inner planets;
- fifth is the need to explain the existence and nature of the swarms of meteorites, asteroids and comets and their relationship to the major planets and to each other;
- and finally one has to take note of the very subtle chemical clues contained within the most ancient material at our disposal, in meteorites.

Carbonaceous chondrite meteorites such as the one that fell at Pueblito de Allende in 1969 have proved to contain a treasury of data on the extremely early history of the Solar System. The evidence is not obvious, but is locked up on complex ratios of elemental isotopes. Parts of the Allende meteorite were found to be extraordinarily rich in an isotope of magnesium known as ^{26}Mg. Now this isotope can only be formed by the radioactive decay of an isotope of aluminium, an isotope whose half life is so short that it

Comets are so insubstantial that the solar wind sweeps particles from them into a tail that always points away from the Sun (top). Cometary orbits are often even more elongated than those of eccentric asteroids. Some circle the Sun within the orbits of the outer planets, but others sweep in from interstellar space and have never been recorded twice. The comet Kahoutek, for instance (above), which appeared in 1973-4, will not return for 10,000 years.

no longer exists anywhere in the Solar System. Physicists have concluded that the only way that this aluminium isotope could have been produced was in the catastrophic explosion of a nearby supernova, and it is this event which they presume to have provided the trigger to the formation of the Solar System.

The story is even more complex than this, however, because other radiogenic isotopes in meteorites indicate that there may have been another, slightly earlier supernova explosion as well, which seeded the meteorites with quite different isotopic species. Both these stupendous explosions seem to have taken place a couple of hundred million years before the actual formation of the Solar System 4,600 million years ago.

The isotopic data is the last firm evidence that we have to go on. The rest is speculation. The present consensus is that the supernova explosions sent great pressure waves through a diffuse cloud of dust and gas that constituted the primitive solar nebula, the site of the formation of the Solar System. Perhaps there was already a local accumulation of gas beginning to glow under the heat of its own collapse – a proto-Sun. The pressure waves may have initiated the condensation of the cloud of dust and gas into a more confined space, where individual particles could begin to interact with one another gravitationally. The explosions also sprayed unmistakable radiogenic isotopes throughout the nebula.

But what were the constituents of this great cloud of dust and gas? It seems certain that most of the cloud was made up of hydrogen gas, the most abundant element in the universe. The solid particles were probably ice, and frozen gases, with small quantities of rocky and metallic material. They may have resembled tiny, fluffy, rather dirty snowflakes.

With the passage of time, these small particles interacted with one another, and began to clump together to form larger and larger bodies, perhaps several hundred metres in diameter. At this stage, the bodies were homogeneous all through: they were simply large, dirty snowballs in space. The analogy with comets is obvious – the comets we see today may be merely relics of this early stage of the Solar System.

Subsequently, things became more complicated. For one thing, as the proto-planets continued to accrete together, their masses became such that impacts between them began to destroy some of them, breaking them down once again into smaller fragments. For another, the Sun now began to exert its influence, and this soon became the most important single feature in explaining the differences between the planets. Those nearest the Sun accreted under high temperatures; those furthest from the Sun under progressively decreasing temperatures. Thus the proto-planet that was to become Mercury was quite unable to accrete volatile materials such as water, gases and even low melting point metals. Uranus, by contrast, was formed in a very chilly part of the Solar System, and was thus able to incorporate into itself even the most highly volatile elements. There is therefore an obvious increase in the volatile content of the planets outward from the Sun, a progression which is slightly complicated by the fact that Jupiter was so massive that the effects of temperature on it were relatively unimportant.

Thus, the accretion of the planets from discrete grains of dust into solid, individual entities was tightly controlled by temperature. Once the planets had formed, however, a whole series of further changes took place within them.

The process of accretion itself liberated large amounts of energy, which caused the inner planets at least to heat up. The effects of this were compounded by heat liberated by short-lived radioactive isotopes trapped within the planets' interiors. The results of this heating are thought to have caused large scale, perhaps complete melting of the inner planets. The melting itself caused massive segregation processes to take place, with dense iron and nickel components drifting downwards to congregate at the centres of the planets, forming their cores. This process was accompanied by still further heating.

Core formation in the inner planets was probably completed within a few tens or hundreds of millions of years of their initial accretion, but the differentiation and segregation that took place above the cores continued long afterwards. It is these processes which led to the well defined threefold structure of the inner planets: core, mantle and crust.

The Earth's crust contains no record whatever of the accretion events; all evidence has long since been wiped away by continuing geological processes. On the Moon, Mercury, and to a lesser extent Mars, however, we can still see traces of those events. The process of accretion did not stop abruptly 4,600 million years ago. Large bodies continued to plough into the surface, throwing up enormous craters and ring basins on the Moon and Mercury, and they continued to arrive in large numbers until about 4,000 million years ago. After that, the number of bodies arriving seems to have dropped off steeply. The process of planet formation was essentially complete.

So much for the planets. But what of the asteroids and meteorites? As has been emphasised earlier, the carbonaceous chondrites are primitive bodies; little has happened to them. But the stony and iron meteorites have clearly experienced episodes of melting and differentiation. So, by analogy, have the asteroids. It seems that a large number of separate, small planetary bodies formed within the present asteroid belt, large enough for some internal segregation to take place, but not large enough for internal activity to persist more than briefly. Occasional impacts between these segregated bodies may have demolished some of them, so that fragments from both their metallic interiors and the stony exteriors occasionally arrive on Earth, 4,600 million years after their formation.

THE MOON'S ACTIVE PAST

The Moon was a desolate, dead world before the first living cells appeared on Earth some 3,500 million years ago. Yet its remote past is easy to read compared to that of our own planet. The lack of atmosphere and geological activity have preserved surface features that on Earth would have been eroded and buckled out of existence.

The largest lunar features, the so-called *maria* or seas, were blasted out by massive meteorites when the Solar System was young over 4,000 million years ago. The craters – most of which are on the near side, for reasons unknown – were then filled with lava welling up from below. The impacts threw up the great mountain ranges and spread liquid rock into strata that can still be seen. Whatever atmosphere there may have been soon escaped the hold of the Moon's low gravity.

Within 1,000 million years, the Moon had cooled and the larger chunks of interplanetary debris had been blotted up. The lava flowed no more (though signs of its former channels endure) and large impacts became rare. But smaller rocks – size and frequency of impact decreasing with time – have since gouged today's pitted surface of overlapping craters.

In addition, a steady rain of micrometeorites has imposed a kind of erosion unknown on Earth. Over the last 3,000 million years, six feet of rock has been scrubbed off high points and deposited as dust in the lowlands.

A photograph from Apollo 8 shows the distinctive Mare Crisium, or Sea of Crises *(upper left)*. The *maria* – ancient, lava-filled craters – never contained water, but the name remains as a reminder of an early astronomical assumption. To the right, is the lunar far side.

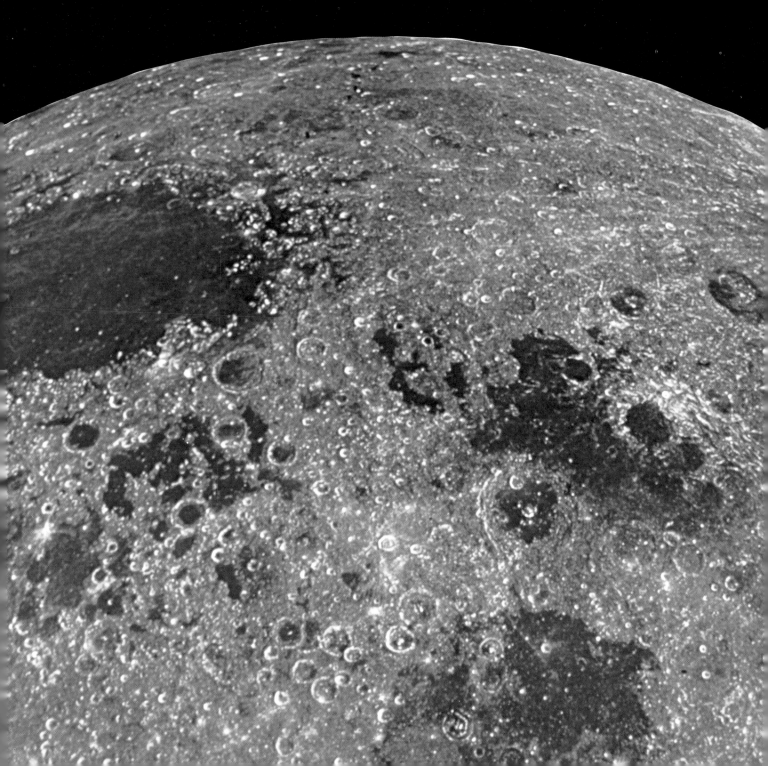

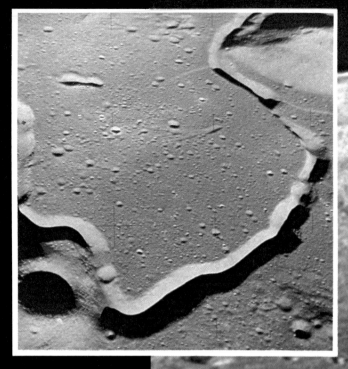

Above: The Hadley Rille is one of the many rilles or clefts that snake across craters and even over mountains. Some are thought to be rift valleys formed in contracting cracking rock. Others may be 'tubes' formed by lava which later collapsed and filled in with dust. Hadley Rille, near which Apollo 15 landed, is 300 metres (1,000 feet) deep. At left is the shadow of Mt. Hadley, with St. George's Crater.

Right: Mt. Hadley shows what looks like lunar stratification across its slope (upper right to lower left). At right is St. George's Crater.

Centre: A far-side view shows how ancient craters have been rounded by micrometeorite erosion, while younger ones retain their sharpness.

Right: Crater Tsiolkovsky, named after the Russian rocket pioneer, is one of the few far-side *Maria*. It is 150 km. (93 miles) across, with the central peak that characterizes many impact craters jutting up like an island from the surrounding sea of lava.

Far right: This tangle of craters on the lunar far side shows how ancient impact craters – like the largest one in this picture – become surrounded, pitted and overlapped by later impacts.

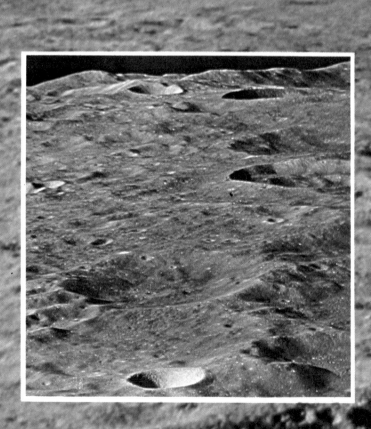

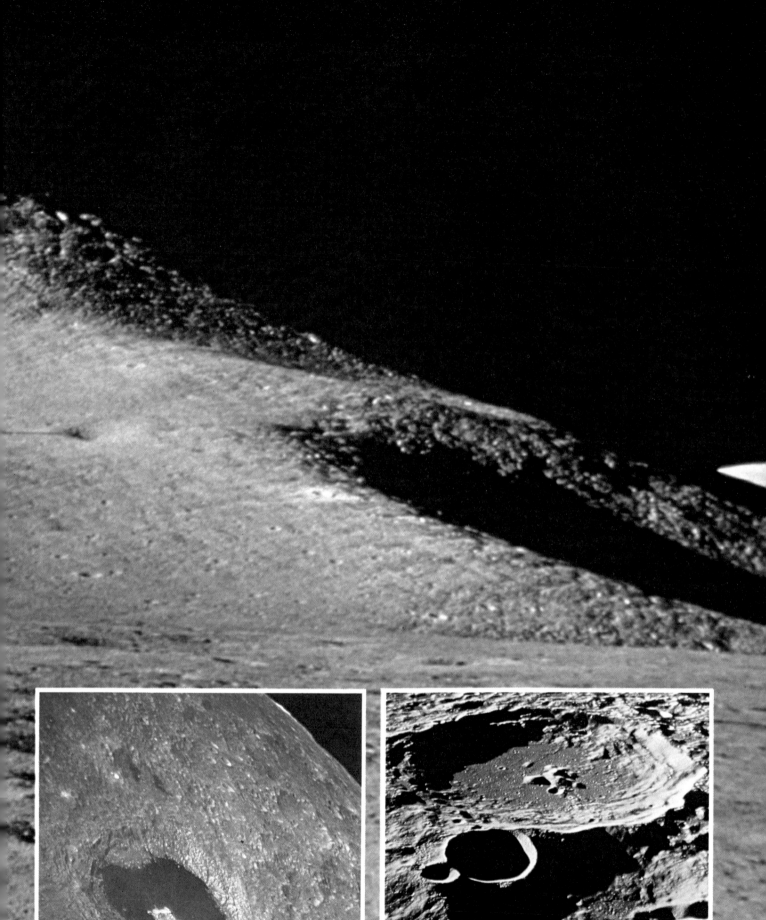
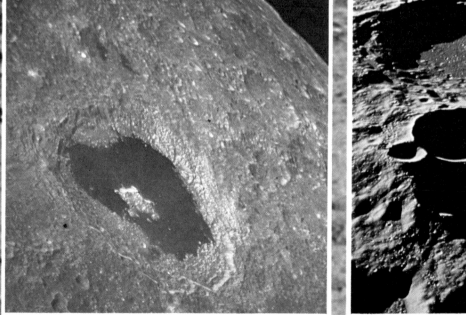

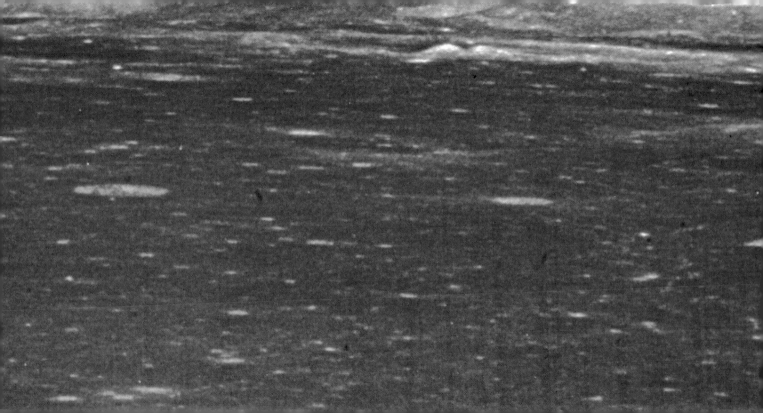

5/PLANET EARTH

PETER OWEN

Our own planet, the third from the Sun, is a peculiar object. It is the only one of the Solar family that lies in the so-called 'zone of life' – the zone within which life-giving water is (generally) neither frozen nor evaporated, but liquid, as even a distant view from the Moon reveals (left). It contains a wealth of life-forms. It is reasonably equable, though the climate does undergo long-term cycles of heat and cold. Yet it possesses a hot interior that keeps its surface features – the continents – in steady, if infinitesimal motion, and provides it with a magnetic field stretching far out into space. In these and in other ways, it is a unique planet.

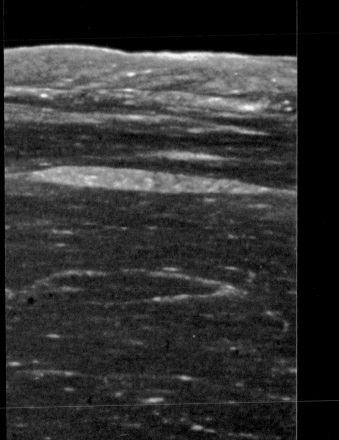

Traditionally, astronomy has been the study of heavenly bodies from the surface of the Earth, and the Earth itself has not usually been considered part of the subject. Of course, the Earth is one of the planets and its annual circuit round the Sun and its daily rotation on its own axis can only be explained in astronomical terms. But beyond this, our planet has been more the concern of geologists, climatologists and biologists than of astronomers.

Since the beginning of the space-age, however, these boundaries have broken down. In recent years, scientists have come to see that there is often no clear distinction to be made between the earth sciences and astronomy. A study of the Earth's unique interior may cast light on the history of the Solar System and thus on the formation of other planets. A study of volcanoes may not only show how our own atmosphere was created, but may also give clues about the composition of the atmosphere of other planets.

We know so much about the Earth that astronomers must draw a line beyond which they can leave research to other specialists. This chapter will look at the Earth from the point of view of an outside observer – an astronomer, say, from an alien civilization, observing Earth for the first time, and seeking to explain those things about it that seem to be unique. (There are many characteristics it shares with other planets of course; these are dealt with in Chapter 4.)

Our alien astronomer would note, for instance, that the Earth is marbled with coloured areas, some of which vary with the seasons; that it is three-quarters ocean; that the continents look like pieces of a jigsaw puzzle; that the poles are ice-covered, but that the ice advances and retreats with the seasons; and that the planet possesses a gaseous blanket capable of sustaining plant life (and possibly, therefore, more complex life forms).

Into the Earth's Interior
The interior of the Earth is totally inaccessible to direct observation. The deepest hole drilled into the surface (an oil well in Oklahoma) penetrates about 10 km. (6.25 miles), but this is only a fraction of the 6,378 km. (3,986 miles) to the centre. Information about the interior must be obtained by, in effect, X-raying it with seismic waves. These are produced by earthquakes and artificial explosions, which make the whole Earth ring like a bell. The waves have different speeds depending on the depth and density of the material they pass through.

The waves are of four types. Two of these, Love waves and Rayleigh waves, only travel near the surface and so tell us nothing about the interior. The other two types, S-waves and P-waves, can travel deep inside the Earth and can be detected at great distances from the earthquakes that produce them. A P-wave is like a sound wave in air and the vibrations are in the direction of motion, whereas they are at right angles for an S-wave. An important consequence of this difference is that S-waves can only travel in solids whereas P-waves can travel in solids and liquids.

Scientists measure the waves with sensitive instruments called seismometers, and from the times at which the waves from a particular earthquake arrive at seismometers in different parts of the Earth the times for them to travel along various paths can be calculated. These times can be many minutes for the longest paths. From the travel times it is possible to calculate the speed of the waves at different depths, and from this the density of the material making up the interior of the Earth.

The most important discovery made in this way was that the density does not vary smoothly with depth but has a number of sudden jumps, showing that the Earth is made of shells having different properties. The top layer of the Earth is called the crust. In 1909 an obscure Croatian seismologist named Andrija Mohorovičić was studying earthquakes in the Balkan Peninsula when he discovered that the speed of P- and S-waves suddenly increased at a depth of a few tens of kilometres. It is now known that this is a worldwide phenomenon, although the depth varies from 30 to 40 km. (19–25 miles) under the continents and about 10 km. (6.25 miles) below the oceans. This depth marks the base of the crust and the top of the layer below, the mantle. The boundary between the crust and the mantle is now called the Mohorovičić discontinuity, or Moho for short.

Mohorovičić's methods can also be applied to study greater depths, although the network of seismic stations must be spread wider than his. In 1906 the British scientist R.D. Oldham showed from an analysis of P-wave data that the Earth has another layer, the core, below the mantle. Later, in 1914, Beno Gutenberg, a German-born geophysicist, showed that S-waves, which travel only in solids, do not penetrate into the core, although P-waves do. This means that the outer layers, at least, of the core must be liquid. Gutenberg calculated the depth of the core-mantle boundary to be 2,900 km. (1,812 miles). In 1936 Inge Lehmann, a Danish seismologist, showed that in the core, only the outer part is liquid. The inner core, from a depth of 4,980 km. (3,112 miles) to the centre, is solid.

The average density of the Earth is 5.518 times that of water, but the seismic evidence shows that the actual density varies with depth. The crust is about three times denser than water on average, while the inner core is about 13 times denser than water. The increase in density with depth is due to the enormous pressure exerted by the weight of the layers above. But the increase is not steady: there is a sudden change at the boundary between the core and the mantle which can only be explained by a similarly sudden change in composition. The outer core is believed to consist mainly of iron, probably with about six per cent of nickel, while the mantle is a mixture of minerals which are mainly compounds of the elements oxygen, silicon, magnesium, calcium

The interior of the Earth, as revealed by the way earthquake waves travel (*right*) and by other, theoretical considerations, consists of four major elements. The inner core, which is solid, and the liquid outer core both consist largely of iron. Movements in the core create the Earth's magnetic field. Outside the core lies the mantle, which consists of solid – but not rigid – rock. Its lower regions are believed to be static, but towards the surface of the Earth, the material of the mantle can flow, very slowly and over long periods of time. Convection currents in this region, the asthenosphere, circulate hot rock from below (red) to a few miles beneath the surface, where it cools (blue) and eventually sinks again. On the mantle's upper region, the lithosphere, is the crust, in which float the lighter continental plates, which are kept in slow but permanent motion by the upwelling rock beneath.

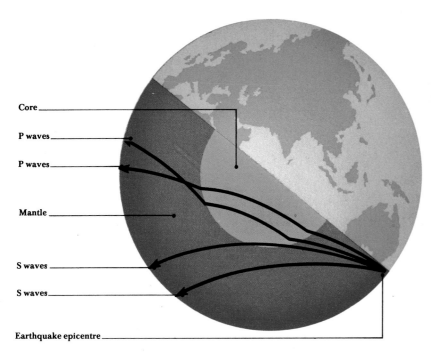

Core

P waves

P waves

Mantle

S waves

S waves

Earthquake epicentre

and iron. Again, the solid inner core is somewhat denser than the liquid outer core, but the difference in density appears to be too great to be just that between a liquid and a solid of the same chemical composition, so there must be some difference in composition also.

The temperature is about 4,000°C at the centre of the Earth and decreases steadily towards the surface. Since heat flows from high to low temperatures there is an outward flow; at the surface it is about 0.06 watts per square metre. Although this is much less than the heat received from the Sun, it is this heat from the interior that supplies most of the energy for volcanoes, earthquakes and mountain building.

It is now generally believed that the Earth formed from small solid particles with a temperature of 200°C or less. Where did the heat come from to raise it to its present temperature? One source is the energy released by the particles that

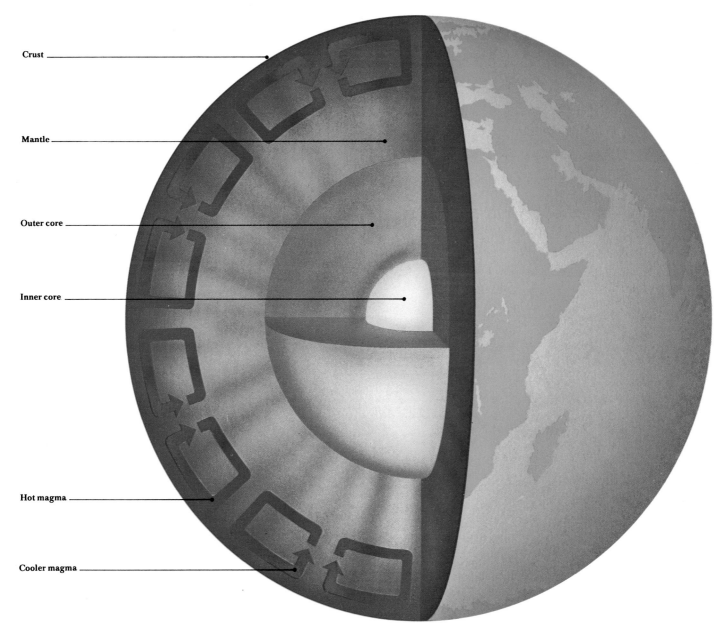

Crust

Mantle

Outer core

Inner core

Hot magma

Cooler magma

collided to form the primitive Earth. Although the total energy released in this way was sufficient to raise the Earth to a temperature of 20,000°C, it was always produced at or near the surface and was probably quickly radiated away. To heat the interior another energy source is required: radioactive decay.

Roughly speaking the radioactive elements responsible for heating the interior of the Earth are of two types: short-lived elements with half-lives of one to ten million years and long-lived elements with half-lives of 1,000 million years or so. These figures should be compared with the age of the Earth, which is 4.6 thousand million years. If there were sufficient quantities of the short-lived elements then the Earth would have quickly heated up. Since all these elements have long since decayed it is not possible to say how important their contribution was. The long-lived elements such as uranium, thorium and the weakly radioactive but abundant potassium are still present. Although the exact amounts inside the Earth are somewhat uncertain there is no question that they can provide the energy necessary to explain the present temperatures.

At some point, the release of energy by radioactive elements must have melted a large part of the Earth since this is the only way known for the separation of the original body of uniform composition into a core and a mantle. A similar process occurs when impure iron is melted in a steelworks and the non-metallic parts separate out to form a low-density slag which floats to the surface.

It was in this way that the primitive crust was formed. The commonest element is oxygen which makes up 46.5% by mass, followed by silicon with 28.9%. Other common elements are aluminium (8.3%), iron (4.8%), calcium (4.1%), potassium (2.4%), sodium (2.3%), magnesium (1.9%) and titanium (0.5%). Everything else together makes up just 0.3%. Only rarely do these elements occur as such; in general they are chemically combined into a vast range of minerals.

The Earth's internal heat has another fundamental significance: it provides power to form the crust – to shift continents, to build mountains, to cause earthquakes and volcanoes.

The Drifting Continents

A glance at any atlas or globe is sufficient to show the apparent fit of the bulge of South America into the bight of Africa. This striking physical feature has caused many laymen and scientists to speculate that the continents were once joined together and have subsequently drifted apart. Ancient myths and legends such as the destruction of Atlantis show that in the past the idea of moving land was quite acceptable. The legends arose because people had experienced volcanic eruptions and earthquakes, and to their minds it was obvious that land moved around, that *terra* was not necessarily *firma*.

The earliest-known written suggestion of continental drift was made by the English philoso-

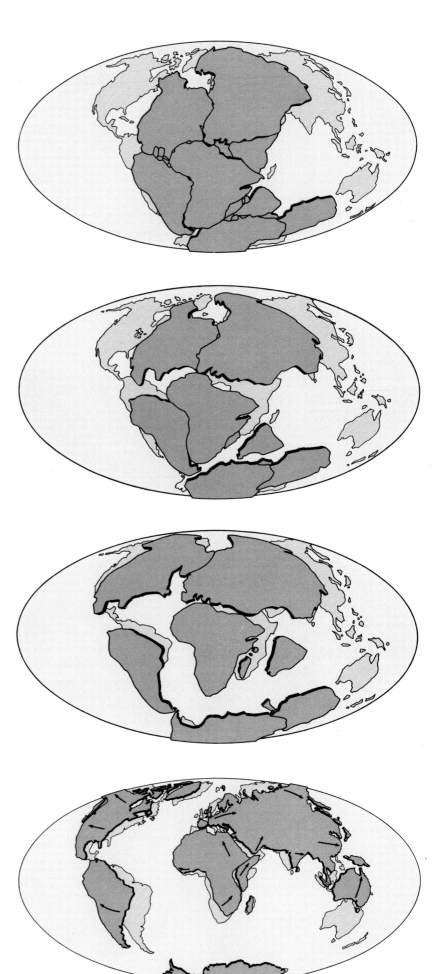

These diagrams show the movements of the continental plates over the last 200 million years, and how they will move in the relatively near future. It is now generally accepted that the plates have been in motion since they were formed over 4,000 million years ago, but the mountain remnants of previous continental collisions have long since been eroded and battered away. A solid geological baseline for 'recent' events is provided by the formation of the super-continent Pangaea ('land everywhere') which was in existence by some 300 million years ago (left).

A hundred and eighty million years ago the northern group of continents, known as Laurasia, had split away from the southern group, known as Gondwana, which itself had begun to break up. India had become an island drifting northwards, but Antarctica and Australia were still joined.

Sixty-five million years ago, North America and Europe were still linked, but the south Atlantic was well formed. Africa had swung to hit the European plate, closing an oceanic inlet to create the Mediterranean, Black and Caspian Seas. Drifting had separated Madagascar from Africa and was beginning to separate Greenland from North America.

Fifty million years hence, the Atlantic will be considerably larger and the Pacific smaller. The San Andreas Fault will have separated, isolating California, and East Africa will have split apart along the Great Rift Valley.

pher Francis Bacon in 1620, but as modern science developed during the 17th century, ancient hearsay such as the legend of Atlantis was deliberately discarded, and with it the ideas of mobile land. For nearly 300 years, the idea of continental drift was rank heresy, though it was supported by a few lone voices such as the German naturalist Alexander von Humboldt around 1800 and the Austrian geologist Eduard Suess a century later. Geology was dominated by the idea that land could rise and fall, but the notion that it could move sideways was regarded as ludicrous.

The man who really established continental drift as a serious scientific proposition was the German geophysicist Alfred Wegener with two articles that he published in 1912. These outlined his wholly original continental drift hypothesis which was by far the most detailed and comprehensive then proposed. He concluded that until 40 million years ago all the continents were joined together in one land mass, which he called Pangaea ('land everywhere'). Pangaea was like a grounded ice floe, which broke up, allowing separate continental blocks to float away on a partially liquefied layer beneath.

Major opposition to continental drift was expressed by the English geophysicist Harold Jefferies in 1924 and his enormous prestige led to initial rejection of the theory. The response may now seem short sighted, but it had scientific validity. Wegener had suggested various forces which would cause the drifting apart of the continents, but most geologists rejected them as too small to produce the required effects. Neither he nor his supporters could come up with a theoretically sound mechanism to explain why the continents move around the globe.

Since then, the climate of opinion has undergone a revolution as profound in its way as the Copernican revolution that set the Sun at the centre of the Solar System. In the 1950's, marine geologists and geophysicists, such as Maurice Ewing and his co-workers at Columbia University, discovered that the crust under the oceans was only a few miles deep, much less than the depth typically found under the continents. They also discovered that a mid-ocean ridge, first discovered in the North Atlantic, was a feature of all the major ocean basins.

In an attempt to explain these phenomena, Professor Harry Hess of Princeton University in 1960 proposed the hypothesis of sea-floor spreading, which led to an exploration of continental drift. He postulated that the heat flow from the centre of the Earth causes convection currents in the mantle and that the crests of the mid-ocean ridges mark the positions of the rising currents. As the currents hit the surface, they spread away from the ridges, and carry the oceanic crust with them. The gap that this leaves is filled with material of the mantle to form new crust.

Hess suggested that along each ridge about 1 cm. of new crust is formed each year. At this rate, it would take a mere 200 million years to form

all the present deep ocean floor. As this time is less than five per cent of the age of the Earth, Hess concluded that old crust is destroyed at the same rate as new crust is generated.

Confirmation of Hess's hypothesis has come from observations of the magnetization of the sea floor rocks, which record the direction of the Earth's magnetic field. Now, the Earth's field is not fixed: it flips directions, for reasons not fully understood. Sea floor observations show that there are stripes roughly parallel to the mid-ocean ridge crests which are alternately magnetized in the same and the opposite direction as the Earth's present magnetic field. As mantle material rises to form new crust, it cools and is magnetized in the same direction as the Earth's field, once its temperature falls below a particular value known as the Curie point. Provided its temperature stays below this value the rock keeps its direction of magnetization, even if the Earth's field reverses. If this is true the pattern of magnetization of the two sides of a mid-ocean ridge crest should be symmetrical, a proposal made by Frederick J. Vine and Drummond Matthews in 1962 and subsequently confirmed by examination of deep-sea sediments.

It follows that if an ocean floor is growing, the continents along its boundaries must be moving apart and so the discovery of sea-floor spreading led to the general acceptance that continental drift does indeed occur.

The theory as now developed is known as the theory of plate tectonics, which differs from earlier theories in seeing the moving units as involving much more than the continental crust. The Earth's mantle has a change in its properties at a depth of about 100 km. (62 miles). The mantle and crust above this depth constitute the lithosphere, and the mantle below it the asthenosphere. The lithosphere is rigid but is broken into seven great slabs (plates) and a number of lesser ones which move about on the asthenosphere below. The movements of the plates are driven by convection currents in the asthenosphere, which, although solid, can creep at a sufficient rate for this to be possible.

As the plates move around they interact with one another along their boundaries in four main ways. One way is for two plates to move apart at a spreading ridge such as the mid-ocean ridge in the Atlantic Ocean. Here, as we have seen, new crustal material is formed and the two plates grow along their common boundary. At other boundaries, two plates collide. If one of them contains oceanic material it is thought that it is forced down below the other in a subduction zone. At the boundary itself there is a trench along the ocean floor. The oceanic plate can be carried down as far as 700 km. (437 miles) before it completely breaks up. The crustal material that is taken down is partially melted and, being less dense than the surrounding mantle, it then rises again towards the surface. Much of it erupts as lava and builds up a chain of volcanic islands, for example the Aleutian, Kurile, Japanese and

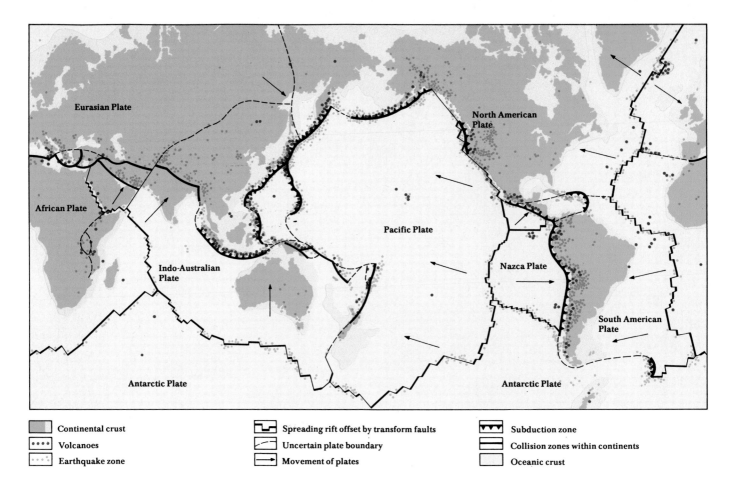

	Continental crust		Spreading rift offset by transform faults		Subduction zone
	Volcanoes		Uncertain plate boundary		Collision zones within continents
	Earthquake zone		Movement of plates		Oceanic crust

The ten major plates that make up the Earth's crust are clearly defined by volcanoes and earthquakes. The plates are in continuous motion, jostled by the motion of the hot, semi-fluid rock beneath. Above upwelling currents, plates move apart, the gaps being filled by hot rock from below. Along other edges – like that marked by the so-called 'ring of fire' circling the north Pacific – the rocks of the lithosphere plunge back into the depths of the Earth, leaving the bulk of the continental rocks unaffected.

Marianas islands. However if the subduction zone directly borders a continent, as along the Pacific coast of South America, the volcanoes form on land, in this case along the Andes mountain range. Friction between the plates at subduction zones causes intense earthquake activity (as in the western parts of South America).

In those areas where two continental plates collide, neither is pushed under the other. Instead, there is a collision zone where the plates crumple and mountain ranges, such as the Himalayas, are formed. Since oceanic crust is being destroyed at a subduction zone the whole ocean is eventually consumed and the zone is converted into a collision zone. Eventually, the Pacific will vanish.

The fourth main type of plate boundary is a transform fault where two plates slide alongside one another with no creation or destruction of plate material. These faults typically occur at intervals along a spreading ridge and provide a series of offsets. A notable example is the San Andreas fault in California. All types of plate boundary are notable for being active zones and almost all the earthquake, volcanic and mountain-building activity of the Earth's crust closely follows them. Almost as a by-product of this activity the continents are carried round as passengers on the plates so that continental drift is a normal part of the Earth's behaviour.

Plates can break and split apart with the formation of a new ocean between them. This has happened in the past; for example South America and Africa split apart when the South Atlantic Ocean began to open about 135 million years ago. The most recent split is thought to have occurred about 40 million years ago when

Australia and Antarctica separated. 200 million years ago the Atlantic, Indian, Arctic and Antarctic Oceans did not exist. The Pacific however is much older and has been closing as the other oceans have opened.

In contrast to the relative youth of much of the ocean floor, most of the continents are much older. A surprisingly large proportion of the continental crust was in existence 2,500 million years ago. Plate movements have since then welded on successively younger extensions.

Predictions of future movements are also possible. In about 10 million years, motion along the San Andreas fault will bring Los Angeles alongside San Francisco. Some geologists have predicted that as Africa rotates anticlockwise, the Red Sea and the Gulf of Aden will widen while the western Mediterranean will swing shut, closing off the Strait of Gibraltar. Finally, the African Rift Valleys will open.

This theory not only neatly explains why the continents are as they are; it also explains why volcanoes occur, which are, as it turns out, vital to the evolution of the Earth in their own right: they are in part responsible for the constitution of the atmosphere.

The Changing Atmosphere
Near the Earth's surface the atmosphere consists of nitrogen (78.1% by volume), oxygen (21.0%), argon (0.9%) and a small amount of carbon dioxide (0.03%). There are also small variable amounts of water vapour and trace quantities of methane, nitrous oxide, carbon monoxide, hydrogen, ozone, helium, neon, krypton and xenon. The pressure falls off with height and at 6 km. (3.75 miles) it has only half its sea-

Eclipses: Shadows of the Earth and Moon

The Earth and the Moon throw long conical shadows into space away from the Sun; when one of these bodies moves into the shadow of the other there is an eclipse. Each shadow has two parts, an *umbra* in which the Sun is totally obscured and a *penumbra* where the obscuration is only partial.

As the Moon goes round the Earth it usually passes above or below the Earth's shadow, but sometimes it goes right through it. There is then an eclipse of the Moon. Such an eclipse starts when the Moon enters the Earth's penumbra. Some of the light from the Sun is cut off so that the Moon becomes darker, but the effect is not generally enough for an observer on the Earth to notice unless he has been forewarned. Later the Moon enters the umbra. There is a partial phase to the eclipse while only part of the Moon is in the umbra but this is followed by a total phase with the Moon completely immersed. No light from the Sun can directly reach the Moon when it is in the umbra but some is scattered by the Earth's atmosphere into the shadow. Different colours are scattered by different amounts and as a result the totally eclipsed Moon is not completely dark but has a distinctive coppery-red hue. The total phase is followed by second partial and penumbral phases as the Moon leaves the Earth's shadow. The total phase can last as long as 1 hour 40 minutes.

Viewed from the Earth, the Moon generally passes above or below the Sun, but it sometimes passes in front and casts its shadow on the Earth. Purely by chance, the Moon and the Sun appear to have very nearly the same size when seen from the Earth and under favourable circumstances the Moon completely covers the Sun and there is a total eclipse. As the Moon moves in front of the Sun its shadow sweeps across the Earth in a narrow path, never more than a few hundred kilometres across.

An observer on the Earth sees a gradually increasing bite taken out of the Sun. As totality is reached, the visible part of the Sun is reduced to a thin crescent. Immediately before totality this crescent is broken by the irregularities of the mountains at the edge of the Moon's visible disc to form what are known as Baily's Beads (after the early 19th-century English astronomer, Francis Baily). The last visible part of the Sun often gives the appearance of a diamond ring. During totality the Sun's atmosphere, the corona, which is normally masked by the glare, becomes visible in all its splendour. After totality the partial phases are repeated in reverse order. At the equator the total phase can last as long as 7 minutes 40 seconds, but this maximum time becomes smaller towards the poles; for example, at latitude 45° it is 6 minutes 30 seconds.

Since their orbits are not completely circular, the distances of the Moon and the Sun from the Earth vary somewhat. Their relative sizes also change. At a total eclipse of the Sun, the Moon only occasionally appears slightly larger. More often, the Sun is visible all round the Moon at mid-eclipse. This is known as an *annular* eclipse.

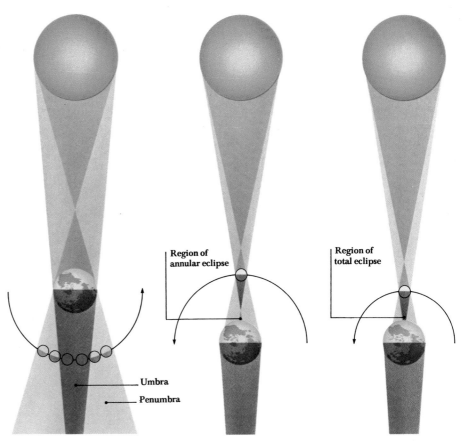

Region of annular eclipse

Region of total eclipse

Umbra

Penumbra

During a total eclipse, as this sequence shot in 1973 shows, the Moon exactly covers the Sun's disc.

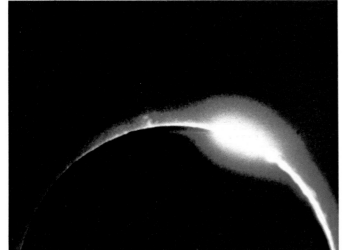

The Sun shining between lunar mountains creates a Baily's Bead.

At totality, the Sun's corona (upper atmosphere) becomes visible.

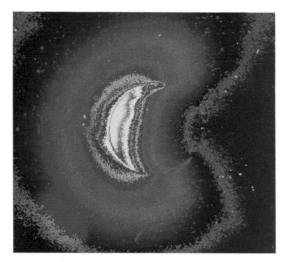

The Earth's atmosphere, which is constructed in four zones and several sub-zones, acts as a blanket that preserves the surface of the Earth from the effects of the most destructive forms of radiation. Atmospheric molecules of hydrogen have been found up to 5,000 miles from Earth, as the ultraviolet picture above, shot from Apollo 16, shows (the different layers of colour reveal temperature differences). But most of the atmosphere, as the diagram at right shows, is held near the surface of the Earth by gravity – 75 per cent is below the height of Everest.

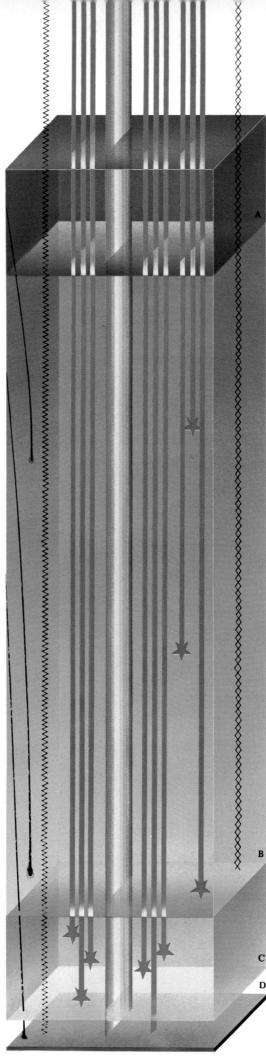

A. The exosphere extends upwards from about 640 km. (400 miles) merging with the interplanetary medium. It consists largely of diffuse hydrogen and helium.

B. The ionosphere, which spans the region from 80 km. (50 miles) to 640 km. (400 miles), is an electrically conducting region. Here the atoms and molecules have had their electrons removed by solar radiation.

C. In the stratosphere, ultraviolet radiation from the Sun converts oxygen into ozone, thus absorbing harmful radiation. This reaction raises the temperature of the stratosphere to about 0°C., acting as a 'lid' on the weather systems of the troposphere below.

D. The troposphere extends from the surface up to about 16 km. (10 miles). This unstable region, subject as it is to the differences in temperature and moisture content of the Earth's surface, is responsible for almost all the Earth's weather. Only radio waves, visible light and a few ultraviolet rays penetrate to this level.

level value. Although the pressure and density drop to very small values at greater heights the atmosphere is still detectable several thousand kilometres above the surface by, for example, the drag it exerts on artificial satellites. There is also a change in composition with height. Below 500 km. (312 miles) oxygen and nitrogen are still the most common elements. Further up helium and then hydrogen are the main constituents.

When we try to discover where the Earth's atmosphere came from and how it has evolved to its present composition we find little direct evidence. It seems unlikely that the Earth captured an atmosphere when it was formed; if it had, then the proportions of heavy gases, such as neon, argon and krypton would have been much greater than they are.

The alternative is that gases were trapped in the rocks and later released. As we have already discussed the Earth must at some time in its past have been almost, if not wholly, melted in order that differentiation into a crust and mantle could occur. The same melting would also cause the release of gases trapped in the rocks to form an atmosphere. This outgassing (as it is called) is also caused by volcanoes and meteoritic impacts. It seems that most of the surface water was released in this way. There is a difference of opinion as to whether carbon dioxide and nitrogen were released in the same way or whether they were produced from methane and ammonia.

Perhaps the most puzzling problem raised by the constituents of the atmosphere is that of oxygen. Oxygen is vital to all forms of animal life; yet it was not part of the primitive atmosphere. Although oxygen is the major constituent of the outer layers of the Earth it was totally locked up in other compounds. Because free oxygen is a very reactive chemical it would have left easily detectable effects, which we do not in fact see. For example, the primary organic molecules necessary for the origin of life would have been destroyed by oxygen, and many old rocks consist of compounds that would have been oxidized if there had been any oxygen present when they were formed. We know therefore, that until about 1,800 million years ago, long after evolution of the first simple life-forms, there was no free oxygen in the atmosphere.

A number of processes (e.g. photosynthesis in green plants) can split up carbon dioxide or water and release oxygen. However, unless some of the products of these reactions are removed they recombine with the oxygen and there is no overall change in atmospheric composition. Only in the case of photosynthesis is it thought that

this occurs. Some of the plant material has been converted into limestone, coal, oil and other minerals leaving the oxygen in the atmosphere. It is a remarkable thought that the Earth was not a ready-made home for creatures like us. It had to be modified by earlier forms of life before it was suitable.

A Magnet in Space
Another peculiarity of the Earth which is also caused by its internal structure is its magnetic field. The Earth's magnetic field is similar to that produced by a bar magnet or a uniformly magnetized sphere, a fact first pointed out by William Gilbert, who was physician to Queen Elizabeth I. It has two magnetic poles which are close to, but not exactly at, the Earth's geographic

The Rhythm of the Tides and Seasons

The tides (right) and the seasons (below) are together the two most obvious consequences of the interaction of the Earth, Moon and Sun.

Although the Moon is so much smaller than the Sun, its proximity to the Earth gives it twice the pull of the Sun on the Earth's waters. The Moon's gravity pulls the oceans into two bulges, one on the side facing the Moon, the other on the opposite side. The Sun's influence is generally ironed out by the greater influence of the Moon. But twice a month, at new and full Moon, the Sun and Moon pull in line and their tidal effects produce tides that are higher than normal (spring tides). When the Moon is at first and last quarter, it pulls at right-angles to the Sun and the tidal effects work against each other producing a small range (neap tides). The Earth rotates beneath the bulges, creating the familiar sequence of high and low tides.

The rhythm of the seasons has nothing to do with any change in the distance from the

Earth to the Sun. Although the Earth's orbit is not a precise circle, the deviation – and thus the change in heat received – is infinitesimal. The true cause of the seasonal change is the tilt of the spinning Earth, about $23\frac{1}{2}°$ away from the perpendicular. Consequently, first one hemisphere and then the other leans towards the Sun.

As the 'summer' hemisphere, for instance, begins to tilt inwards, the Sun, seen from the Earth, climbs higher in the sky. This firstly makes the proportion of daylight hours a greater fraction of the whole 24 hours, reaching an extreme inside the Arctic and Antarctic circles, where the summer Sun never sets. Secondly, it brings the Sun's rays down more nearly vertical at local noon, concentrating their heating effect rather than spreading the warmth over a slanting path.

In the winter, the opposite happens: daylight hours are reduced (the poles lie in permanent darkness for several months), the Sun only rises low above the horizon, and its heat is weakly dissipated over a slanting path.

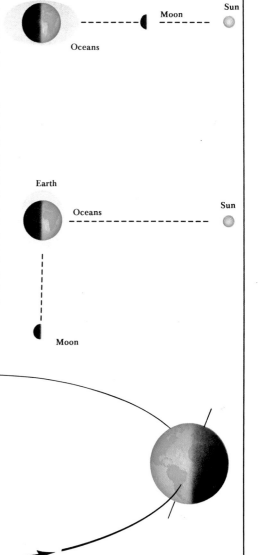

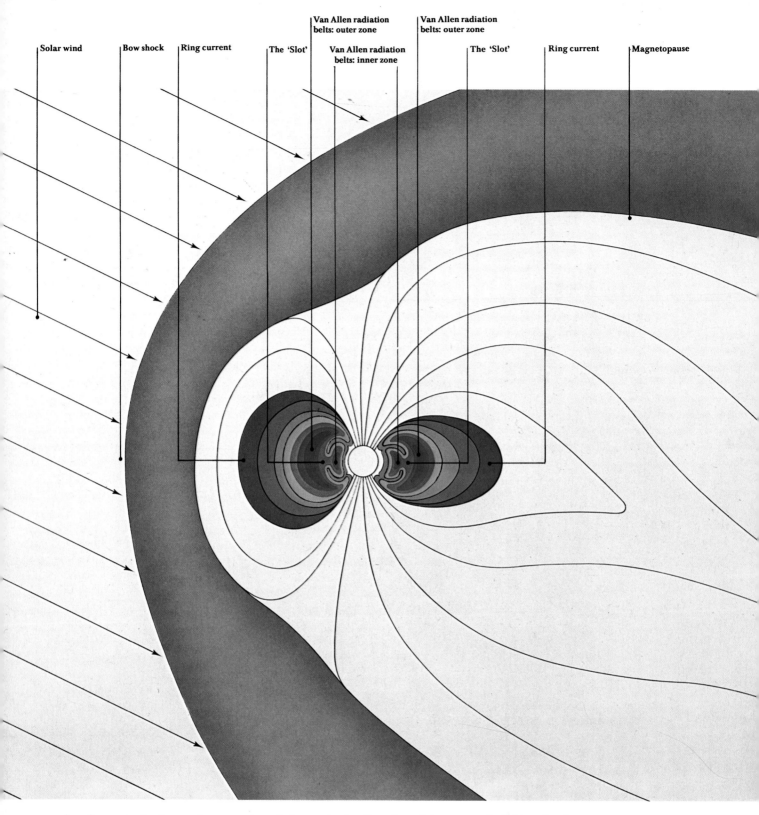

Solar wind | Bow shock | Ring current | The 'Slot' | Van Allen radiation belts: outer zone | Van Allen radiation belts: inner zone | Van Allen radiation belts: outer zone | The 'Slot' | Ring current | Magnetopause

poles; for example the north magnetic pole is at present in Canada 1,300 km. (812 miles) from its geographic counterpart. Scientists have been measuring the strength and direction of the magnetic field for about 400 years, during which time there have also been quite substantial changes in the position of the magnetic poles. The rapidity of these changes shows that their cause cannot be in the solid outer layers of the Earth. This is further confirmed by the irregularities in the field which are quite unconnected with surface features such as continents and oceans.

Some rocks are naturally magnetic (some of them, as we have seen, lie on the ocean floor) and when they were formed they were magnetized in the direction of the magnetic field at the time. This magnetization has been maintained to the present day by many rocks. Paleomagnetism, which is the study of this fossil magnetism, can give scientists information about the magnetic field millions of years ago. One of the most amazing outcomes of such studies was the discovery that about once every 400,000 years the field completely reverses in direction, a phenomenon first observed by the Frenchman Bernard Brunhes in 1906. During the last 71 million years there have been 171 such field reversals, the most recent of which was 700,000 years ago. Why the reversals occur is at present a mystery.

The Earth's influence spreads far beyond the confines of its own surface. Besides its gravitational field, it also possesses a complex magnetic field, the magnetosphere, which begins some 300 miles (500 km.) above the surface overlapping the upper regions of the atmosphere. As in a bar-magnet, lines of force arc out from the north and south magnetic poles, creating a region that can be detected by special photographic film (*left*). But the field is distorted. In its orbit, the Earth is swept by radiation from the 'solar wind', which flattens the leading edge of the field in to about 70,000 km. (40,000 miles) from the Earth, while away from the Sun the field forms a tail hundreds of thousands of miles long. The field deflects most of the solar wind but some particles enter the Earth's atmosphere at the poles causing dramatic electrical displays, aurorae (*right*). Some particularly energetic particles penetrate the magnetic shield and become trapped in two intense zones, the van Allen radiation belts.

As we have seen, the origin of the magnetic field is not to be found near the surface of the Earth. Because of the continuous changes and frequent reversals of the field the obvious place to look for its origin is in the liquid outer core where we can hope for motions that are much more rapid than those that can occur anywhere else.

The first problem is to find out the mechanism that maintains the field. Permanent magnetism is not believed to be possible in a liquid, and even if it were the different parts would soon be mixed up and there would be no general magnetism. The other way of producing a magnetic field is by electric currents. The Earth's core is mainly iron and nickel, so being metallic, it is a good conductor of electricity. However, calculations show that if an electric current were started in the Earth's core it would die away after several tens of thousands of years. This is such a small fraction of the age of the Earth, which is 4,600 million years, that there must be some force driving the currents.

The general view today is that the field is caused by what scientists call a self-exciting dynamo. In a man-made dynamo, when an electrical conductor is driven round in a magnetic field it generates an electric current. But part of the electric current can in its turn be used to generate the magnetic field, which then creates more electricity. (This is what happens in the generators at a power station.) It is possible to connect two dynamos in such a way that, with the same direction of rotation, the electric currents and magnetic fields can flow in either direction and can even spontaneously reverse from time to time. Such a system has a complex structure, which is not true of the Earth's core; but there could be complex motions of fluid in the core that compensate for the simplicity of structure to produce the field as we measure it. Scientists are

now trying to discover if there are any such motions in the core.

Whatever its cause, the Earth's magnetic field extends well above the surface. The Sun continuously emits a stream of charged particles known as the solar wind. As the stream flows past the Earth, the magnetic field – like a big, soft ball in a current of water – is flattened on its leading side and stretched on the far side. The region within which this occurs is called the magnetosphere. The boundary of the magnetosphere, the magnetopause, is about 70,000 km. (43,750 miles) from the Earth on the sunward side, but on the other side the magnetosphere's tail extends beyond the Moon's orbit.

The magnetic field strongly affects the paths of the charged particles, such as cosmic rays, that approach the Earth. Those of low energy, such as most of the protons in the solar wind, cannot enter the Earth's field at all. More energetic particles can enter but more easily near the poles than at the equator. As they enter the upper atmosphere, they produce an electrical display as if in a giant neon light – curtains of colour known as aurorae. These Northern and Southern Lights (Aurora Borealis and Aurora Australis) are common in polar regions, and sometimes seen in temperate zones as well.

The paths of the particles are generally very tortuous but there are certain regions where particles can follow fairly stable orbits. These regions have crescent-shaped cross-sections. Particles are trapped in them and they are called Radiation Belts or Van Allen Belts. Evidence for the existence of these belts came from the earliest Russian and American satellites, and in 1958 the American Dr. J.A. Van Allen and his colleagues provided the first picture of their shape and distribution.

The Rhythm of the Ice Ages

If our alien astronomer observed the Earth for a very long time – long enough to measure the continents moving – he would also notice that the polar ice would sometimes spread over a much larger area, and that glaciers would flow down from high mountains to blanket the surrounding countryside in ice. He could, in other words, notice the recurrence of ice ages, the last of which ended some 10,000 years ago.

The causes of ice ages have not been identified with certainty. But *within* major ice ages the ice ebbs and flows – we are, for instance, at present in an 'interglacial' period within a major ice age, and the ice may return to northern latitudes in another few thousand years. The ebb and flow of ice probably owes a good deal to the peculiarities of the Earth's spin and orbit, peculiarities which affect the amount of heat we receive from the Sun.

This idea appeared in its classic form in the work of Yugoslav astronomer Milutin Milankovich in the 1930's, but became established in the eyes of professional meteorologists only in the 1970's. The theory has been summed up by

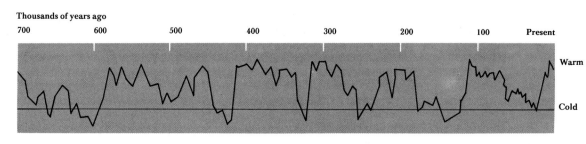

Thousands of years ago

700   600   500   400   300   200   100   Present

Warm

Cold

Though the causes of ice ages remain obscure, one theory currently in favour – the so-called Milankovich model – relates the minor recurrence of ice ages to small irregularities in the Earth's spin and orbit. Every 26,000 years, the Earth wobbles like a spinning top; every 40,000 years, its tilt varies slightly; and about every 100,000 years its orbit becomes more elliptical. These effects combine to reduce the amount of heat the Earth receives from the Sun, triggering a slow but inexorable spread of polar ice.

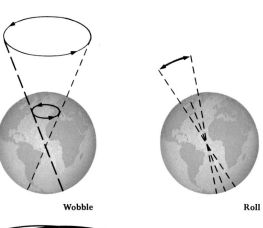

Wobble   **Roll**

Stretch

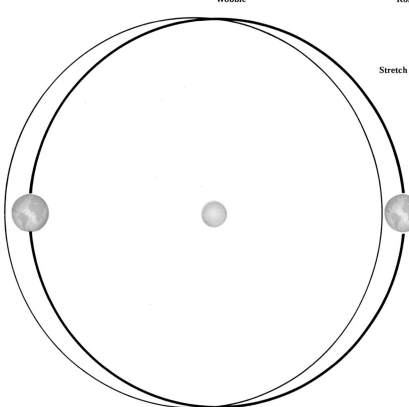

Professor B.J. Mason, the Director-General of the UK Meteorological Office.

The Milankovich Model, as it is still called, explains detailed changes in ice cover *within* a full ice age in terms of changes in the orbit of the Earth, and the orientation of our planet relative to the Sun. Three separate cyclic changes in the Earth's movements combine to produce the overall variations in the solar radiation falling on the Earth which are the key to the theory.

The longest is a cycle of between 90,000 and 100,000 years over which the orbit of the Earth around the Sun 'stretches' from more circular to more elliptical and back; next, there is a cycle of some 40,000 years over which the tilt of the Earth's spin axis – the cause of the seasons – varies as the Earth 'nods' up and down relative to the Sun; finally, the combined pull of the Sun and Moon on the Earth causes our Planet to wobble like a spinning top as it orbits around the Sun, with a rhythm 26,000 years long. These effects combine to produce changes in the amount of heat arriving at different latitudes in different seasons.

Professor Mason has calculated just how much 'extra' heat is needed to melt ice at high latitudes when an interglacial begins, and just how much summer heat must be 'lost' to account for the advance of the ice nearly 100,000 years ago. A variety of pieces of evidence, such as the record of changing sea levels and the scratches left by glaciers in the rocks, tell geologists how much ice was around in each millennium over the past 100,000 years, and the Milankovich Model tells how much heat was arriving from the Sun each season. When Mason put the two sets of figures together, the agreement was impressive.

Between 83,000 and 18,000 years ago, the 'deficiency' of northern summer heat (insolation) added up to a staggering 4.5×10^{25} calories (45 followed by 24 zeros); dividing this vast number by the equally vast weight of the ice sheets which built up in the northern hemisphere at that time, Mason arrived at a more every-day figure; about 1,000 calories for every gram of ice formed. And, when a gram of water vapour is cooled all the way down to a gram of ice, the amount of heat involved in the change is 677 calories. The Milankovich effect is more than enough (but only *just* more than enough) to account for the great freeze-up from 83,000 to 18,000 years ago.

What happens when the ice melts? Much less 'extra' summer insolation is needed, because the ice only has to melt to water and run off into the sea, it doesn't have to be evaporated all the way back into vapour from its frozen state. From 18,000 years ago to the present, 4.2×10^{24} 'extra' calories were available from the changes of the Milankovich cycles; and the amount of heat needed to melt the ice that geologists know was melting over that time turns out to be 3.2×10^{24} calories. Again, the Milankovich effect is just a little bit bigger than the minimum needed to do the job.

The model also gives us a long-range forecast – the present interglacial is just about over and we are heading back into a period of colder northern and southern winters, with a new full ice age looming up a few thousand years ahead.

SPACE AGE VIEWS OF CLOUD AND LAND

Satellite photography has already transformed major Earth sciences. Even without the benefits of work done previously on Earth, views of the Earth from space give a vivid impression of conditions below. Scientists of an alien civilization, reporting on the Third Planet, would immediately conclude they had found a dynamic world whose wealth of vegetation and life-giving water would make it ideal for higher life forms. They might even comment (as Earth scientists often do when confronted by pictures like these) that the new-found world, with its combination of soft, marbled colours and jigsaw puzzle shapes, was one of startling beauty.

Alien geologists would soon guess that continents now separated by hundreds of miles of ocean were at one time joined. Meteorologists recording the variety of cloud forms – from swirling hurricanes to fluffy cumulus – could picture the weather systems, temperatures and wind-speeds of the atmosphere. Biologists would point out that clouds bear water from ocean to land, creating an efficient hydrological cycle. Ground cover, from lush green jungles to arid, brown deserts are easily visible.

By day, no signs of higher life forms can be seen. But an orbiting telescope – or a night time view of the glow of major cities – would soon provide evidence that at least one species on the world beneath had already evolved a technological capacity.

Above a scattering of cumulus and nimbus clouds, Skylab 1, solar panels extended, hangs in orbit.

Above: Moonrise from Earth orbit produces an image distorted by the lens of the Earth's atmosphere.

Above right: At sunset, the upper atmosphere scatters the Sun's blue light more than the red, producing a band of cerulean blue familiar from the surface.

Right: An extraordinary view taken just after sunset reveals turbulence in upper-atmosphere clouds.

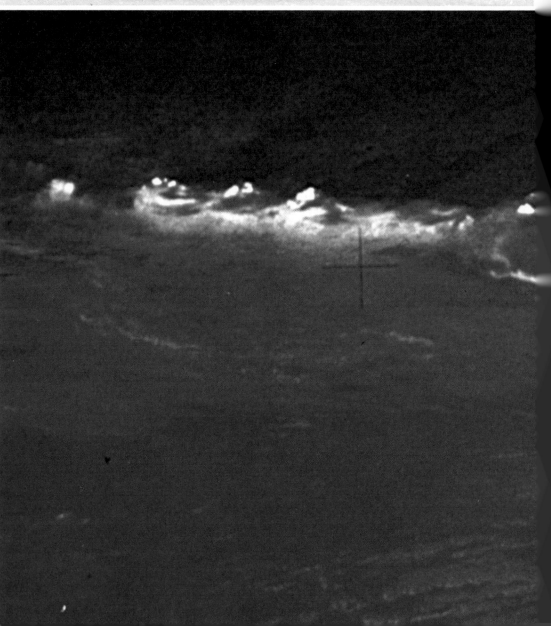

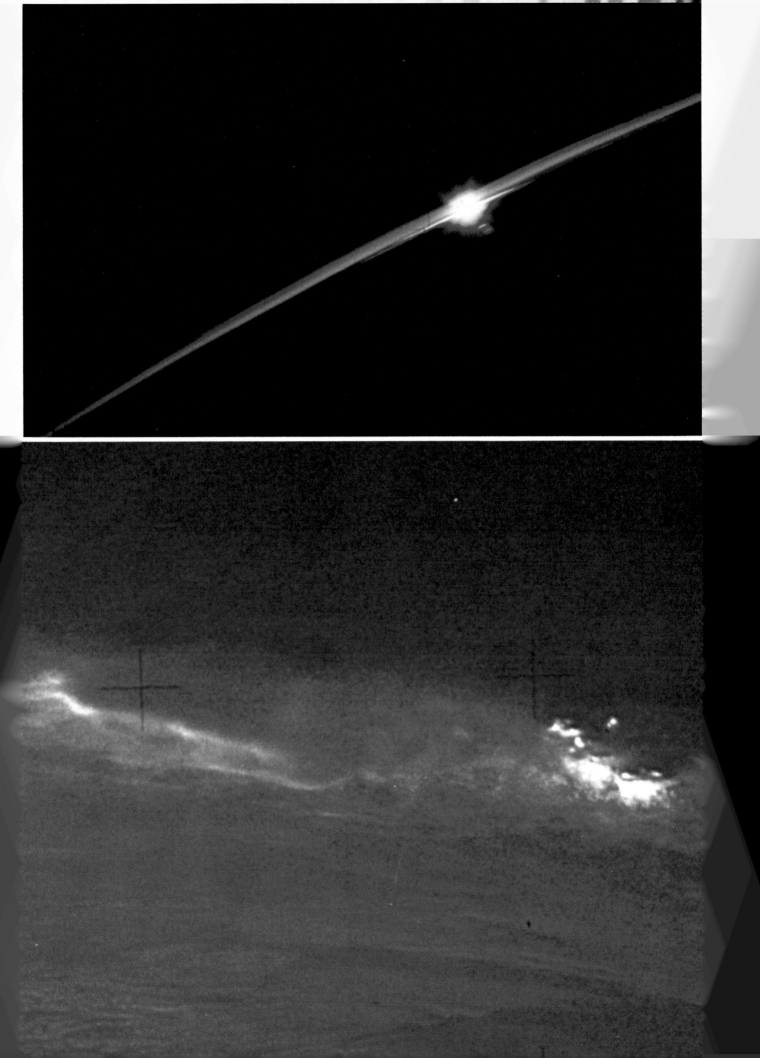

Left: A panorama looking south down the Nile, with the Red Sea on the left, shows the green of well irrigated areas contrasting with the desert aridity to the west (*right*) and south (*top*).

Below: Even at a distance of several thousand miles, the sandy wastes of the Sahara and Saudi Arabia are clearly visible.

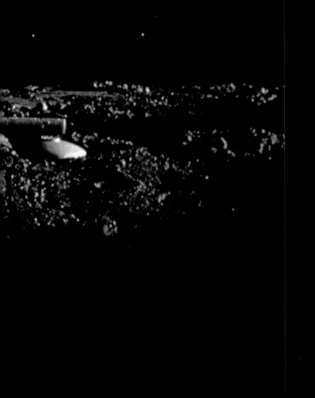

6/THE LIVING VOID

IAN RIDPATH

The possibility that other life-forms may exist in the
Universe – and that we may one day contact them – has
been a staple of science-fiction novels for a century or
more. Now, however, the subject has moved out of the
realms of fiction. By asking questions about the formation
of planets, astronomers hope to learn more about the
nature of life and the chances of its recurrence
elsewhere. By searching the skies for other planets from
Earth and later with telescopes based in space (left),
astronomers hope to answer for the first time the
question: Are we alone? Intuitively, many in the past
have said that we cannot be. Perhaps we shall soon
know. News that the Earth is populated by
technologically-advanced beings has already reached the
nearest stars. If there is anyone listening, we could make
contact with an alien civilization within a generation.

On 8 April, 1960, an American radio astronomer, Dr Frank Drake, made mankind's first deliberate attempt to pick up radio messages coming from another civilization in space when he turned the 26-m. radio telescope of the National Radio Astronomy Observatory at Green Bank, West Virginia, towards the nearby stars Tau Ceti and Epsilon Eridani. These stars, 12 and 11 light years away respectively, are similar to the Sun, and thus likely candidates to have planetary systems that might support life like our own. Drake's interstellar listening attempt was whimsically named Project Ozma, after the princess of the fictional land of Oz, a place 'populated by strange and exotic beings'.

After three months of listening to Tau Ceti and Epsilon Eridani, no alien signals had been heard and the experiment was abandoned. Drake confessed that he was not particularly surprised at the result. Another civilization that close to us would be celestial overpopulation: if there were alien civilizations *that* close together, we would certainly be aware of the fact, for messages would be cluttering the interstellar wave-bands.

Project Ozma ended, but was not forgotten. It caused astronomers to consider seriously the science-fiction notion of alien intelligence and – if it exists – the problems of establishing contact. As a result many astronomers have come to accept the once-controversial possibility that life is widespread throughout space. This belief rests not upon speculations about UFO's or alleged alien visitations in times past, a subject which most scientists frankly reject, but comes from an assessment of the various factors involved in the origin and evolution of life.

The problems of making contact are rather different, and may be depressing to anyone hoping for immediate enlightenment from advanced civilizations. If there are intelligent beings circling the nearest star to the Sun, it will take eight years to say 'Hello' and receive a reply. For radio messages to be exchanged across the Galaxy would take 200,000 years. Nevertheless the possibility of life elsewhere now seems so high and the implications so startling that most astronomers think it worthwhile to search for clues to its existence, even if the civilizations concerned may have vanished long before their radio signals arrive on Earth, and even if communication (assuming the senders are still around) must be made on a rather extended time-scale.

Life: A Universal Phenomenon?

One strong reason for believing in the widespread existence of life is the abundance of organic chemicals, both on Earth and, more surprisingly, in space.

Stanley Miller, an American biochemist, in 1953 conducted a classic experiment on the origin of life on Earth when he assembled a mixture of hydrogen, methane, ammonia and water vapour – gases that are believed to have surrounded the early Earth. This atmosphere has since been replaced by a new one made of the gases released from volcanoes and plants (but the giant planet Jupiter still retains a primitive atmosphere to this day). Theorizing that the primitive atmosphere could have been made to synthesize more complex chemicals and that a suitable power-source was readily available in the form of lightning, Miller passed an electric spark (to simulate lightning) through his artificial atmosphere. At the end of the experiment he had produced a wide range of complex organic molecules, including a number of amino-acids – molecules that are vital to life on Earth, for they form chains that make up proteins, the structural material of living things.

Since then, scientists have found that amino-acids are formed from constituents of the Earth's early atmosphere under a wide range of conditions, such as when shock-waves and ultra-violet light are used as the energy sources, artificial equivalents of thunderclaps and radiation from the Sun. With so many ways to make organic molecules, the seas of the primitive Earth must have been in part a complex chemical soup. From this soup, life on Earth arose. Remains of simple organisms called blue-green algae are known from rocks over 3,000 million years old. For well over half its existence, our Earth has been a host to life. Presumably, a similar pattern would hold for any Earth-like planet.

But life may not be simply the product of Earth-type environments. Space itself is a chemical crucible. Radio astronomers have found that in the dense clouds of gas where stars are forming there exist a number of complex molecules, many of which contain carbon. These molecules give off characteristic long-wavelength radiation which radio astronomers can detect. So far, about 40 molecules have been detected in interstellar space, including ammonia, water, formaldehyde,

In this stony meteorite, which fell near Murchison, Australia, researchers found amino acids, the building blocks of protein and essential ingredients of terrestrial life. This is direct evidence that complex organic compounds were synthesized in space, perhaps even before the formation of the planets. This startling finding has suggested the theory that the basic chemistry of life could be established in space and then 'sown' on the Earth (and, presumably, other suitable planets in the Galaxy) as part of the normal development of solar systems.

A Solar System Full of Life

It seems to many astronomers today unlikely that Earth has a monopoly on life. The idea is not original. Here, the French 18th-century philosopher Fontenelle argues that life is common in the Universe:

'We find that all the Planets are of the same nature, all obscure Bodies, which receive no light but from the Sun, and then send it to one another; their Motions are the same, so that hitherto they are alike; and yet if we are to believe that these vast Bodies are not inhabited, I think they were made but to little Purpose; why should Nature be so partial, as to except only the Earth? But let who will say the contrary, I must believe the Planets are peopled as well as the Earth....

'I cannot help thinking it would be strange that the Earth should be so well peopled, and the other Planets not inhabited at all: For do you believe we discover (as I may say) all the Inhabitants of the Earth? There are as many kinds of invisible, as visible Creatures; . . . there are an infinity of lesser Animals, which would be imperceptible without the aid of Glasses. We see with Magnifying Glasses that the least Drop of Rain Water, Vinegar, and all other Liquids, are full of little Fishes, or Serpents, which we could never have suspected there . . . Do but consider this small Leaf; why, it is a great World, inhabited by little invisible Worms, of a vast extent. What Mountains, what Abysses are there in it?

The insects on one side, know no more of their fellow creatures or the other, than you and I can tell what they are now doing at the Antipodes; does it not stand more to reason then, that a great Planet should be inhabited . . . imagine those Animals which are yet undiscovered, and add them and those which are but lately discovered, to those we have always seen, you will find the Earth swarms with Inhabitants, and that Nature has so liberally furnished it with Animals, that she is not at all concerned for our seeing above one half of them . . . Why should Nature which is fruitful to an Excess here, be so very barren in the rest of the Planets, as to produce no living thing in 'em?'

---

hydrogen cyanide, and others whose tongue-twisting names testify to their chemical complexity. Among these are formic acid and methylamine, which can react together to give the amino-acid glycine. Other complex molecules may await discovery in space, perhaps including amino-acids themselves.

Already amino-acids and other molecules have been found in meteorites, lumps of rock which crash to Earth like the one which landed at Murchison, Australia, in 1969. Biochemists who analysed the meteorites, using sensitive equipment originally designed for analysing Moon rock, determined that the organic chemicals had been formed in space, and were not the result of terrestrial contamination. Most recently, Professors Sir Fred Hoyle and Chandra Wickramasinghe of University College, Cardiff, have proposed that life actually originated in the dust and gas clouds of interstellar space, and was seeded onto the Earth by comets. If so, then life must be both abundant and built along similar chemical lines throughout the Universe.

This is a highly controversial theory: other scientists believe that conditions for the formation of organic molecules would have been more favourable in the rich atmosphere of the early Earth than in the gas clouds of space. But wherever life is formed, either on Earth or out between the stars, there is clearly no shortage of the right raw materials in the Galaxy.

In theory, any star could be the Sun of a life-bearing planet. The key to life is liquid water. Each star is surrounded by a so-called life zone, a kind of interplanetary green belt in which conditions are suitable for life. The Earth, for instance, lies at the heart of the Sun's green belt. Temperatures on Earth are not so hot that water boils, as has happened on Venus, nor so cold that water freezes, as on Mars.

But in fact stars like our own Sun seem particularly favourable as centres for advanced life, for a combination of reasons: they have a sufficient energy output to warm any surrounding planets, and they are long lived enough for any life on those planets to become highly evolved.

Stars smaller and fainter than the Sun live far longer, but they have much more restricted life zones around them. A red dwarf such as Barnard's star lives an astounding 100 times longer than the predicted 10,000 million years of the Sun. Yet it glows so feebly that only the very closest planets to it would be warm enough for life. Larger stars than the Sun are much hotter and have correspondingly larger life zones, but they emit searing amounts of dangerous short-wavelength radiation, and they burn out much more quickly than the Sun. For instance, the bright star Sirius is over twice as massive as the Sun, and has approximately one-tenth the predicted lifetime, which means that it can live for no more than 1,000 million years. Our Sun, which is an average star, seems to have a suitable compromise between energy output and lifetime, and so astronomers have concentrated their search for intelligent extraterrestrial life on stars of similar type to the Sun.

What are the chances of intelligent life emerging from the raw materials, the chemical soups of dust-clouds and planets? And what are the chances of the emergence, successively, of advanced civilizations and of contact between them and us? Frank Drake has originated a rough-and-ready method to help answer these questions. His approach has been discussed at various scientific meetings, most prominently in 1971 when Soviet and American specialists met for an international scientific summit on extraterrestrial life at the Byurakan Astrophysical Observatory in Soviet Armenia.

Drake devised an equation that multiplies seven factors basic to the existence of intelligent extraterrestrial life:
  - the average rate of star formation in the Galaxy;
  - the fraction of stars with planets;
  - the number of planets suitable for life;
  - the percentage of planets on which life actually does arise;
  - the likelihood of intelligent life;

– the desire of that life-form to communicate;
– and, finally, the average longevity of civilizations, like our own, which satisfy all the above requirements.

The average rate of star formation over the history of the Galaxy can be easily estimated. There are at least 100,000 million stars in the Galaxy, and the Galaxy is approximately 10,000 million years old. Simple division gives an average birth-rate of 10 stars per year.

After this, however, the figures become mere guesswork. Astronomers have estimated that at least one star in ten may have planets, but the number of those planets that will be suitable for life is even less certain. In our own solar system,

only one planet is ideal for life (though Mars is nearly favourable). It seems a fair guess that other solar systems have an average of one habitable planet.

We can only guess that life is likely to arise wherever conditions are right, and that in many cases it will develop into an intelligent, technological civilization, given sufficient time. But the number of other civilizations around at present depends on how long each survives and this is the factor that has caused most disagreement among scientists. If each new group of beings in space rapidly wipes itself out, there will be very few around for us to talk to. On the other hand, if each civilization never dies out, then the Galaxy

A Nearby Sun, A Nearby Planet

No planets of stars other than our Sun are visible to Earth-based telescopes. Yet one star – Barnard's Star, a faint red dwarf a mere 5.9 light years away in the constellation of Ophiuchus – is known to possess at least one planet. Life is at least a possibility elsewhere.

Barnard's Star, discovered by Edwin Barnard in 1916, has the fastest 'proper motion' of any star – it is moving across laterally at 55 miles a second and approaching us at 67 miles a second.

Astronomers realized the star offered a unique opportunity to test for the existence of other planets. Any large companion would be detectable from the slight irregularity imparted to the star's path as both bodies revolved around their common centre, the barycentre (far right). The rapid motion of Barnard's Star would show up any irregularities more rapidly.

The task was formidable. One man, Peter van de Kamp, of Sproul Observatory, Pennsylvania, has devoted over 20 years to it. In 1956, after 18 years of studying thousands of photographs, he identified an irregularity with a cycle of 25 years, then spent another seven years verifying the results. Seen from the Earth, the irregularity is minute – the equivalent of detecting a one-inch movement at 100 miles.

Van de Kamp concluded one planet a little larger than Jupiter or two smaller ones could be responsible for the irregularity. If one, it would have to be in a highly elliptical orbit. It seemed more likely that any planets would have nearly circular orbits, and van de Kamp suggests Barnard's Star has two planets, in orbits 4.7 and 2.8 times the Earth–Sun distance respectively and with periods of 26 and 12 years.

His theory is summarized in the diagram at right, which shows the straight-line path of the system's barycentre. Barnard's Star, its deviations exaggerated, is shown at 25-year intervals, when the planets might be expected to line up.

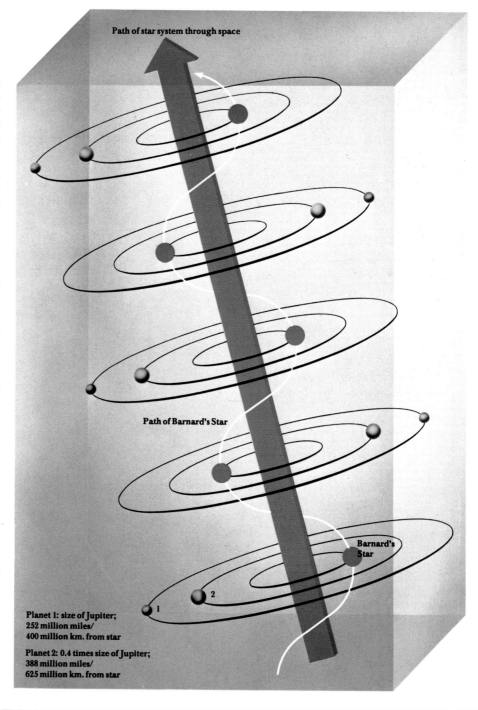

Path of star system through space

Path of Barnard's Star

Barnard's Star

Planet 1: size of Jupiter; 252 million miles/ 400 million km. from star

Planet 2: 0.4 times size of Jupiter; 388 million miles/ 625 million km. from star

should be literally crawling with aliens. Apparently, it is not.

The scientists at the Byurakan conference estimated that the average lifetime of galactic civilizations is 10 million years (though this seems to many a very optimistic figure, given the speed with which we have developed enough nuclear hardware to blast ourselves out of existence). Inserting this figure into the Drake equation along with plausible values for the other factors gives us an estimate of the frequency of other civilizations in space. And it turns out that there could be a total of one million civilizations like ourselves or more advanced throughout the Galaxy. This means that there could be one civilization between every 100,000 stars, and that the nearest neighbours with whom we may communicate by radio are several hundred light years away. In other words, the chances would be that we have no near neighbours.

Some scientists naturally think this figure of one million civilizations could be too low, while others regard it as a gross overestimate. To get a better answer to the problem, scientists are seeking to refine their knowledge of the various factors in the Drake equation. Chief among these factors is the frequency of planetary systems around stars. Earlier this century, planetary systems were thought to be rare. At that time, the standard theory said that the planets of our own solar system were formed from a filament of gas pulled out of the Sun by a passing star, and such a close encounter between stars would be extremely unlikely.

But astronomers now believe that the births of stars and planets are intimately linked. According to modern theories, stars are surrounded by a ring of material left over after their formation. In many cases this material may form a second star, but in others it will give rise to a system of planets. Therefore, if a star is not a member of a binary system we should expect it to have planets.

Exactly what proportion of stars is double or multiple? Two astronomers at Kitt Peak Observatory, Arizona – Helmut Abt and Saul Levy – have examined this question. They studied 123 stars like the Sun in our stellar neighbourhood, finding that two-thirds of them were double, while the rest had no detectable stellar companions. Abt and Levy predicted that these stars also had companions that were too faint to be seen.

A true star needs a mass of at least seven per cent the mass of the Sun, otherwise conditions in its interior do not become extreme enough to spark nuclear reactions. Objects which lie between this limiting mass and the mass of Jupiter are called degenerate stars; they are nevertheless too large to be regarded as planets.

Some of the stars studied by Abt and Levy will have degenerate stellar companions. But, according to the two astronomers' calculations approximately 20% of the stars they sampled should be accompanied by genuine planets. If this figure is correct, and if it applies to all stars as well as the small sample actually studied, there must be a vast number of potential homes for life in space. The chances of our having neighbours from whom we could expect a reply within a century increase accordingly.

We cannot see planets around other stars with existing telescopes on Earth because they are too faint. But while no direct evidence exists of other planetary systems in space, there are a number of indirect methods by which planets might be detected, one of which has already produced encouraging results. This method involves measuring the slight position changes of a star as it moves through space (its *proper motion*). If the

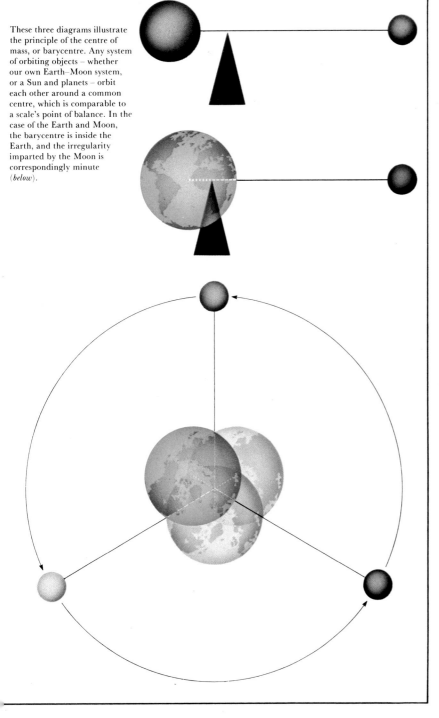

These three diagrams illustrate the principle of the centre of mass, or barycentre. Any system of orbiting objects – whether our own Earth–Moon system, or a Sun and planets – orbit each other around a common centre, which is comparable to a scale's point of balance. In the case of the Earth and Moon, the barycentre is inside the Earth, and the irregularity imparted by the Moon is correspondingly minute (*below*).

star is single, its proper motion should be in a straight line. But if it has a companion, the gravitational pull of the companion will cause the star to wobble, the amount of the wobble depending on the companion's mass and distance.

At Sproul Observatory in Pennsylvania, Professor Peter van de Kamp has been photographing the motion of Barnard's star since 1937. Barnard's star, a red dwarf six light years away, is the second closest star to the Sun. Van de Kamp has announced that the star shows a slight wobble in its motion which indicates it is accompanied by two planets similar in size to Jupiter and Saturn, orbiting the star every 12 and 26 years respectively. Smaller planets like Earth might also exist, but would produce too small an effect to be detected. Astronomers at other observatories are photographing Barnard's star and other stars in an effort to confirm van de Kamp's results.

Another method of detecting planets is to measure the slight Doppler shift in a star's light as it swings around in orbit with its companions. Very sensitive measurements are required to detect this motion, and are now being undertaken by astronomers at Kitt Peak observatory. Results are not expected for several years yet. Further in the future, it might be possible to see planets directly by special optical systems attached to telescopes in space.

The factors in the Drake equation governing the evolution of intelligence and technology are difficult to assess because we have only the example of ourselves to go on. Anthropologists can trace the dawn of man to three or four million years ago on the plains of Africa. Then, several species of ape-man lived side by side, but only one species survived to become true man. *Homo sapiens* as we know him today has existed for perhaps 40,000 years, and there have been no physical changes in mankind since that time.

We are very recent arrivals in the history of the Earth, the product of a long and complex evolutionary chain. It seems inconceivable that the same evolutionary path could be duplicated exactly anywhere else, so how can we hope to find anyone in space remotely resembling our-

The Canals of Mars

One of the oddest controversies in the history of astronomy is that concerning the canals of Mars. The idea was first popularized by the Italian astronomer, Giovanni Schiaparelli, after observations in 1877. Schiaparelli suggested that the *canali* were systems of rivers that distributed melt-water from polar ice. His ideas were extended by Percival Lowell at Flagstaff, Arizona, who claimed the canals must be artificial, an idea that inspired countless science fiction stories until the Viking missions revealed Mars's dusty, rock-strewn and cratered surface. Schiaparelli and Lowell's work, which reflects the difficulties of observing accurately through the Earth's turbulent atmosphere, was purely wishful thinking. The vague, shadowy patches were joined in the mind's eye, as beads seen from a distance seem to form single lines.

It is startling now to see how highly developed their theories were and how little supported by the reality of the Martian surface. As Schiaparelli wrote:

'The polar snows of Mars prove in an incontrovertable manner, that this planet, like the Earth, is surrounded by an atmosphere capable of transporting vapor from one place to another. These snows are in fact precipations of vapor, condensed by the cold, and carried with it successively. How carried with it, if not by atmospheric movement? . . .

'All the vast extent of the continents is furrowed upon every side by a network of numerous lines or fine stripes of a more or less pronounced dark colour, whose aspect is very variable. These traverse the planet for long distances in regular lines, that do not at all resemble the winding courses of our streams.

Some of the shorter ones do not reach 500 kilometers (300 miles), others on the other hand extend for many thousands, occupying a quarter or sometimes even a third of a circumference of the planet . . .

'As far as we have been able to observe them hitherto, they are certainly fixed configurations upon the planet. The Nilosyrtis [a canal Schiaparelli estimated at 200–300 km. (120–180 miles) wide] has been seen in that place for nearly one hundred years, and some of the others for at least thirty years. Their length and arrangement are constant, or vary only between very narrow limits. Each of them always begins and ends between the same regions. But their appearance and their degree of visibility vary greatly, for all of them, from one opposition to another, and even from one week to another . . .

'Every canal (for now we shall so call them), opens at its ends either into a sea, or into a lake, or into another canal, or else into the intersection of several other canals. None of them have yet been seen cut off in the middle of the continent, remaining without beginning or without end. This fact is of the highest importance.

'The canals may intersect among themselves at all possible angles, but by preference they converge towards the small spots to which we have given the name of lakes . . . That the lines called canals are truly great furrows or depressions in the surface of the planet, destined for the passage of the liquid mass, and constituting for it a true hydrographic system, is demonstrated by the phenomena which are observed during the melting of the northern snows . . . Such a state of things does not cease, until the snow,

reduced to its minimum area, ceases to melt. Then the breadth of the canals diminishes, the temporary sea disappears, and the yellow region again returns to its former area . . . We conclude, therefore, that the canals are such in fact, and not only in name. The network formed by these was probably determined in its origin in the geological state of the planet, and has come to be slowly elaborated in the course of centuries. It is not necessary to suppose them the work of intelligent beings.'

Upon which theory Lowell built his own:

'To review, now, the chain of reasoning by which we have been led to regard it probable that upon the surface of Mars we see the effects of local intelligence. We find, in the first place, that the broad physical conditions of the planet are not antagonistic to some form of life; secondly, that there is an apparent dearth of water upon the planet's surface, and therefore, if beings of sufficient intelligence inhabited it, they would have to resort to irrigation to support life; thirdly, that there turns out to be a network of markings covering the disk precisely counterparting what a system of irrigation would look like; and, lastly, that there is a set of spots placed where we should expect to find the lands thus artificially fertilized, and behaving as such constructed oases should . . .

The fundamental fact in the matter is the dearth of water. If we keep this in mind, we shall see that many of the objections that spontaneously arise answer themselves. The supposed Herculean task of constructing such canals disappears at once; for, if the canals be dug for irrigation purposes, it is evident that what we see, and call by ellipsis the canal, is

selves? Fortunately, evolution has the knack of reaching the same result by several different pathways. For instance, several different types of creature have independently invented flight: insects, mammals, reptiles, and even fish. Therefore there seem to be good reasons to suppose that intelligent, technological beings similar to ourselves have arisen on many planets around solar-type stars throughout the Galaxy.

Of course, local conditions will impose certain differences. On a high-gravity planet, for example, creatures (if they had muscles and bones like ours) are likely to be squat and heavy; whereas on a low-gravity planet the beings might be tall and slender, with large noses to breathe the thin air. On planets with vast expanses of flat land, some creatures may have developed wheels rather than legs. If the planet is cold, they could be hairy, and possibly white like polar bears. Some aliens may even look like centaurs, with four legs and two arms.

Such speculations go beyond the strict numerical predictions of the Drake equation. But what that equation does show is that the chances of other life existing in space are sufficiently good for it to be worth our while making the effort to look.

The Search for Life Nearby

Beginning on our own doorstep, are there any signs of life in the solar system?

Mercury, the closest planet to the Sun, is an airless and waterless body of extreme temperatures. Life as we know it could not possibly exist there. Venus is not much better. It has a dense atmosphere, but this traps the Sun's heat to produce roasting temperatures, and the atmosphere at the surface bears down with 90 times the pressure of the atmosphere on Earth. Suitable conditions for life may exist higher in the atmosphere, but it's a slim hope.

We know that the Moon is lifeless, having sampled its rocks, but Mars has always looked a likely home of life. At the turn of the century the American astronomer Percival Lowell drew a whole network of so-called canals crossing the planet's surface. He believed that the canals

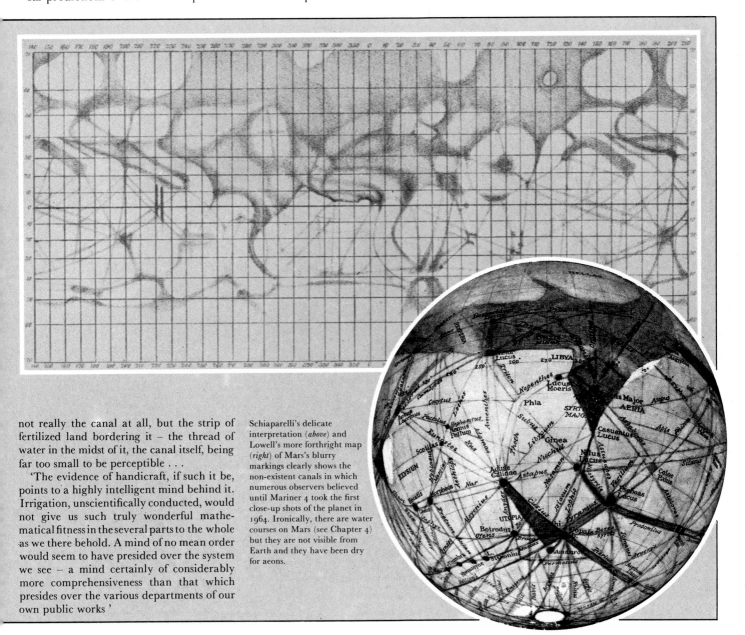

not really the canal at all, but the strip of fertilized land bordering it – the thread of water in the midst of it, the canal itself, being far too small to be perceptible . . .

'The evidence of handicraft, if such it be, points to a highly intelligent mind behind it. Irrigation, unscientifically conducted, would not give us such truly wonderful mathematical fitness in the several parts to the whole as we there behold. A mind of no mean order would seem to have presided over the system we see – a mind certainly of considerably more comprehensiveness than that which presides over the various departments of our own public works '

Schiaparelli's delicate interpretation (*above*) and Lowell's more forthright map (*right*) of Mars's blurry markings clearly shows the non-existent canals in which numerous observers believed until Mariner 4 took the first close-up shots of the planet in 1964. Ironically, there are water courses on Mars (see Chapter 4) but they are not visible from Earth and they have been dry for aeons.

were dug by Martians to bring melt water from the planet's polar caps to their crops near the equator, and he wrote books describing his vision of an inhabited Mars.

Other astronomers failed to see the canals he so confidently drew, but that didn't prevent the idea of an inhabited Mars catching on in the public imagination and in science fiction. While Lowell's ideas of Martian civilization were officially frowned upon, many astronomers were at least prepared to concede that the dark areas visible on Mars could be areas of vegetation, such as mosses or lichens, and this view was held into the 1960's.

Then, space probes began to change our view of Mars. Mariner 4 in 1965 showed that Mars is more hostile than had previously been supposed, with thin air and a crater-strewn surface. Of large expanses of vegetation there were no signs.

This gloomy picture of Mars persisted until 1971, when Mariner 9 made a complete photographic map of Mars from orbit and ignited new hope among the life-seekers. Most important among the features spied by Mariner 9 were what appeared to be dried-up river channels, as though water had once flowed freely across the surface of Mars, possibly in times of more clement climate. Giant volcanoes were discovered which might in the past have erupted sufficient gases

to form a dense atmosphere, along with liquid water. Therefore some kind of micro-organisms might cling to existence in the soil of Mars. The only way to find out was to go down and look.

In 1976, two Viking spacecraft were sent to search for life on Mars. Each of the identical craft carried twin cameras for photographing the surface, and each was equipped with an on-board biological laboratory to analyse samples of soil. A long mechanical arm reached out to scoop up soil for analysis.

Before the Vikings landed on Mars, optimistic biologists had speculated that organisms large enough to be visible to the cameras might exist on the surface. These organisms, the speculation ran, would obtain their water from the rocks or from thin nightly frosts, and would be protected from the Sun's ultra-violet radiation by parasols made of silica.

Both Vikings were targeted to lowland areas on Mars. One probe landed in a basin called Chryse where water was believed to have flowed during the wet times on Mars, and the second made landfall in an area called Utopia, which is covered with frosts in winter. Colour photographs from the Viking landers showed that the surface of Mars is a red, stony desert. Dust blown into the thin atmosphere makes the sky pink. Despite careful scrutiny, not a single cactus nor the

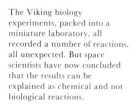

The Viking biology experiments, packed into a miniature laboratory, all recorded a number of reactions, all unexpected. But space scientists have now concluded that the results can be explained as chemical and not biological reactions.

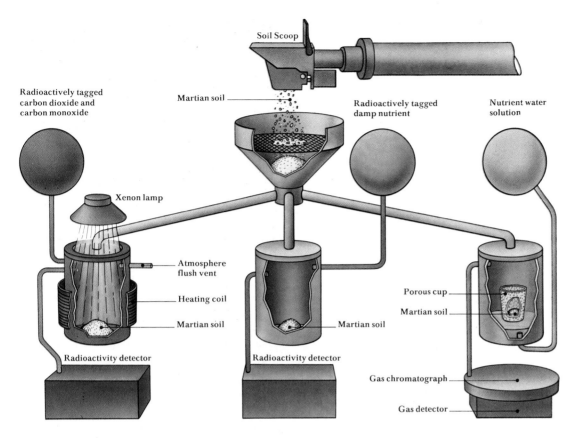

Soil Scoop

Radioactively tagged carbon dioxide and carbon monoxide

Martian soil

Radioactively tagged damp nutrient

Nutrient water solution

Xenon lamp

Atmosphere flush vent

Heating coil

Martian soil

Radioactivity detector

Martian soil

Radioactivity detector

Porous cup

Martian soil

Gas chromatograph

Gas detector

The pyrolytic release experiment was designed to detect photosynthesis, which would reveal the existence of plants, algae or bacteria. Radio-active carbon was introduced to a sample of soil. This was incubated and heated and the gases given off measured to see whether any carbon had been assimilated.

The labelled release experiment assumed that any Martian micro-organisms would require water. A number of nutrients – eg. formate and lactate – labelled with radio-active carbon were introduced with water to the sample. Any micro-organisms present would ingest the nutrients and release the radio-active carbon.

The gas exchange experiment also assumed that Martian bio-chemical reactions were based on water. It was designed to measure an exchange of chemicals between any organisms and the atmosphere. The sample was placed in a porous cup and a nutrient-rich water solution, introduced first round the cup and then directly into it. Gases given off could then be measured by the chromatograph.

smallest patch of vegetation was visible. Not even a single cockroach or a sandfly hopped through the pictures. In the two locations where the Vikings landed, Mars looks depressingly dead.

Some idea of the hostile conditions on Mars was given by the Viking meteorology instruments. Even on a warm summer's afternoon, the air at the Chryse site reached −29°C, and at dawn had fallen to −85°C. At neither Viking site does the ground become warmer than 0°C at any time of the year. Combined with an atmospheric pressure similar to that at a height of 32 km. (20 miles) above Earth, Mars is indeed a harsh place.

Better news for the life-seekers came from the instrument that analysed the Martian atmosphere. In addition to carbon dioxide it detected a few per cent of nitrogen, the existence of which was previously unknown on Mars. Therefore the three main ingredients for life have been found on Mars: carbon, nitrogen and water. Viking's instruments also confirmed that the atmosphere of Mars was denser in the past. So biologists could hope that life had once arisen on Mars and still clung to existence in the form of micro-organisms in the soil.

Viking carried three biological experiments to find out. Together, they could detect traces of life in soil many times more barren than Earth's most barren deserts. Each experiment incubated the soil in a different way in an attempt to make Martian micro-life grow. A number of samples were scooped up by the craft's mechanical arm and tipped into the biology laboratory for analysis.

First, let's look at the 'gas-exchange experiment'. This fed a sample of Martian soil with a rich nutrient broth, and analysed the gases given off. When the experiment was run at each Viking site, puffs of carbon dioxide and oxygen were released. Experimenters concluded, though, that chemical processes in the soil were responsible, not Martian bugs. (Water vapour in the nutrient solution displaced the carbon dioxide from the dry Martian soil, and chemical reactions with oxides in the soil released the oxygen.)

The other experiments were not so easy to interpret. In the 'labelled-release experiment', a nutrient containing radioactive carbon was injected into the samples. If organisms existed in the soil, they might be expected to give themselves away by releasing gas containing the radioactively labelled carbon. And indeed, sudden surges of radioactive gas were recorded when this experiment was run at both Viking sites. Radioactive gases were not emitted from soil samples that had been sterilized at 160°C, while certain amounts of gas were emitted by soil that had been heated to 50°C.

Results like this in terrestrial soils would be a clear indication of life, but Mars is a very different world. Any life there experiences continual sub-zero temperatures, and heating to 50°C would be expected to have a more adverse effect than actually observed. Once again, it could not be ruled out that chemical reactions with the highly oxidized soil were responsible for the results of the labelled-release experiment.

Third of the biology tests was the 'pyrolitic-release experiment'. This worked in the opposite way to the labelled-release experiment, by seeing if carbon was taken up by the soil sample. Each sample was incubated in a test chamber with an

A probe seen from Callisto, Jupiter's fifth moon, heads into a low orbit round the gaseous giant planet to assess the molecular structure of the upper atmosphere, which – some scientists believe – may be a possible habitat for life.

artificial atmosphere containing carbon dioxide and carbon monoxide labelled with radioactive carbon. After incubation, the sample was heated (pyrolysed) to drive off any carbon that had been assimilated, which was then measured.

Several soil samples were tested at each Viking landing site, with positive results. Something in the soil was combining with radioactively labelled carbon from the atmosphere. But the uptake of carbon was unaffected by heat treating the soil at $90°C$, and even sterilization to $175°C$ did not completely eliminate the reaction. The experimenters concluded that biological processes could not be responsible, and that soil chemistry must be the answer.

To interpret the biology experiments we must also take into account the results from a related instrument, known, somewhat forbiddingly, as the 'gas chromatograph mass spectrometer' (GCMS). This device searched for organic compounds by heating soil samples and analysing the gases driven off. It would reveal the existence of living things even if they had failed to respond to the biology experiments. To most people's surprise, there was no sign of any organic compounds in the soil of Mars, not even those molecules detected in meteorites on Earth.

This negative result is strong evidence against the possibility of life on Mars. Although the GCMS was not sensitive enough to detect a mere scattering of Martian micro-organisms in the soil sample, there should have been plenty of organic material in the form of food and dead bodies for it to register. Without a detectable level of organic molecules in the soil, it is difficult to attribute the results of any of the Viking biology experiments to life.

The possibility of Martian organisms cannot yet be completely ruled out. Some scientists believe that life may exist on Mars in certain favoured oases, or a small population of Martian organisms might exist that did not respond to the Viking experiments. Whatever the case, a more detailed examination of Mars is required to settle the matter finally. (Some plans for the future exploration of Mars are discussed in the concluding chapter.)

Jupiter, the giant planet of the solar system, appears totally unlike the Earth, yet it may be the most promising other site for life in the solar system. Its atmosphere is made of the same gases which surrounded the primitive Earth, and from which life is believed to have arisen: hydrogen, ammonia, methane, carbon dioxide and water vapour.

Organic molecules are formed in abundance from simulations of the atmosphere of Jupiter, as in simulations of the atmosphere of the primitive Earth. Typical of the molecules formed are reddish-brown tarry substances. The red and brown colouring in the swirling clouds of Jupiter is probably caused by organic molecules.

Jupiter has no solid surface beneath its clouds; instead, the gases of which it is made keep on getting denser. At its cloud tops, the temperature

of Jupiter is $-120°C$, but deeper down in its atmosphere Jupiter is warmer. Liquid ammonia exists below the visible clouds of Jupiter, and this region of Jupiter's atmosphere could be a likely habitat for life using liquid ammonia as a solvent instead of water. Deeper still, about 80 km. (50 miles) below the visible cloud tops, temperatures have risen sufficiently for clouds of water vapour to exist, making this an even more favourable locale for Jovian life.

Some scientists have speculated that living organisms may float in the clouds of Jupiter, in similar fashion to fish floating in the seas of Earth. These creatures would be gas bags, filled with helium, and moving around by expelling gas. They would feed off organic molecules in the clouds around them. Possibly the Jovian organisms may themselves be coloured red. Their favoured site may be the great red spot, an updraught of warm air from deep within Jupiter. Probes diving into the atmosphere of Jupiter will be needed to discover whether or not there is life there.

Saturn is a similar planet to Jupiter except that its atmosphere is less colourful, and so there is less chance that it contains complex organic molecules. Saturn's largest satellite, Titan, is a fascinating body. With a diameter of 5,800 km. (3,625 miles), Titan is larger than our own Moon, and has a denser atmosphere than Mars. Titan may be the most favourable site for life in the outer solar system.

Beyond Saturn the solar system becomes icily cold. Uranus and Neptune are so remote that sunlight reaching them is too weak to nurture living organisms. There seems no hope of finding life that far from the Sun.

Listening for Messages from the Stars

Whether or not there is life on Mars, Jupiter, or any other planet of the solar system, we will find advanced beings only by looking much further afield. As yet, we cannot send probes to other planetary systems (though that day may not be too far away – see Chapter 12) but we do have the ability to exchange radio messages with beings around other stars (though such an exchange would take a decade or two for close neighbours and many centuries for more distant civilizations). It was this realization that sparked Frank Drake's Project Ozma attempt to pick up interstellar messages in 1960. Improvements in electronics since then mean that current radio telescopes are far more sensitive. The world's largest individual radio astronomy dish, the 305-m. radio telescope at Arecibo in Puerto Rico, is powerful enough to exchange messages with an instrument of similar size anywhere in the Galaxy.

The first task is to find out if anyone is transmitting. Two main problems hamper the search for radio messages from the stars: where to look, and on which wavelength. If we assume that Sun-like stars are the most favourable for the emergence of advanced life, then we can easily assemble a list of nearby targets – although if we

wish to look further afield than a few hundred light years the number of stars increases so rapidly that it becomes best to scan the whole sky.

Choice of wavelengths is much more difficult. In 1959, as plans for Project Ozma were being hatched, two American physicists, Giuseppe Cocconi and Philip Morrison, proposed that the natural channel for interstellar communication would be 21 centimetres. This is the wavelength emitted naturally by hydrogen gas in space, to which radio astronomers all over the Galaxy would – it could reasonably be expected – already be tuned. Project Ozma listened for artificial signals from aliens at 21 cm. wavelength, as have many other searches since then.

But other scientists engaged in the search for extra-terrestrial intelligence (SETI) have argued that the existing noise from hydrogen makes 21 centimetres the last wavelength to choose for interstellar signalling. Discoveries of emissions at specific wavelengths from various other molecules in space have complicated the issue, and radio astronomers are now abandoning the idea of a 'preferred' wavelength for interstellar communication. They think we will need to scan the entire radio window observable from Earth, which adds greatly to the search effort.

A number of searches have been made to date, concentrating on specific wavelengths. First to follow the lead of Project Ozma was a Soviet team led by Vsevolod Troitsky of Gorky State University, who in 1968 began using a 15-m. radio dish to listen to 11 nearby stars, plus the Andromeda galaxy, at a wavelength of 30 cm. Troitsky and his colleague Nikolai Kardashev subsequently set up across the Soviet Union a network of low-sensitivity aerials which remains in operation in the hope of receiving intensely powerful transmissions from advanced civilizations in the Galaxy.

In 1972, Gerrit Verschuur at the National Radio Astronomy Observatory, Green Bank, reckoned the time was ripe for a sequel to Project Ozma. He used the observatory's giant 91-m. and 43-m. radio telescopes to listen on 21 cm. to ten nearby stars, including the original Ozma stars, Tau Ceti and Epsilon Eridani. Such was the improvement in equipment that Verschuur calculated his search was over 1,000 times as sensitive as Project Ozma. But still nothing was heard.

A more comprehensive search has been made at Green Bank by Ben Zuckerman and Patrick Palmer. Again using the 91-m. and 43-m. telescopes at 21 cm. wavelength, Zuckerman and Palmer surveyed more than 600 Sun-like stars out to a distance of 80 light years. A 40 megawatt transmitter beaming through a 100-m. dish around any of those stars (equivalent to terrestrial capabilities) would have been detected. Nothing positive was heard.

Following these pilot schemes, the search for extra-terrestrial intelligence is widening out. At Ohio State University Radio Observatory, Dr Robert S. Dixon has been scanning the entire sky at 21 cm. wavelength since 1973, and plans to continue indefinitely. At the Algonquin Radio Observatory in Canada, Paul Feldman and Alan Bridle are engaged in surveying several hundred stars at a wavelength of 1.35 cm., which is the wavelength emitted by water molecules – and might thus be a natural-seeming communications channel for water-based beings.

Frank Drake and Carl Sagan are using the Arecibo radio dish to survey several nearby galaxies in the hope of finding some super-civilization capable of transmitting over inter-galactic distances. A number of other small-scale searches have also been undertaken at Arecibo, Green Bank and elsewhere.

There have been several false alarms in the search for radio signals from space. In 1965, Soviet radio astronomers thought they had detected signals from a super-civilization deep in space. But subsequent investigation showed that they were picking up natural variable radio emission from a quasar, CTA-102. Radio astronomers at Cambridge got a bigger scare in 1967 when they found the first of the regularly

Close Encounters: Myth and Reality

Every age spawns its fashionable myths, and the myths of the space age are naturally concerned with alien spaceships and astronauts. since the start of the modern 'flying saucer' era in 1947, countless people have believed that the Earth is being frequently visited by alien beings. It is also maintained that alien astronauts visited Earth many times in the past.

Such ideas are not absurd in principle. At present, it seems to most scientists almost inconceivable that Earth should be the only planet in the Galaxy to produce a technological civilization. But the distances and time-scales involved make it seem extraordinarily unlikely that we should be the objects of special – and strangely secretive – study. Besides, there is no convincing evidence to support the notion of alien visits.

Most UFOs (unidentified flying objects) are misperceptions and do not remain unidentified for long. Major culprits are bright stars and planets, aircraft, meteors, satellites and weather-balloons. Tales of abduction by extraterrestrials have turned out to be hoaxes, or at best bizarre delusions. Several large financial offers have been made for proof that at least some UFOs are extraterrestrial spacecraft; though a hard-core of UFO cases so far defies explanation, firm proof of alien visitations has not been forthcoming. Photographs, like those at right, are typically out of focus and easily faked.

Supposed evidence of alien visits in the past is equally weak. Massive stone structures built by remote civilizations are exhibited as examples of alien technology. The fact that even more grandoise structures were built by Greek, Roman and medieval engineers in Europe without outside help is ignored. In many cases, our own ignorance about our terrestrial past is paraded as evidence to justify extraterrestrial explanations. The reasoning is spurious: ignorance cannot be evidence of anything.

But the possibility of alien visitation is a gripping one, even to the most sceptical, for a genuine extraterrestrial spaceship or a real alien artifact would be of unparalleled scientific importance. The search for evidence continues.

ticking radio sources known as pulsars. They wondered if these might be pulsed communications from other civilizations, until it became clear that the signals were of natural origin.

Soviet radio astronomers made another false announcement in 1973 when they claimed to have picked up transmissions from an alien probe in the solar system. They subsequently realized that the probe was an American one, spying on the Soviet Union. In the Zuckerman and Palmer search, ten stars showed so-called 'glitches' or unexplained spikes of energy. These have been attributed to terrestrial interference, as none of the glitches reappeared when the stars were re-surveyed.

What sort of signal should we be on the look-out for? Probably the signals will be confined to a very narrow-frequency range, perhaps only one Hertz (cycle per second) or less. (For a given transmitter power, a narrow-frequency signal goes further than a broad-band one.)

Most normal objects in space emit over a wide range of frequencies. Even the 21 cm. radiation of hydrogen is smeared over a certain frequency range by the internal motions of the hydrogen clouds. Consequently radio astronomy receivers are tuned to accept a wide-frequency band, usually several kiloHertz, which degrades their ability to detect narrow-band signals. Therefore standard radio-astronomy receivers are far from ideal for SETI purposes.

To undertake a comprehensive search of the radio window for possible messages from the stars, we need to build special multi-channel receivers capable of examining a wide range of frequencies simultaneously with high sensitivity. And this is just what NASA plans to do in a major SETI endeavour due to begin in the early 1980's.

NASA scientists have designed a device known as a multi-channel spectral analyser, which can examine a million radio frequencies at a time. Along with a device for decoding any signals that may be received, this will be attached to radio telescopes in the U.S. and elsewhere for a five-year search for messages from the stars.

Two groups are engaged in the project: the Ames Research Center, and the Jet Propulsion Laboratory, both in California. The JPL group plan to scan the entire sky visible to them using their 26-m. dish at Goldstone in the Mojave desert usually used for tracking space probes. Their receivers will cover all the so-called radio window from 30 cm. to 1.2 cm., which is the range of radio wavelengths reaching the surface of the Earth. Longer wavelengths than 30 cm. are swamped by background noise from the Galaxy, while wavelengths shorter than 1.2 cm. are hampered by noise from the Earth's atmosphere.

By contrast, the Ames group plan to concentrate on individual stars out to 1,000 light years from the Sun, and to examine them in a region of the spectrum known as the water hole. This region lies between the 21 cm. line of hydrogen and the 18 cm. line of the substance known as hydroxyl, OH. It takes its name from the fact that H and OH go together to make water, H_2O. The water hole is a particularly attractive slot for interstellar communication since it coincides with a natural minimum in background noise from space and also because of the importance of water to all life as we know it. Three radio telescopes will be employed for the Ames survey: Arecibo, one at Green Bank, and one in the southern hemisphere.

Soviet radio astronomers have also announced plans for a major survey for interstellar messages, using antennae on the ground and attached to space stations. Among the radio telescopes to be used is the RATAN 600, a ring-shaped instrument 600 metres in diameter. In addition to individual stars, the Soviet researchers intend examining globular clusters and the rich star fields at the centre of our Galaxy, plus galaxies of our local group.

What if the current searches using existing radio telescopes fail? Perhaps no one is sending deliberate signals, or the signals are too weak for us to pick up. In that case, larger collecting arrays are necessary – the bigger the telescope, the fainter the signals it can hear.

NASA scientists have hatched a plan for a major radio array called Cyclops, a big eye on the Universe. Cyclops would be an unprecedentedly vast and expensive undertaking, growing to full size over a number of years. At its maximum, it

SETI PROJECTS	Year	Observatory and instrument	Wavelengths	Targets
	1960	Green Bank 26 m.	21 cm	2 stars
	1968–69	Gorky 15 m.	30 cm	11 stars, Andromeda galaxy
	1970–present	Eurasian network. Omnidirectional.	Several	All sky
	1972	Green Bank. 91 m, 43 m.	21 cm	10 stars
	1972–76	Green Bank. 91 m, 43 m.	21 cm	659 stars
	1973–present	Ohio State University. 175 m.	21 cm	All sky
	1974–present	Algonquin. 46 m.	1.35 cm	500 stars
	1975	Arecibo. 305 m.	Several	Nearby galaxies
	1980's	Various US instruments.	1.2–30 cm	All sky and selected stars

would consist of 1,000 or more antennae each 100 m. in diameter spread over 20 square km. (8 square miles). Cyclops could detect interstellar communications beamed in our direction from as far away as the centre of the Galaxy, and it could also eavesdrop on leakage noise from domestic transmissions, such as TV broadcasts, around nearby stars. Therefore Cyclops could detect other civilizations by their radio noise even if they were not deliberately hailing us.

Cyclops is intended to be built on Earth, but radio interference from satellites, radars, aircraft, and other terrestrial sources may eventually force SETI receivers into space. Cyclops on the Moon, or even dishes suspended in lunar craters have been considered; these would be expensive, but on the far side of the Moon they would be insulated by over 3,000 km. (1,875 miles) of solid rock from the Earth's radio chatter. Best of all might be giant orbiting antennae several kilometres in diameter constructed by the space industrial facilities that will be in operation next century (see final chapter). One such dish would be as powerful as the complete Cyclops system, and would be able to scan the entire sky from its vantage point in space. Each dish would be screened from radio noise on Earth. Most of the interstellar communication in the Galaxy is probably between large antennae in space.

How will we recognize that first interstellar signal and what will it say? Radio noise confined to a narrow-frequency band should by itself be a giveaway, but nature has played tricks on us before, and it will be necessary to examine suspicious signals carefully before announcing that they are the products of alien intelligence rather than natural emission. An interstellar 'Hello' may be a simple, continuous tone, or more likely it may come as a string of on-off pulses – a kind of celestial Morse code, like the binary code used by computers. Evidence of a message coded on the radio signal would clinch the case for an intelligent origin.

What the message might say is limited only by the imagination of the senders, and our ingenuity in decoding it. Simple mathematics is one suggested possibility, perhaps including a translation programme that will enable us to decode

more complex messages to follow. But the most favoured idea is a form of interstellar television, in which the message pulses can be arranged to give a dot picture showing the aliens, their home planet, and such simple scientific facts as their biochemical make-up and aspects of their lives and technology. Surprising amounts of information can be conveyed pictorially, and this form of communication avoids all language problems.

Though most scientists believe that radio is the best way to communicate over interstellar distances (because less energy is needed to send a radio message than any other form of signal), radio is not the only way for interstellar civilizations to communicate. Laser beams at infra-red, optical and ultra-violet wavelengths have been suggested, and the Copernicus satellite has been used to search for ultra-violet laser emissions from the vicinity of the stars Tau Ceti, Epsilon Eridani, and Epsilon Indi, though without success.

Roland Bracewell, an American radio astronomer, has gone as far as to suggest that the best way of establishing communication between civilizations 100 light years or so apart is by sending automated probes to the target stars. The probes would take many centuries to get to their destinations, but once there they would orbit the stars indefinitely, awaiting the radio noise which would signal the emergence of technological civilization. The probes could then converse with them while messages announcing the discoveries were on their way to the parent star. Such a strategy cuts out the otherwise inevitable delay in establishing communication over vast distances in space, and Bracewell urges radio astronomers to be on the alert for strange radio signals coming from within our own solar system which might actually be the communication attempts of an alien probe.

Messages from Earth

To look at the problem from the other side, what could another civilization learn of us?

Britain's Astronomer Royal, Professor Sir Martin Ryle, has been among those who have urged that we should keep our existence quiet for fear of attracting the attention of potentially

Project Cyclops, named after the one-eyed giant of Greek mythology, foresees the construction of hundreds of 100 m. (300 ft.) radio-telescopes, which would in collecting power be many hundreds of times more powerful than any single dish. Suggested by Frank Drake of the Arecibo Observatory, Puerto Rico, the array could ideally be built on the Moon's far side (*left*), where it would be shielded from the radio-interference that would disrupt the performance of an Earth-based system (*right*).

hostile extraterrestrials. But already it is too late. For the past 30 years or so, high-frequency radiation from TV and radar has been streaming through the ionosphere into space, and it will inevitably be picked up by anyone listening nearby.

Most of the radio noise we are emitting today comes from UHF television stations in the United States and western Europe. Extraterrestrials of a similar technological level to ourselves could pick these transmissions up at a distance of 25 light years, a distance encompassing about 300 stars, and would receive beams from our ballistic missile early warning radars at up to 250 light years, a range encompassing several hundred thousand stars.

Inhabitants of Alpha Centauri and beyond would not be glued to their screens for weekly episodes of *Star Trek* because programme information itself would be too weak to decipher over interstellar distances. But the aliens would be able to pick up the powerful carrier waves on which the programmes ride.

Careful analysis of Doppler shifts and other characteristics of the signals would reveal the size of the Earth, its orbit around the Sun, and other information we might expect to extract from leakage noise from another civilization that Cyclops or a large space antenna could pick up.

One short radio message has been sent to the stars. On 16 October 1974, the Arecibo radio telescope transmitted a three-minute burst of pulses towards the globular cluster M 13 in the constellation of Hercules. If the 1,679 pulses are arranged into a grid 23 characters wide by 73 deep, they form a pictogram telling the aliens

about ourselves. But even if there is someone among the 300,000 stars of M 13 who will pick up the message and decode it, the cluster is 24,000 light years away so an answer cannot be expected until around AD 50,000.

Our radio messages may never have listeners, but we have sent messages of other kinds across space and time that may, millennia hence, be picked up. First to leave were small engraved plaques attached to each of the Pioneer 10 and 11 probes which will eventually exit the solar system and drift outwards into the Galaxy where some advanced civilization with an interstellar early warning radar might detect and recover them. If aliens do find the Pioneers, they will be able to tell from the plaques something about the probe's makers and their location (see box).

A second pair of probes, called Voyager, are following the Pioneers deep into the solar system. After their planet-probing missions are over they, too, will end up floating dead and dark between the stars. They carry a different form of message in a bottle for other civilizations to decode: long-playing records. Records were chosen because large amounts of data can be squeezed onto them.

Each record begins with 115 pictures which are encoded electronically into the grooves. These pictures include views of the solar system, the Earth and its inhabitants, and human technology. Then follow spoken greetings in a wide variety of languages, and a so-called 'sound essay' consisting of sound effects of water, volcanoes, living beings, and technology such as a rocket launch. The record concludes with Earth's greatest hits, a musical selection ranging from tribal songs to Bach, Beethoven and Chuck Berry. One wonders what the hypothetical extraterrestrials will make of it all.

How might we first learn that we are not alone? One day we might intercept a probe-borne 'culture capsule' from another civilization, perhaps in the form of the interstellar messenger probes suggested by Ronald Bracewell. We might find a commemorative plaque at the site of an alien landing on another planet, or the Moon, similar to the plaque attached to the Apollo 11 lunar module. There are, of course, those who think contact has already been made with UFO's and that visits have been frequent; the hard evidence, however, is still lacking. Probably the first real evidence of life elsewhere will be in the form of radio waves, the easiest way of sending information over great distances.

Whoever we contact in space (assuming the message does not come from too distant a star) is almost certain to be more advanced than we are, so we stand to learn a great deal from the extraterrestrials. Secrets of new energy sources, medical breakthroughs, and even the key to peaceful coexistence are some of the suggested benefits from communication with advanced beings. They may also be able to tell us how and where to get in touch with more civilizations, so that we become plugged in to some kind of galactic telephone network.

This pattern represents a string of 1,679 on-off pulses, which in 1974 were beamed by Puerto Rico's vast Arecibo radio-telescope towards the globular cluster M.13 in the constellation of Hercules. When arranged in a grid 23 × 73, the pulses form a pictorial message of numbers, atoms basic to life, and formulas for DNA (the molecule which encodes genetic instructions). Curved lines, representing DNA's spiral structure, point out a human shape. Other information – Earth's population, human height, the Solar System with the Earth displaced towards the human figure, the Arecibo telescope – would perhaps be baffling. But at least a recipient civilization would, if it could break the code, know there was intelligent life elsewhere.

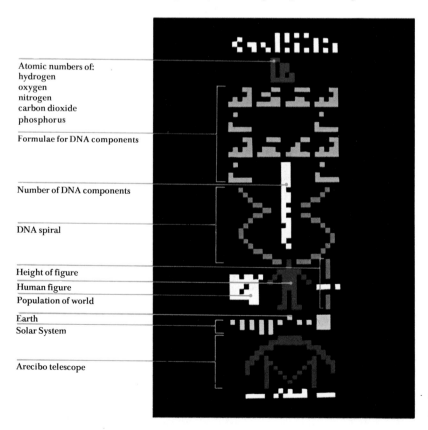

Atomic numbers of:
hydrogen
oxygen
nitrogen
carbon dioxide
phosphorus

Formulae for DNA components

Number of DNA components

DNA spiral

Height of figure

Human figure

Population of world

Earth

Solar System

Arecibo telescope

It is hard to imagine the sensation that the first announcement of contact with extraterrestrials will cause. Even if the message is incomprehensible, or if it says nothing of interest at all, its very existence will have answered the question: Are we alone? And knowing that advanced civilizations exist in space will be encouraging news to a young civilization such as ourselves, embattled as we are with the problems of technological development. It will demonstrate that such civilizations can survive without destroying themselves. Perhaps we will be able to emulate them. Contact with extraterrestrial life could mean the end of our childhood.

A Message from Earth

This engraved metal plaque was attached to both the Pioneer 10 and 11 space probes to Jupiter. Devised by the astronomer Carl Sagan, it is written in the only universal language: science. It should be comprehensible to any advanced technological civilization. It shows, at top left, a representation of the hydrogen atom, which produces the characteristic 21-cm radiation familiar throughout space. This is the plaque's unit of measurement, as indicated by the stroke beneath (the binary number 1). Below that is a starburst pattern which illustrates the direction to 14 pulsating stars as seen from the Sun, with their periods indicated in binary code. Any civilization clever enough to intercept the probe would know about pulsars, their frequencies and their decay rates, and so would be able to work out roughly where the probe came from (by mapping the point of origin relative to the pulsars) and when it was sent (by comparing the observed period of each pulsar with the period indicated on the plaque).

With the point of origin pinpointed to within several hundred stars, the receiving civilization could then – in theory at least – go further.

At bottom is our Solar System – an almost certainly unique arrangement of planets – showing a Pioneer flying out into space. The tote mark at the right indicating the height of the female figure is tagged '8' in binary; 8 × 21 cm is the height of the woman, whose size is related to the probe itself, which provides a double check that the basic unit is understood correctly. The two strange creatures may be the most difficult part of the message for an alien civilization to understand, but their very oddity should suggest a biological, rather than technological, interpretation.

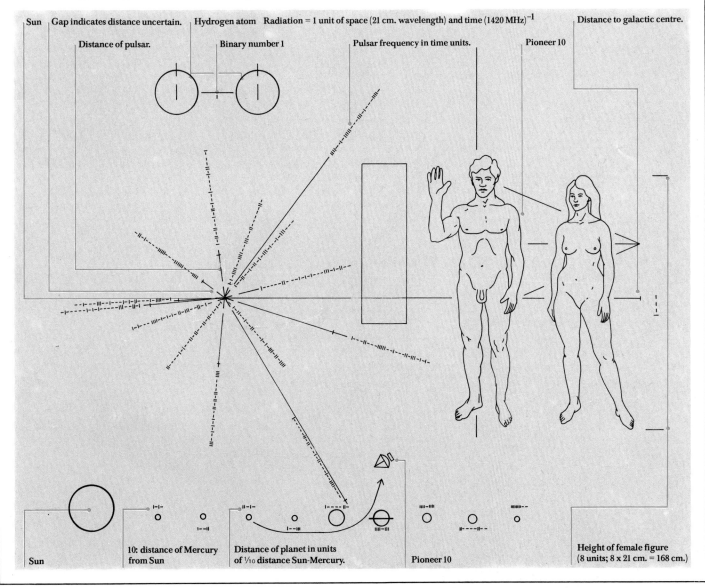

Sun | Gap indicates distance uncertain. | Hydrogen atom | Radiation = 1 unit of space (21 cm. wavelength) and time (1420 MHz)$^{-1}$ | Distance to galactic centre.

Distance of pulsar. | Binary number 1 | Pulsar frequency in time units. | Pioneer 10

Sun | 10: distance of Mercury from Sun | Distance of planet in units of 1/10 distance Sun-Mercury. | Pioneer 10 | Height of female figure (8 units; 8 x 21 cm. = 168 cm.)

7/THE EXPANDING UNIVERSE

JOHN GRIBBIN

Mankind has always been awed by its insignificance in the framework of the Universe. But just how insignificant we are emerged only in the 1920's, when astronomers learned for the first time that all the stars we see at night – and millions more – are virtually in our own backyard, members of our own Galaxy, the Milky Way (seen in the reconstruction at left as if from a planet on the edge of a distant globular cluster). But beyond our Galaxy, 20 times its diameter away, lies another Island Universe, 300 billion stars wheeling like a vast Catherine-wheel in space. And that galaxy, too, is a next-door neighbour compared with the distant communities of galaxies that stretch out thousands of millions of light years in all directions. Moreover, this whole vast system of galaxies is expanding. The most distant known galaxies are receding from us (and we from them) at thousands of miles per second.

Once astronomers had realized that the stars are separate Suns, and not little lights fastened to a crystal sphere rotating about the Earth, our planet was relegated to a relatively minor position in the Universe. But for hundreds of years astronomers failed to appreciate just how minor. They were generally prepared to say that the Universe was 'infinite', but they had no idea of its actual scale or structure. Until well into the 20th century, the established picture was one of a Universe full of stars – the Milky Way we see in the sky – with, perhaps, some of those stars also possessing their own planets, replicas of our Solar System.

In the past few decades, we have discovered that the Universe is almost unimaginably grander than this. We now know that our Milky Way is but one system of stars, an island galaxy separated by great stretches of empty space from other island galaxies, many of them far bigger than our own Milky Way system, which reach across the sea of the Universe to the edge of infinity; and we know that the whole Universe is expanding, the result of a giant explosion at a time we can date with some degree of certainty. In other words, it seems we know a good deal about the ultimate limits of time and space (as far as *this* Universe is concerned).

The Mystery of Darkness

The story of these staggering discoveries begins back in the 18th century, when one astronomer stumbled across a profound puzzle concerning the nature of the Universe. As such, it marks the beginnings, in a modern sense, of the science of cosmology – a strange devotion more like philosophy than everyday science, in which theorists struggle to explain, on the grandest of all scales, how the Universe in which we live got to the state it is in today, and where it is going in the future.

The key question – a deceptively simple one – was asked in 1744 by a man who we might justly regard as the first cosmologist: de Chesaux. It was he who first wondered why a supposedly infinite Universe full of bright stars should be dark at night.

The reasoning behind the question is as follows: if we are living in a Universe which is literally infinite in extent, and if that Universe is full of stars, then wherever we look in space there must be a star, no matter how far away. Every bit of the night sky, however small, should contain a star. There should not be any gaps in the dome of blazing light.

Alas for de Chesaux, no one else at the time seemed able to realize the significance of his question, which only gained notoriety when it was thought up again, independently, by a German astronomer, Wilhelm Olbers, in 1826. For this reason, the puzzle is known to this day as 'Olbers' Paradox'. The paradox, of course, applies equally well in a supposedly infinite Universe full of island galaxies. It is just as real today as ever, and its resolution – as we shall see –

tells us something very basic about the nature of the Universe.

It is possible to set out the paradox rather more scientifically. The best way to look at the problem is to imagine thin layers, or shells, of stars (though the argument applies equally to galaxies) surrounding the Earth *ad infinitum*. The intensity of the light from all the stars in any one shell can be easily worked out, once you accept the assumption that the Universe is full of a more or less even spread of stars. Each shell of stars contributes its own share of light. The share should increase with the greater number of stars; but in fact, since each shell is proportionately further away, the light is dimmed by distance and all the shells with the same thickness contribute exactly the same amount of light at the surface of the Earth, no matter how far away they are. The distance cancels out the increase in star-light. Nevertheless, in an infinite Universe, there would be an infinite number of shells (whatever thickness of shell we chose) so that we might guess that the Universe should have an infinitely bright sky. In fact, though, nearby stars block out some of the light from further away. When this is taken into account, we can say that the night sky should be 'only' as bright as the surface of an average star – in total, just 40,000 times the brightness of the Sun at noon. The puzzle is not, therefore: why is the sky dark at night? It is: why is the sky so 'dark' even in the daytime?

How do we resolve the paradox? Three basic assumptions are hidden in the preceding argument. These are:

– that the Universe is infinite – there is no 'edge' beyond which no more stars or galaxies are to be found;

– that it is uniform – the density of stars doesn't decrease the further away you get from the Sun;

– that it has always been the same – the stars from long distances away haven't 'switched off' or moved away.

One at least of these assumptions must be wrong

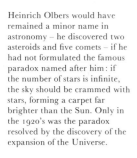

Heinrich Olbers would have remained a minor name in astronomy – he discovered two asteroids and five comets – if he had not formulated the famous paradox named after him: if the number of stars is infinite, the sky should be crammed with stars, forming a carpet far brighter than the Sun. Only in the 1920's was the paradox resolved by the discovery of the expansion of the Universe.

if we are to resolve Olbers' Paradox. But which?

Olbers' Paradox caused no revolution in astronomical thought even in the 19th century, being regarded as no more than a curious puzzle. With all the advantages of hindsight, however, we can say that it *should* have caused a revolution, because the options by which the paradox can be resolved are very limited.

The first guess might be that the Universe isn't uniform. All the best observations have shown that – despite local clumping effects – the Universe is indeed uniform. This seemed true in the 19th century, and is accepted as true today: the number of galaxies in a particular volume of space is the same everywhere. Nor have astronomers – then and now – ever seen any sign of an 'edge' to the Universe; it is so big that even if it is not literally infinite in extent then the sky should still be a blaze of light. So only one assumption is left to be challenged – the assumption that the Universe is unchanging. There seems little escape from the theoretical conclusion that the whole Universe is evolving in some sense; and, since we see no signs that the more distant stars and galaxies have 'gone out', the *only* way to resolve Olbers' Paradox is to say that the Universe must be expanding.

Expansion solves the problem. The light and heat from all the objects in the sky has to spread ever more thinly into the void to fill the increasing space between them. This effectively weakens the light, lowering the density of the energy of starlight – the brightness of the sky – everywhere. The faster each light source recedes, the weaker its contribution will be, and in a Universe with the more distant shells receding more rapidly than those nearby, only the very nearest are left making any noticeable contribution to conditions here on Earth.

Today, this theory is an established and fundamental part of cosmology. It is hard to imagine the difficulty with which cosmologists convinced themselves, almost against their better judgement, that the Universe really is expanding in this way. By the time Einstein came along to provide cosmology with a new mathematical framework, the theory of relativity, the standard theory was still that the Universe was a static place. Einstein's equations were developed into a cosmology that predicted an expanding Universe, by de Sitter, but even Einstein himself, while overturning the whole of our understanding of physics, seemed almost to bend over backwards in his efforts to mutilate his own theory into a form which could accommodate a static Universe by inserting a 'cosmological constant' into his equations. But no one seemed too bothered by the paradox at the time, for no one as yet had a clear idea of the scope of the Universe.

It took a series of observations by the great American astronomer Edwin Hubble to show that the Universe actually is expanding. Only then did the astonished cosmologists, desperately seeking a model to explain the observations, realize that de Sitter had already found one. Then

The 18th-century German thinker, Immanuel Kant, though best known as one of the most significant philosophers ever, also contributed to cosmological theory by proposing that stars were grouped in a lens-shaped system – our own Galaxy – and that fuzzy clouds (nebulae) like the Andromeda nebula were other galaxies. As it turned out almost two centuries later, he was right.

the Einstein-de Sitter model became the classic standard description of the real Universe; and only much later still did cosmologists start to point out to each other that they should have realized all along, from Olbers' Paradox alone, that the Universe really is expanding.

Hubble's observations – which established the existence of other galaxies, the scale of the Universe, and the fact of expansion – mark the beginning of the second phase of modern cosmology, the factual kick in the pants that put the theorists on the right track.

The Great Expansion

But Hubble, of course, did not produce his discoveries out of his hat. He was the intellectual descendant of a breed of telescopic observers going back to Galileo, who, amongst other things, observed some of the other galaxies in the Universe without realizing their true nature. Galileo's great work, pioneering the use of telescopes in astronomy, concerned the nature of our Solar System and the realization that the Sun is one star among many in the Milky Way. But – as anyone with a small telescope or even binoculars can do today – Galileo also noticed many small clouds of light, which were known as '*nebulae*' (Latin: clouds). He even found, with the aid of his telescopes, that some of nebulae were made up of individual stars packed tightly together. Fuzzy blobs, though, could not compare with the dramatic discoveries about the Solar System being made at that time, or with the study of stars and other phenomena over the next 300 years.

It was the burst of interest in comets which followed Halley's work in the early 18th century that led several people to catalogue the positions of these fuzzy nebulae. Comets offered the prospect of instant fame, thanks to the tradition that a new comet is named after its discoverer. Observers found that they could be fooled by one of these odd nebulae into thinking that they had found a comet, only to be disappointed when the fuzzy blob stayed in one place instead of sweeping in towards the Sun. The Frenchman Charles Messier helped to sort out the confusion by cataloguing 103 of the brightest nebulae and recording their positions, so that comet hunters would know the wood from the trees. These nebulae are identified by the letter M and a number corresponding to their position in Messier's catalogue, a system still used to this day. Some of them are literally clouds of glowing material inside our own Galaxy; but many are whole galaxies in their own right, including our near neighbour the Andromeda Galaxy, M31 in the catalogue.

In the second half of the 18th century, the German philosopher Immanuel Kant observed both the Milky Way and the Andromeda Nebula. Noticing that the latter was made up of uncountable stars, he suggested that it might be a separate system like the Milky Way, and that the Milky Way must be a mass of stars similar to the Andromeda Nebula. By the end of the 18th century, therefore, the concept of island galaxies had already appeared; but the idea was a philosophical one, and not as important to contemporaries as the great puzzles of the time, such as the nature and origin of the Sun.

Still, the catalogues continued to grow, with William Herschel and his daughter Caroline locating more than 2,000 previously unclassified objects of various kinds. William's son John continued this work into the 19th century and moved to South Africa in 1833 to begin cataloguing the nebulae of the southern skies. During the 19th century, as telescopes got bigger and better (the greatest was the Earl of Rosse's 72-inch reflector), astronomers began to identify the different kinds of nebula, distinguishing some of the starry island galaxies of Kant from gaseous blobs (which we now know are within our own Galaxy). They even began to unravel the structure of some of the external galaxies, notably the great spiral nebulae like the classic archetype M31.

At the end of the 19th century, a powerful new tool was harnessed to the long-range vision of the telescope with the invention of spectroscopy, which gave astronomers the ability to investigate the nature of the light from stars in our own Galaxy or from more distant sources. In cosmological terms, this was where the story *really* began.

The spectroscope showed, first of all, which nebulae are definitely groups of stars, since their light is characteristic of the light from many stars added together. Glowing clouds of hot gas in the Milky Way have quite different spectra, and from now on the two could never be confused again. But at the beginning of the 20th century, one of the key problems in astronomy was to determine just how far away these various objects were. Only with some idea of the distance to an object like the Andromeda Galaxy could astronomers work out, from its apparent size in the sky, how big it really was; and as well as its

intrinsic interest, that measurement would provide a model that would indicate how big our own Milky Way is, on the assumption that it is roughly the same kind of object.

One key to distance is provided by the pulsating stars, the Cepheids. We saw in Chapter 3 how the Cepheids show a regular variation in brightness which depends on their absolute brightness. This discovery gave astronomers a kind of standard by which they could measure distances – measure the period of a Cepheid, and you know how bright it really is; measure its *apparent* brightness and you can discover how far away it is. This discovery had already led to a revolution in the understanding of our own Galaxy. Then, in the 1920's, the combination of these new techniques with the completion of what was at the time the largest telescope in the world, the 100-inch reflector on Mount Wilson, California, produced the vital cosmological breakthrough, thanks to the patient work of Edwin

The Great Spiral in Andromeda is the most distant object in space visible to the naked eye and also the only extra-galactic object visible to the naked eye in the northern hemisphere. Without a telescope, it looks like a blurred, oval cloud, but it is now known that it is a spiral galaxy, larger than our own, containing some 300 billion stars, all bound together to form an object 20 times further away than any star in our own Galaxy.

John Herschel was a scientist with an astounding range of interests. He was a pioneer photographer and made a name for himself in the fields of biology, botany and meteorology. He was also a co-founder of what is now the Royal Astronomical Society. In the 1830's, he took the 18-inch telescope built by his father, William, to Feldhausen, at the Cape of Good Hope (*left*) to observe the southern skies. Partly as a result of his work, a catalogue of some 10,000 cloudy objects had been built up by the end of the 19th century. But there was no way of knowing at the time that many of these objects lay beyond the confines of our own Galaxy.

Hubble.

The new telescope showed more clearly than ever before that the Andromeda Nebula is indeed a galaxy, composed of swarms of stars. Any lingering doubts that the spiral nebulae might be mere swirling gas clouds on the periphery of the Milky Way vanished. The next step was obvious – to look for Cepheids in the An-

dromeda Galaxy to measure its distance. This Hubble did. The answer came out to be more than a million light years – far further than any star.

Now, at last, the Milky Way chauvinism of the past 300 years was put in its proper place, as the true enormity of the Universe began to be realized. The science of extragalactic astronomy

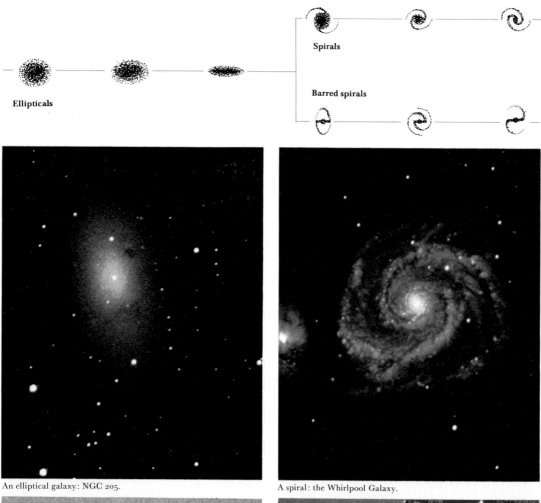

Ellipticals

Spirals

Barred spirals

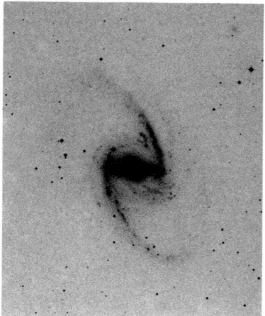

An elliptical galaxy: NGC 205.

A spiral: the Whirlpool Galaxy.

A barred spiral: NGC 1365.

An irregular: the Large Magellanic Cloud.

Edwin Hubble, in the space of just five years (1923–9) made two discoveries comparable in their significance to the Copernican revolution. He proved that some nebulae were systems of stars far beyond the Milky Way and he discovered that the galaxies seemed to be receding with speeds that increased with distance. He also proposed a system of classifying galaxies that is still in use. Hubble classified galaxies into ellipticals, spirals, barred spirals and irregulars. Elliptical galaxies are oval in shape, some of them three times the size of our own galaxy. The spirals have a structure like our own galaxy, with two arms swinging out from a nucleus. Some have a bar across them, with the arms beginning at the end of the bar. As the diagram (top) shows, these can be neatly related, but it is not known whether one class evolves into another. The irregular galaxies, not classified on the diagram, are usually smaller bodies, without any particular shape.

– objects outside the Milky Way – soon gained a prominence it has never relinquished. At once, two burning questions demanded to be answered: what are the different kinds of galaxies beyond the Milky Way? And how far away are they, remembering that the Andromeda Galaxy seems to be a relatively near neighbour? This second question becomes a vexed one when a galaxy is so far away that individual Cepheids can no longer be identified.

Hubble studied hundreds of galaxies and classified them into a simple scheme, based on their appearance, which remains invaluable today. Three 'branches' of galaxy provide the basis of the scheme, but it is important to remember that there is no suggestion that the branches represent evolutionary paths. Hubble did not say that each type of galaxy develops into the neighbouring type. He just put similar ones together.

The three chains are roughly as follows:

First, the spirals that we have already heard about. These range from tightly wound spirals in which the 'arms' marked by bright stars twist sharply around the bright nucleus, through various more loosely-wound spiral types in which the whole pattern is more spread out.

Alongside this family of 'ordinary' spirals is an equivalent family of 'barred' spirals, which have a bright bar of stars across their centres, with the spiral arms twisting off from the ends of the bar, not from the centre of the galaxy proper.

And then, quite different from the spirals in appearance, there is the family of elliptical

This cube, of which each side represents four million light-years, contains a three-dimensional map of the 'local group' of galaxies. Besides our own Galaxy (*centre*) only the two Magellanic Clouds and the Andromeda Galaxy are visible to the naked eye. The group contains a number of dwarf galaxies of perhaps a million stars each, which may be globular clusters that have escaped from the larger spirals.

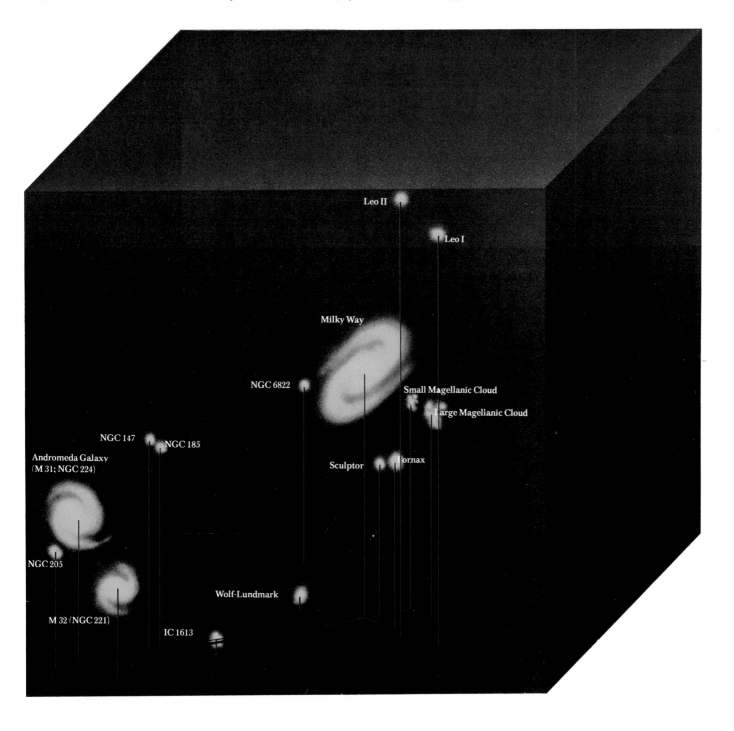

galaxies. These overwhelmingly dominate the Universe, making up about 80 per cent of all observed galaxies. They are uniform in appearance, look elliptical in photographs but actually range from long thin cigar shapes to completely round spheres of stars. The smallest are scarcely distinguishable from the large globular clusters of our own and other galaxies; the largest contain the equivalent of 10 million million Suns.

Last of all, since not quite everything can be fitted into this neat picture, Hubble invented a grab-bag category of irregular galaxies for everything else.

Now, more than 50 years later, with even bigger and better telescopes available, there is still debate about how the different kinds of galaxy relate to one another, and whether some kinds do actually evolve into others. In particular, the discovery in the 1950's and 1960's of the violence associated with so many galaxies provided yet another new insight into the nature of the Universe (see Chapter 8).

Hubble also set out to determine the distances to galaxies out into space beyond the range of the Cepheid yardstick – beyond a few million light years, in the region where individual Cepheids are just too faint to be seen against the background of the millions of other stars in a galaxy.

The first step was to make the reasonable guess that there is some limit to the brightness any star can achieve, and that the brightest stars in any galaxy must be close to this limit. This guess can be checked for the nearest galaxies, by comparison with the Cepheid method, and seems to hold true. With this aid, looking at the very brightest individual stars in the external galaxies and using their apparent brightness as a guide to distance, Hubble pushed his measurements out to about 10 million light years. But even that proved to be a tiny step into the Universe. Most galaxies are so far away that *no* individual stars can be picked out, the whole object, many millions of stars, is just a fuzzy blob on a photographic plate. Already it was clear that the Universe we can see was hundreds of millions, possibly thousands of millions, and perhaps even millions of millions of light years in extent. How could Hubble measure the distances to these fuzzy blobs?

As a very rough guess, he tried reasoning that all galaxies are equally bright – that they all contain the same number of stars – so that the dimmest are simply the furthest away. All the galaxies of one type are nearly the same size, so the rule can be applied with some effect. Another approach was to look at clusters of galaxies and to reason that just as the brightest star in any galaxy has the same absolute brightness, so the brightest galaxy in any cluster will be about the same brightness and size as the brightest in any other cluster. (Notice, by the way, how matter shrinks in the mind's eyes as we step out into the vastness of space – whole galaxies become too small to take into account, and we must puzzle about clusters of galaxies, perhaps hundreds in

number, wheeling together through space like a swarm of insects.)

These tactics were very rough-and-ready. There seemed to be no reliable way to get to grips with the distances to the furthest galaxies, or even to those beyond our immediate back yard, the 'local' region of space a mere 10 million light years or so across. Then, in 1929, came another big breakthrough.

This centred on the application of spectroscopy to the study of galaxies. One of the most interesting features of spectroscopy is that it provides a way for astronomers to measure the velocity with which a distant star is moving. This is because of the Doppler shift, a displacement or shift of the characteristic lines in a stellar spectrum towards either the red or blue end of the spectrum, depending on whether the star is going away from or coming towards us. Astronomers at the Lowell Observatory had been applying this technique to galaxies in the 1920's and by 1929 had measured the shifts of 40 or so; almost all showed a shift to the red.

In principle, light shift – the distortion of light that indicates the retreat or approach of stars and galaxies – is similar to the way in which sound waves are compressed by an approaching source to sound higher, and stretched by a receding source to sound lower. Whether applied to changes in sound or light, the effect is named after the 19th-century Austrian physicist, Christian Doppler. A stationary source (*top*) spreads its light evenly across the spectrum. If the source is receding, the spectrum – and any absorbtion lines it contains – is 'shifted' towards the red end (*centre*). The amount of the shift – which applies to all radiation, not just visible light – indicates the speed of recession. If an object is approaching us – a rare event – the information is shifted towards the blue end of the spectrum (*bottom*).

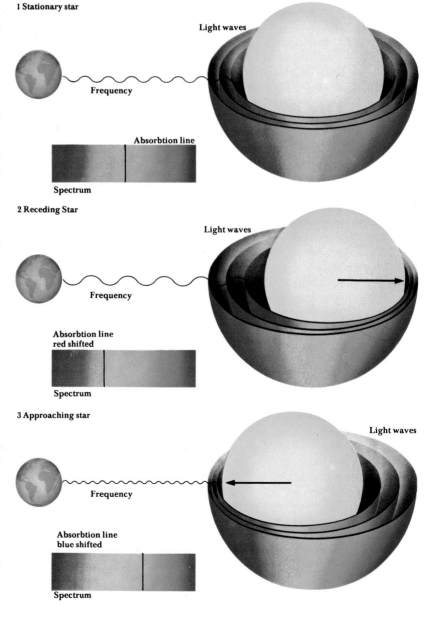

1 Stationary star

Light waves

Frequency

Absorbtion line

Spectrum

2 Receding Star

Light waves

Frequency

Absorbtion line red shifted

Spectrum

3 Approaching star

Light waves

Frequency

Absorbtion line blue shifted

Spectrum

At the same time, Hubble had been measuring the distances to some of the same galaxies (quite nearby ones, of course) and in 1929 he put the two sets of results together to arrive at a dramatic conclusion. Not only did it seem that all of the galaxies he had measured except our very nearest neighbours were receding – otherwise they would not show a red shift – but the velocity of recession was exactly proportional to their distances. Hubble guessed this was a universal effect; if so, the whole of the visible Universe could be measured, with the distance of every galaxy determined from a measurement of its redshift (the concept is now so common that astronomers tend to write it as one word).

The discovery went far beyond providing a tool to measure distances to galaxies. For, remember, astronomers were still happy with a 'static' universe in the 1920's. Now, with Hubble's discovery they found they had been wrong all along, that the Universe is really expanding – and in just the way required to explain the curious paradox that the sky is dark at night.

Hubble's discovery also suggested a curiously egocentric picture in which all the galaxies seemed to be receding from our own. Were we, after all, at the centre of the Universe? Alas, no. A *uniform* expansion, with each galaxy moving further away from every other, produces just the same kind of picture wherever you happen to be in the Universe. The best analogy is that of a balloon on which a row of evenly spaced dots is painted. When the balloon is blown up, all of the dots move further apart from each other; and the further away the dots are from the observer, the faster they move. Yet no dot is 'central'.

Over the years since 1929, improved measurements of distances to nearby galaxies have given a more accurate value to the speed of recession as related to distance – the Hubble constant, as it is known; it is estimated at between 50 and 100 kilometres per second per megaparsec (one million parsecs). In other words, for every 75 km. (47 miles) per second of measured recession velocity, a galaxy is about 1 megaparsec ($3\frac{1}{4}$ million light years) away from us.

But this was almost a minor technical achievement compared with the turmoil in cosmology created by the discovery of the expansion of the Universe. For, as many astronomers were quick to see, if the Universe is expanding now, then it must have started from some great original explosion. This is the Big Bang model of the Universe, described by Einstein without his spurious cosmological constant. Hubble's constant tells us how quickly the Universe is expanding, which tells us in turn how long it is since the Big Bang itself. The age of the Universe turns out to be about 15,000–20,000 million years, roughly comparable with the calculated ages of the oldest known stars.

Big Bang vs. Steady State

Some cosmologists simply didn't like the idea of a definite beginning to the Universe as we know it.

Indeed, some still dislike it: it forces them to face the philosophical – even religious – question of what existed before the Universe started up in this way.

In the 1940's, several cosmologists asked themselves the question of whether an expanding Universe could be explained without reference to a Big Bang. Out of this renewed interest in the nature of the Universe, fired by the Einstein–de Sitter equations and Hubble's observations of an expanding Universe, Fred Hoyle and Thomas Gold, working together, and Hermann Bondi on his own, came up with what was known as Steady State model.

Their reasoning ran as follows:

The most fundamental assumption of cosmology is that the Universe is essentially the same everywhere we look, and that conditions in our immediate neighbourhood are indeed typical of the whole vast Universe. This assumption has to be made before puzzling over the nature of the Universe; after all, if Einstein's equations, say, only apply over the nearest 10 million light years then we have no way of guessing how things work further away, and no basis to make mathematical descriptions – models – of the Universe. This idea has become enshrined as the 'Cosmological Principle', which states that the Universe, averaged on a suitable scale, is the same everywhere.

But why not go a step further in philosophical terms? The Steady Staters argued that the simplest assumption of all is that the Universe is not only the same everywhere now, but also that it always *has* been the same as it is now. This idea of a Universe unchanging in both space and time they dubbed the 'Perfect Cosmological Principle', and set out to reconcile it, if they

In explaining the expansion of the Universe, the recession of the galaxies is often compared to the way markings on a balloon recede from each other when the balloon is blown up. All points seem to recede from each other with a speed that is directly proportional to their distance apart. In this example, the 'balloon' of the Universe undergoes two doublings over two equal time periods. In that time, galaxies A and B have increased their distance by three units, but galaxies B and C have increased their distance by nine units.

Among Sir Fred Hoyle's numerous contributions to astronomy was his Steady State theory according to which new matter is formed in space to fill the increasing gap left as the galaxies recede from each other.

could, with the observation that the real Universe expands.

This is simple enough to do in theory. Remember we are talking about averages – it is the *average* appearance of the Universe which has to stay the same for all time. Now, if you take a large chunk of the Universe, say 100 million light years across, and count the number of galaxies in it you will get a figure for the density of galaxies in the Universe now. In a few thousand million years, those galaxies will have moved further apart as the Universe expands, reducing the density and violating the Perfect Cosmological Principle. But what if new galaxies are being created all the time to fill in the gaps as the old galaxies move apart? In this theory the Universe looks the same all the time; although at each time the galaxies in our chosen volume of space may be different, the density is the same.

Of course, this Steady State model does require the continuous creation of new matter to fill in the gaps – not whole galaxies popping complete into existence, but a few atoms of hydrogen in the vastness of space, atoms that slowly form giant gas clouds, which condense into galaxies of stars as the gaps between the old galaxies open up. This is quite a leap for anyone brought up on the very clear evidence from experiments here on Earth that matter can neither be created nor destroyed (except through interchange with energy in line with Einstein's prediction); but it is no worse than having to accept, as the Big Bang theory requires, that *all* of the matter now in the Universe was created at one instant.

These questions come close to those of theology, let alone philosophy, and the passions they roused in the 1950's were reminiscent of some great religious debate or the controversy over Darwin's theory of evolution. Because the observations could not distinguish between the two theories, Big Bang and Steady State, astronomers divided into two camps solely on the basis of which one they liked best. At the highest level the leading proponents of each theory were faced with a situation in which their deepest cosmological beliefs were no more than that – beliefs. Lacking solid evidence to settle the issue one way or the other, they were left with philosophical arguments that relied as much on emotion as on reason.

This was an unsatisfactory state of affairs for scientists in the mid-20th century, especially astronomers in a generation that had grown used to the idea of exploring the depths of the Universe with giant telescopes, and which now had available the dramatic prospect of opening up the first new window on the Universe for 350 years, in the form of radio astronomy. So the rivalry between the two cosmological schools of thought helped to foster the very rapid growth of observational astronomy, and especially radio astronomy, through the 1950's and into the 1960's. If no one had ever proposed the Steady State theory, and one established view of the Universe had held sway (in the way that the static Universe concept had dominated for centuries) then cosmology would hardly have developed as an observational science with the same speed.

Ironically, the specific observations encouraged by the Steady State–Big Bang rivalry never did resolve the issue; it was settled by a completely different sort of observation, made more or less by accident, which involved a whole new revolution in cosmology, and which demands a chapter on its own (Chapter 8).

Basically, the cosmological rivalry required one thing of the observers, whether they were using the big optical telescopes or the then new tools of radio astronomy: they had to look deeper and deeper into the Universe. While the Steady State theory held that however far the optical and radio telescopes reached they ought to find the same overall appearance, the Big Bang theory required that the Universe must be evolving. The vital point was that, because of the time it takes for light and radio waves to cross the enormous spaces between the galaxies, the further away we look, the longer ago the radiation we now see set off, the further back in time we see and the closer our information will be to the beginning of the Universe (if it had a beginning).

Look again at the idea of successive spherical shells surrounding our own Galaxy. From one sphere, say 100 million light years in radius, we are receiving radiation (light, radio and other electromagnetic waves) that set out 100 million years ago. From another shell, 1,000 million light years in radius, we receive information that is 1,000 million years old; and so on. Each layer, like the skins of an onion, can be peeled by our detectors to provide a picture of what conditions in the Universe were like at different times in the past. In a Steady State Universe, each layer should look the same; but in an evolving Big Bang Universe, the further layers, where we see the Universe as it was long ago, should show signs of the greater density with which galaxies were packed together at the Big Bang.

The effort to probe deeper and deeper into the Universe, driven by the need to resolve the Big Bang–Steady State conflict and aided by ever improving observational techniques, threw up a whole zoo of new astrophysical phenomena. Once again – at the dawn of the 1960's – astronomy was in turmoil.

ISLANDS IN THE VOID

On photographic plates, stars seldom emerge as anything but pinpricks of light. Clouds of gas, though beautiful, are often unstructured. But the galaxies – island universes, each of billions of stars – show a beauty of form and colour that make them the most startling objects to be revealed by large telescopes.

The most impressive shapes are those of the spiral galaxies, which make up about three quarters of the known galaxies in the Universe – there are about 1,000 million within the range of our telescopes. Spirals have central rounded nuclei, composed of old stars, while young stars and gas which has not yet been formed into stars, lie in a disc (as in our own Galaxy) about 100,000 light-years in diameter and a mere 2,000 light-years thick.

This material orbits the centre, like a Solar System on a cosmic scale, with the more distant stars travelling more slowly to form a whirlpool-like structure with two spiral arms. Oddly, the spiral arms do not consist of the same stars all the time. They are more like waves flowing through the galaxy, Catherine-wheel versions of ripples in a pool.

NGC 6946, a spiral galaxy in the constellation of Cygnus, shows the loose, clumsy arms surrounding the small nucleus that is typical of its class.

Overleaf: The sprawling, loose spiral of the Whirlpool Galaxy (M51) is accompanied by a small, irregular companion. One arm of the Whirlpool seems to cross in front of its companion, and astronomers assume the two are connected.

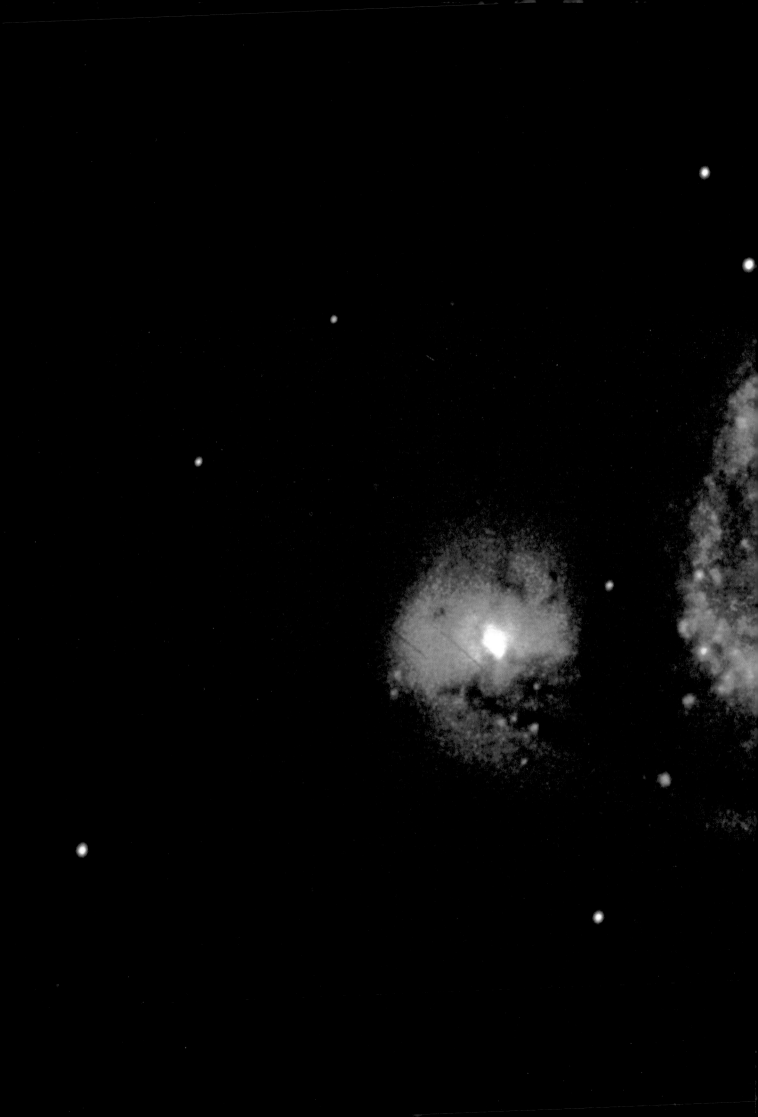

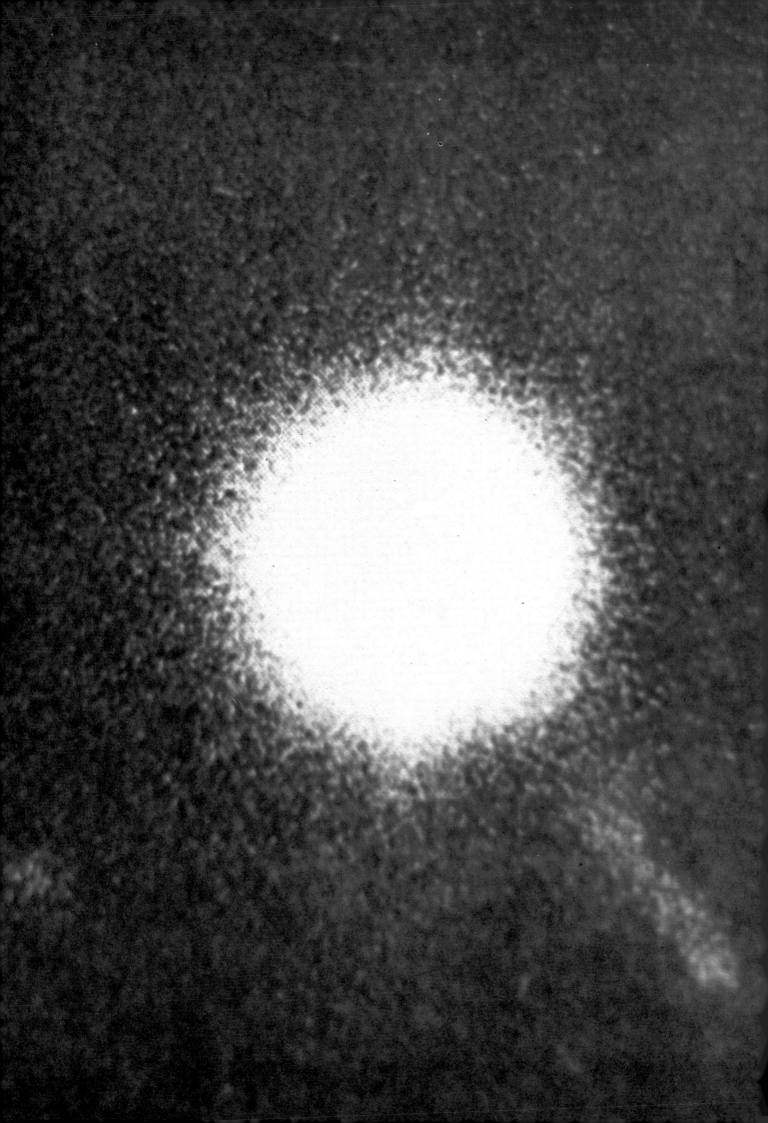

8/TO THE EDGE OF ETERNITY

JOHN GRIBBIN

By timing the expansion of the Universe, we can now say that it came into existence no longer than 20,000 million years ago, in a monumental fireball that provided the raw materials from which galaxies, stars and planets are still formed. We see around us the effects of that violent birth. The galaxies – as we know from studying energetic, dense quasars like 3C 273 (left) – are fleeing from

In the early 1960's, Hubble's law of the red-shift-distance relationship for galaxies was almost completely accepted by astronomers, and the evidence accumulating from studies of distant objects was beginning to tip the balance slightly in favour of a Universe that is evolving – changing with time. The Big Bang began to look stronger as a theory than the Steady State, although the case was far from proven.

Then, in 1963, there came a bombshell. A combination of radio astronomy and optical techniques revealed a completely new and un-expected kind of object in the Universe, the quasars. And these objects had such remarkable properties that, for a time, it seemed that their behaviour cast doubt on the validity of Hubble's law, removing the very foundation stone of the science of observational cosmology. For a time astronomers actually had to ask themselves not whether either Steady State or Big Bang theory was a better description of the Universe, but whether they really knew anything fundamental about the workings of the Universe at all.

The Coming of the Quasar

By 1962, radio astronomy had already begun to change our pictures of the Universe, revealing that many distant galaxies are intense emitters of radio energy, and must be associated with power-fully energetic astrophysical processes. But it remained very difficult to identify the exact optical counterparts of many radio sources, because the radio techniques then available gave only a rough indication of the direction of such a source, identifying a patch of sky in which there might be dozens or hundreds of visible galaxies, any one of which might be the source of the radio emission. (Today, more precise radio observa-tions are possible, see Chapter 9).

The key to the discovery of quasars was the Moon, which astronomers use to locate distant radio sources. In effect, the radio astronomers used the whole of the Earth-Moon system, 250,000 miles long, as their 'sighting tube'. During its orbit around the Earth, the Moon passes in front of all the objects on a narrow band of sky. It happens that one interesting and un-identified powerful radio source lay in this band, and that the Moon would pass in front of it during 1962, blocking off its radio noise temporarily. By observing the source with a radio telescope and waiting for the moment of cutoff, the astronomers knew that the source must lie somewhere along the 'front' edge of the Moon's disc at that second; when the signal reappeared, the source must be somewhere along the back edge of the disc of the moving Moon. Observers could time the gap between the signal's 'off' and 'on'; they also knew the time it took the various parts of the moon's disc to pass – the lunar equator hides stars for longer than the poles do. By fitting one measurement to the other, two particular spots could be identified – where the 'front' and 'back' arcs crossed. As it happened, only one of these points marked the site of a visible optical source.

The observers thus identified the optical counter-part of the radio object known as 3C 273. But to their surprise this turned out to be not a galaxy of stars, but a single star-like object.

This was puzzling enough, for it suggested that some stars inside our own Galaxy might be strong sources of radio noise. But worse was in store. When the redshift of 3C 273 was measured, it was found to be far too big to fit into any known pattern of star movements; and, if interpreted in line with Hubble's law, the redshift placed the object well outside the Milky Way, at galactic distances. 3C 273 didn't seem to be a star at all, but rather a star-like bright source as far away as a galaxy – hence the name, 'quasi-stellar' source, from which 'quasar' is derived.

To be so small that it looked like a star, yet so bright that it was visible from as far away as a galaxy, 3C 273, and the other quasars that began to be identified in the wake of this discovery, must produce as much energy as a galaxy but all from within a region no bigger than the nucleus of a 'normal' galaxy like our own.

As more and more quasars were discovered, many with redshifts so big that they made 3C 273 seem like a near neighbour, the problem got worse. Could there really be objects in the Uni-verse which were much smaller than galaxies but radiated as much power as all of the stars of a large galaxy put together? Naturally, astrono-mers began to look for alternative explanations. But these were just as worrying. Could quasars be fairly ordinary stars, much closer to home? That meant that Hubble's law might not apply to them, and once doubts were cast on Hubble's law as applied to quasars, there seemed to be room to doubt the law as universally applicable to galaxies.

Some astronomers argued that quasars might be 'local' objects, shot out from the centre of our own Galaxy in some vast explosion hurtling away from us so fast that their light is redshifted. But, if so, where was the source of all that energy?

Others argued that the redshifts might be nothing to do with velocity at all, but might really be produced by very strong gravitational fields, in line with Einstein's theory, which tells us that light struggling away from a very massive object will be redshifted in the same way as light from a rapidly receding object.

It took a good ten years for the debate to be resolved. Throughout most of the 1960's, there was no accepted picture of the Universe which could fit quasars into the same framework as galaxies, and while most astronomers clung to the Hubble law for galaxies, with its implications of an expanding Universe, the heretics could argue from a strong position that whatever strange process produced quasar redshifts also produced galaxy redshifts, and that the Universe might after all be static, as Einstein once thought.

It would simply be confusing to describe here all the different ideas that were thrashed out over the 15 years following 1963; but it is important to realize how the dramatic shock of the discovery of quasars opened up astronomical thinking.

It is really only since 1963 that we have come to appreciate just how violently active a place the Universe is. The discoveries of pulsars and X-ray sources (see Chapter 3) encouraged theorists to develop their ideas of what happens to matter when it is compressed to high densities. What they came up with – the modern understanding of black holes – carries across into the study of quasars to explain the origin of their energy.

Quasars come in a variety of shapes and sizes, sometimes with a radio structure spreading across millions of light years, but all of them deriving their energy from some tiny central source, which may produce bursts of energy up to ten thousand times the energy of all the stars in our Milky Way Galaxy put together. From Einstein's most famous equation, $E = mc^2$, which tells us how much energy could be obtained by converting *all* of a mass into energy (c is the speed of light); and assuming that only a fraction of this mass-energy is being liberated in quasars, we know that the *total* mass involved must be much more than a million times the mass of the Sun. Yet it is typically squeezed within a volume of space no larger across than our Solar System – the size of the largest stars. Such an object can only be a giant black hole. But black holes, in theory, *absorb* radiation; they cannot shine as the brightest phenomena in the Universe.

But there is no paradox. It is not the quasar that shines; it is the matter around it. A quasar is a supermassive black hole, formed as the result of a lot of matter collapsing together in one place. Gravity pulls the matter into a compact state, and as more matter piles on top this becomes more and more compressed, so that the gravity field surrounding the object becomes stronger and stronger. Eventually, nothing, not even light, can escape its grip. Inside the black hole, the mass must, according to the equations, continue to collapse into a mathematical point, the mirror image of the explosion outwards from a mathematical point that is at the heart of the Big Bang

model – but that bizarre pattern of events is beyond our observing. All we can know about directly is what is happening at the edge of the black hole, where any passing material will quickly be swallowed up in the maelstrom, a kind of gravitational whirlpool in space.

If the 'hole' were very big, then matter could quietly slip into it without a large release of energy. But if the mass of several million Suns is compressed into the space of the Solar System at the heart of a galaxy like the Milky Way, then there will be a massive pileup of material – gas, dust and even whole stars – sucked in by the intense gravity field but unable to squeeze immediately into the tight 'throat' funnelling down into the hole.

With the effects of rotation and magnetic fields added in, we now have a very efficient cosmic energy source – a central black hole surrounded by a swirling mass of material which is constantly being fed from outside and heated up by collision; as much as 20% of this whirlpool mass can be turned into radiation at wavelengths that span the electromagnetic spectrum – and into escaping energetic particles (cosmic rays). This is how a quasar can shine for a hundred million years or more while devouring the heart of a galaxy.

Such a process also explains the spread of radio emission into a typical 'double lobe' pattern on either side of a quasar or radio galaxy. As the central mass spins, winding its magnetic fields around itself, it spreads a blanket of infalling material in the plane of rotation, like the rings of Saturn but on a vastly greater scale. The heat generated in this mass produces energy which powers the fast particles of cosmic rays. Some of the energetic particles escape along the 'lines of least resistance' at the poles, producing jets pushing out on either side of the central black hole.

The whole picture hangs together in a thoroughly satisfactory way, given the reality of black holes and the efficiency with which gravity

According to current thinking, the presence of all matter curves space-time to create gravity. A black hole distorts the fabric of space-time to such an extent that beyond a certain point – the 'event horizon' – electromagnetic radiation itself (blue arrows) is locked in by the intense gravitational field, which steadily consumes nearby matter (red arrows). What happens to matter beyond the event horizon – at the hole itself – we can never directly know. But the presence of such objects can be deduced by the radiant energy of matter as it crowds into the lip of the hole.

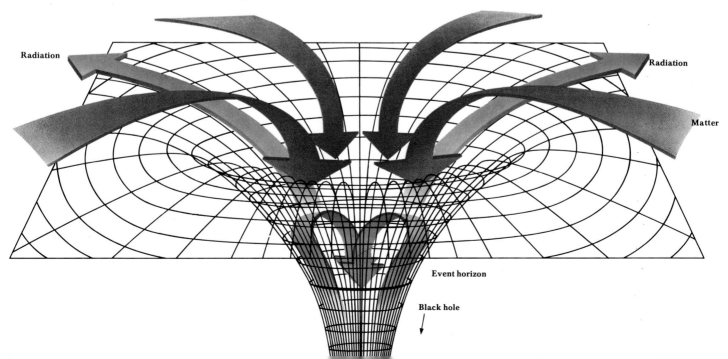

Radiation

Radiation

Matter

Event horizon

Black hole

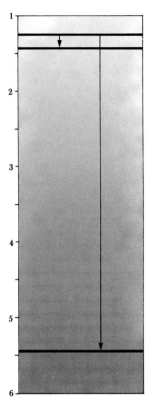

Since the discovery of quasar 3C 273 in 1962, the speeds – and distances – of remote objects have stretched out considerably. This graph shows one hydrogen absorbtion line, known as Lyman Alpha, redshifted by two objects. The scale, in Angstrom units (1,000's) shows observed frequencies. The top line is 'static' – it represents absorbtion in our own atmosphere. The other two lines, which record radiation emitted at precisely the same frequency, have been redshifted to much higher frequencies. The second line is of 3C 273 and indicates a recession speed of 15 per cent of the velocity of light. The bottom line is of quasar oH 471, receding from us at 80 per cent of the speed of light.

can turn a million or so solar masses into energy. Of course, the energies are like nothing we experience on Earth, or even in our own Milky Way Galaxy; but why should they be? The Universe is not only much bigger than we can really imagine, it is unimaginably more violent than anything conceived of only a few years ago.

The energetic activity harks right back to the Big Bang; for, with the energy source of quasars no longer a mystery, there is no need to invoke any bizarre explanations for their redshifts, and we can once again have faith in the fact of universal expansion and in the validity of Hubble's law.

Now, quasars provide a bonus for the cosmologist. Because they are so bright, many of them can be detected from much further away across the Universe than any ordinary galaxy; with the aid of redshift measurements, they provide us with the best cosmological probes of all, and scope to test the evolutionary theory of the Universe against the rival Steady State.

The techniques mentioned at the end of Chapter 7 depend on counting the number of sources we detect at different redshifts, equivalent to counting the numbers active at different times through the history of the Universe. Some evidence in this direction comes from 'ordinary' radio sources, but the best comes from quasars. Quasars are not, as it turns out, evenly distributed along the redshift scale. There are more fast-moving ones than slower ones. Since the faster moving ones are further away and older, the source counts show that there were more quasars long ago than there are now.

This evidence is on its own almost enough to settle the issue, but there is still a loophole. Can anyone *prove* that the chain of logic by which we reason out cosmic distances does not contain a flaw? Can we be sure that our instruments are not fooling us in some way, and that the source counts are not being 'biased' by some kind of consistent error?

There is one piece of evidence which seems to have settled the issue once and for all. It is the most remarkable scientific observation made in modern times, equalled in philosophical importance only by Olbers' paradox. This observation too was made by radio astronomers, and it was again a breakthrough of the 1960's, although with hindsight it is difficult to understand why it wasn't made at least ten years earlier. The discovery was of cosmic radiation that permeates all of space and is a distant cosmic echo of the Big Bang itself. With that as evidence, who could remain a Steady Stater?

The Echo of the Big Bang

Back in the 1940's, many theorists were grappling with the puzzle of just what the Universe must have been like in the period immediately after the Big Bang. In particular, they were eager to explain how the Universe ended up containing the observed proportion of basic material, about 80% hydrogen and 20% helium, from which stars and galaxies condensed, and which has

since been partly built up into other elements through nucleosynthesis inside stars (see Chapter 3 for details).

Because we have a good idea of how much helium was 'made' from hydrogen when the Universe was young, and because we know a good deal about nuclear reactions and the conditions required to allow this kind of fusion, by 1950 it was possible to get a rough idea of what happened to the Universe as it expanded away from the initial singularity. In particular, it was clear that unless the Universe had contained some inhibiting factor during the first minutes of its life, nuclear reactions would have proceeded so rapidly that many heavy elements would have been produced right away, leaving very little hydrogen left over for stars as we know them ever to get started. The inhibiting factor was intense electromagnetic radiation.

The proportion of helium actually made tells us how intense this inhibiting background radiation was early in the history of the Universe; and, allowing for the expansion of the Universe since the beginning, that makes it possible for astronomers to predict that the Universe today should still be full of radiation, but in a much weaker form. An image often used to explain the problem is that of a box full of gas at high pressure, a box that has expanded hugely. However large the box becomes, it will still contain the same amount of gas, but spread ever more thinly, at ever lower pressures. Radiation's equivalent of pressure, in this context, is temperature. The expansion of the Universe spreads the radiation more and more thinly and the temperature drops. The prediction, even from the first calculations of this kind, was that there should still be a cosmic background radiation, with a temperature equivalent to a few degrees above absolute zero ($-273.2°C$ is absolute zero, the theoretical point at which all molecular motion ceases.) Weak, to be sure; but radiation that ought to be easily detectable with simple radio telescopes, such as were available in the 1950's, at microwave frequencies. Yet no one even looked for such radiation then, and when it was found in the mid-1960's the discovery was by accident – the predictions had been largely forgotten and ignored.

In 1978, though, that discovery turned out to have been a particularly lucky accident for Arno Penzias and Robert Wilson, who made the first observations of the cosmic background, since they received the Nobel Prize for their work. Their story begins in 1964, when as young radio astronomers working at the Bell Telephone Laboratories in New Jersey they happened to have access to an unusually sensitive radio antenna-receiver system, designed and built for communications using the Echo satellite. Penzias and Wilson were interested in measuring the background radio noise coming from the Milky Way, and were considerably puzzled when they found, in 1964, a faint radio hiss, a universal background, coming from all directions in space.

Einstein's Universe: Strange but True

Albert Einstein, whose ideas were published over 60 years ago, is still regarded as the greatest theoretical physicist of this century, and will probably remain so until well into the next one. His theories of relativity – the Special Theory (1905) and the General Theory (1916) – were so complex, and so revolutionary that their conclusions took decades to assimilate and are still debated.

The Special Theory is founded on the principle that the velocity of light is the same for all observers, no matter what their motion relative to each other. The velocity of light is thus the ultimate speed limit of the Universe. At speeds approaching that of light, time begins to run more slowly, objects contract along their own length and they become increasingly massive. In Einstein's famous 'Twins Paradox', one fast-moving space-travelling twin would age more slowly that his stay-at-home brother. Einstein also showed the possibility of changing matter into energy, a theory summarized by his famous equation $E = mc^2$ (i.e. the energy of an object is equated with its mass times the square of the speed of light).

These bizarre predictions, which demand we abandon our common-sense notions of space and time, have been confirmed in many ways by experiment. For instance, orbiting cesium clocks, accurate to one part in many millions, have been measured running slower than equivalent clocks on Earth.

The General Theory extends relativity to gravitational fields. Einstein concluded that the presence of matter distorts space and time – space-time must be regarded together as curved. No-one can visualize curved space but the distortion is something like the 'distortion' of the Earth's surface when mapped in Mercator projection. The two-dimensional flatness of the map's surface represents a 'flatness' curved in three dimensions. In the same way, according to Einstein, we normally view the Universe as a three-dimensional map of a four-dimensional reality. The presence of matter creates additional local curvitures (as mountains do on the Earth's surface). To make another comparison, a star is like a metal ball placed on a rubber sheet. Other objects like stars and planets tend to 'roll' towards it down a kind of gravitational slope created by its presence in the Universe.

The effects of space curviture can be calculated and checked. A massive star's gravitational field should slightly reduce the energy of radiation leaving it. Starlight is bent as it passes the Sun. Mercury's eccentric orbit should revolve at a different speed than that predicted by Newton. Such effects have been verified.

Exactly *how* the Universe is curved, no-one yet knows. If it is positively curved, it may be finite in size, yet have no boundary, like the surface of the Earth. If a person could travel in an apparently straight line far enough, he would find himself returning to the point from which he began. Alternatively, it may be negatively curved, in which light follows an open path, like a parabola.

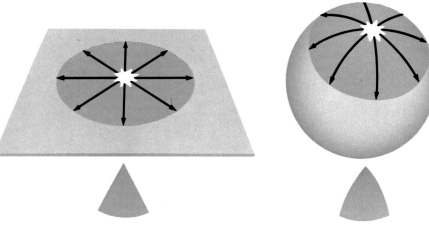

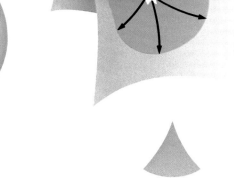

Flat Universe   Positively-curved Universe   Negatively-curved Universe

Arno Penzias (*left*) and Robert Wilson stand on the antenna with which they measured the temperature of the Universe in 1965 – a temperature of 2.7°K (−454.8°F), consistent with a cosmic 'Big Bang' 10,000–20,000 million years ago and the consequent expansion of the Universe since.

Over several months, they repeated the observations and tested their equipment to make sure the observations were genuine and not the result of faults in the system, but they ended up with the same result. It was only in 1965 that Penzias heard, at second or third hand, about new work by Princeton University theorist P.J.E. Peebles, who had done some new calculations predicting that the Big Bang origin of the Universe should have left a residue of radio noise detectable today as radiation with a temperature of a few degrees – no more than 10° above absolute, which is usually referred to as 0° on the Kelvin scale, named after the British physicist Lord Kelvin.

This immediately explained the observations, and caused enormous excitement as the news spread throughout the world's astronomical community, not least because the detection of the 'echo of the Big Bang' firmly nailed the lid on the coffin of the Steady State theory. But among all those excited astronomers there was one team that must have felt a deep sense of frustration. For two students, P.G. Roll and D.T. Wilkinson, had already been working at Princeton under the guidance of an older hand, Robert Dicke, to build a small radio telescope intended to test Peebles' prediction. They did this, confirmed the discovery made by Penzias and Wilson, and were in fact so hot on their heels that their publication of their own first observations of the background radiation appeared alongside the announcement of the discovery, the scientific paper by Penzias and Wilson, in the specialist *Astrophysical Journal*.

It seems more than a little harsh that the team which actually predicted the existence of the background, then went out and found it and explained exactly what they had found ended up with only a footnote in the history books, while the team that found the background by accident and couldn't explain it at all until prompted by others ended up with world acclaim and recognition from the Nobel Committee.

But there is another ironic footnote to the story. One of the first groups to make any prediction of the nature of the cosmic background radiation had been working on the problem as long ago as 1946. Their work, however, wasn't known even to the Princeton group of the mid 1960's – which is surprising, since one of those theorists of 1946 was the same Robert Dicke who was on the Princeton team in 1964! Steven Weinberg, discussing this curious tale in his book *The First Three Minutes*, comments: 'This is often the way it is in physics – our mistake is not that we take our theories too seriously, but that we do not take them seriously enough. It is always hard to realize that these numbers and equations we play with at our desks have something to do with the real world.' Had any radio astronomer of the 1950's accepted the existing Big Bang calculations as really the best description of the Universe, he could have made the relatively simple observations involved with equipment existing at that time, killing the Steady State theory before its battle with the Big Bang cosmology really got under way – and, incidentally, writing his own name into the history books.

Since 1964, observations of the background radiation have been improved until its 'temperature' can be set very accurately at 2.7K; that in turn provides a fine tuning so that cosmologists can, in their mind's eye, wind the clock back 15,000–20,000 million years to interpret very accurately what happened to the Universe, in terms of the all-important balance between matter and radiation, in the first few minutes after the Big Bang.

A Brief History of the Universe

We have become accustomed in this book to immense spaces of times. Now we have to slow the clock right down, for the story of that first immense explosion takes longer to tell than the events themselves.

We can pick up the story from the time just after 'the beginning' when the temperature of the fireball was a million million degrees. At such temperatures, particles of matter such as protons and electrons interacted continuously with their sub-atomic mirror images – their 'anti-particle' equivalents – and with the high temperature (that is, highly energetic) radiation background in a maelstrom of reactions. All the while, as the seconds passed, radiation was being turned into matter and matter into radiation.

As the Universe expanded, as seconds became minutes, the radiation cooled (became less energetic), and things became a little more orderly. The mass-energy increasingly stayed locked up in material particles, with the more massive particles settling out from the maelstrom first, as expansion continued and the energy density of the background radiation declined. At

100,000 million degrees, the protons and neutrons destined to make up virtually all the matter in the Universe as we know it had stabilized, but electrons and positrons continued to interact with the fading radiation left over from the primeval fireball.

At 1,000 million degrees, the weakening background radiation lost its capacity to make electron-positron pairs, and the left-over matter for the Universe as we now see it was fully settled. At about this temperature, too, the background allowed some of the protons and neutrons to 'cook' into helium, just like the nuclear fusion which operates in the Sun today.

Everything from the 'beginning' to the end of the era of nucleosynthesis – from a million million degrees down to 1,000 million degrees – took just over three minutes (hence the title of Weinberg's book). From then on, the time-scale began to stretch out.

The thousands of millions of years since the fireball have been taken up with the birth of stars, the condensation of galaxies, appearance of solar systems and – almost certainly not just on Earth – the development of life.

After the turmoil of the first million years or so following the Big Bang, the Universe was left full of swirling clouds of material, chiefly hydrogen and helium gas. At this time, and ever since, two conflicting forces were acting on the material. One was the universal expansion, tending to stretch the gas thinner and spread molecules and atoms apart from one another; the other was gravity, tending to hold things together so that if once a group of atoms combined in a swirling eddy they would attract others and grow into a bigger and bigger irregularity.

In a perfectly uniform Universe, there is no way in which large concentrations of matter could ever occur; indeed, it is quite difficult to explain how concentrations of matter as big as the galaxies can have formed in the time since the Big Bang, no more than 20,000 million years. Something in the initial fireball, it seems, must have produced density fluctuations, so that after the first million years the Universe was already 'lumpy' enough for galaxies to be able to grow. How this happened is not known; but it is straightforward to calculate how a galaxy would form from a 'proto-galaxy', one of these clouds of gas, held together by gravity, and containing enough material to form thousands of millions of suns.

The standard picture of galaxy formation envisages the gas collapsing first into a roughly elliptical shape under the influence of gravity, with stars forming out of irregularities in the collapsing cloud. At first, large hot stars, composed just of hydrogen and helium, will form in the young galaxy, run through their life cycles quickly and explode, scattering heavier elements into the interstellar medium. From these materials, 'second generation' stars can form, stars like our Sun, which have enough heavy elements for the leftover debris of star formation to produce planets and even life.

Heavy elements mix with gas into a dusty band and the whole galaxy settles into a spinning spiral – a picture that neatly explains how the Milky Way evolved, and why the stars of the halo are old and lack metals, while those of the disc are young and metal-rich. It is harder to explain why some galaxies stay as ellipticals, although it may be that this is related to how fast

The Sombrero Hat Galaxy (M 104; NGC 4594) is a spiral some 10 times the mass of our own Galaxy. It lies 41 million light-years away in Virgo. It may be a 'young' galaxy with a core of hot stars and a bank of gas from which second generation stars will later form.

the proto-galaxy was rotating, and to the power of the initial stellar explosions. Without much rotation to slow the collapse, very large early stars might form in the nucleus of an elliptical galaxy, then explode with such violence that they sweep gas and dust right out of the galaxy, leaving no scope for second generation stars to form. This would also, incidentally, leave a large mass at the nucleus, ready to coalesce into a black hole.

Although the pieces of this picture hang together fairly well, there are astronomers who do not believe that 'irregularities' as big as galaxies could have grown out of the initial fireball at all, unless something much more 'lumpy' was present to start with. They argue that from the very beginning there may have been compact, massive irregularities in the Universe, 'seeds' on which the material of the galaxies could condense by gravitational attraction.

The story is far from complete and astronomers are still puzzling over the differences between spiral and elliptical galaxies, and over the origin and evolution of both.

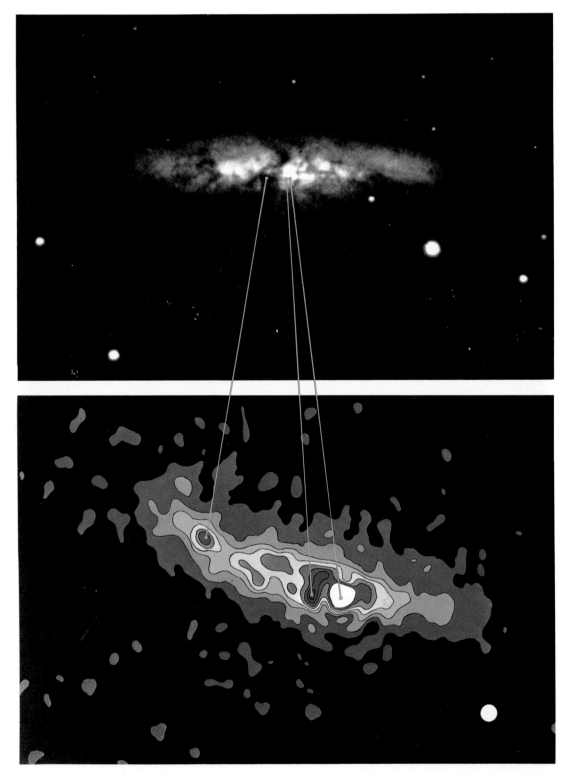

The irregular galaxy M82 (NGC 3034) consists mainly of billowing clouds of exploding gas and dust. Its central region is an active radio source (*below*, with colours showing increasing intensity towards the white centre). The galaxy is the nearest radio source to us, at 10 million light-years. What caused the explosion, which occurred between one and 10 million years ago, is not known

The Violent Galaxies

They have a mass of extraordinary objects to study, all of which could provide additional pieces in the puzzle of galactic evolution. Quasars are not the only violent active phenomena in the Universe. There are, for instance, exploding galaxies of many different kinds. For a while, astrophysicists were faced with the daunting prospect of having to find a different explanation for every kind of exploding galaxy. But today a coherent picture has emerged, with quasars seen as simply the most extreme version of a kind of activity that spreads right across the scale down to such ordinary galaxies as our own Milky Way. *All* galaxies, it now seems, may be active from time to time during their lives, and *all* galaxies including our own, may harbour black holes at their centres.

The classic examples of exploding galaxies are called Seyfert galaxies, which show activity somewhere in the middle range between a galaxy like the Milky Way and a quasar. This type of galaxy was first identified in 1943 by Carl Seyfert. after whom they are named. He found that about one per cent of all spirals have very insignificant spiral arms but very bright central regions (nuclei). Such systems have now been studied across the spectrum – from X-rays through ultra-violet and optical light into the infrared and radio regions; the observations show that the radiation of the bright nucleus is produced by hot gas in violent motion, with a central 'condensation' concentrating the mass of millions of suns in a small volume. This, of course, is just the kind of system now thought to provide the energy for quasars, the difference being that it is operating on a slightly smaller scale and that we can see (just) the spiral arms of the surrounding galaxy.

On either side of the Seyfert phenomenon in the chain of energies, there are active but apparently more 'normal' galaxies (intermediate between Seyferts and galaxies like our own), and even more compact and energetic objects, the N-galaxies (intermediate between Seyferts and quasars). As far as physical characteristics are concerned, there seems to be a continuous gradation from the most compact energy sources, quasars, right through to quiet galaxies. It may very well be that the differences are simply due to the different amount of matter that is present in the central black hole in each case. As more and more material is swallowed up in the nucleus, any particular energetic galaxy may slowly be converted into the next step up the chain.

During its active life, a system might undergo several spasms of activity, rather than shining brightly all the time. It may swallow matter up in bursts, then 'rest' while more matter falls in towards the black hole. Instead of one per cent of spirals being Seyferts, it is more likely that *all* spiral galaxies are Seyferts for one per cent of their lifetimes. Even our Milky Way Galaxy shows signs of a peculiar source at the galactic centre, modest by some standards, but perhaps a black hole with a mass of five million suns, sufficient to produce energetic activity from time to time, although not quite enough for our Galaxy ever to have been a quasar.

Eventually, though, any black hole of this kind will sweep the central regions of its parent galaxy clean, so that very little material is left to fall in, although stars may continue happily in their orbits far away from the nucleus. The quasar or Seyfert activity will fade away, and what we think of as an 'ordinary' galaxy will be left. But the fading away may take millions of years – in the meantime forming giant radio sources, the last pieces in the puzzle of galactic evolution.

Radio galaxies were first identified in the 1950's, when radio observing techniques became accurate enough to pinpoint the optical counter-parts of some of the many strong celestial radio sources that were already known then. One of the first identifications made was of the brightest radio source in the direction of Cygnus, dubbed Cygnus A. This is the second brightest object in the sky at radio wavelengths, and it had seemed natural to guess that it must be fairly nearby. Yet the identification, made in 1954, showed that Cygnus A is actually associated with a galaxy, shown by Hubble's law to be almost 650 million light years away from us. For the radio energy of the galaxy to be so bright at such a distance, it must be 10 million times more powerful, as a radio source, than a galaxy like Andromeda or the Milky Way.

Progress in identifying radio galaxies was slow, and by 1970 only a couple of hundred positive identifications of radio sources with galaxies had been made; but now the figure is well above 1,000, and the broad features of this class of cosmic phenomenon can be discerned. It turns out that Cygnus A, the first to be identified, is indeed typical of its kind.

Once again, radio galaxies typically show signs of active central nuclei; the most powerful radio galaxies are generally associated with giant elliptical galaxies; and, most intriguingly and significantly of all, most powerful radio galaxies show a characteristic 'double lobe' pattern, with the most intense radio noise coming from *either side* of the central galaxy, with only a weak radio emission from the nucleus itself.

There is very little doubt that these powerful radio sources are the result of gigantic explosions in the nuclei of the galaxies with which they are associated. Energetic charged particles are squirted out in opposite directions, and produce radio emission as they interact with the magnetic fields around the galaxy. But these moving particles can only move out at speeds less than the speed of light, and in the most extreme case, the giant radio source 3C 236, the ends of the radio lobes are 20 million light years apart. In other words, the explosion which produced the radio source we see now occurred at least 10 million years ago, since it must take that long for the material in each lobe to get where it is today. To

Centaurus A (NGC 5128), a giant radio galaxy in the southern hemisphere some 16 million light years away, emits radio waves from two lobes about 100 times the size of the visible galaxy.

put this in perspective, the distances involved are roughly 10 times the distance to our near neighbour, the Andromeda galaxy.

On the other hand, the fact that no bigger sources are found also tells us something about the activity of galactic nuclei, for it seems that these giant explosions never proceed for more than a few tens of millions of years without dying away, and certainly not for anything like the lifetime of a galaxy, which is around 10 *thousand* million years. What happens when they do fade away?

We can get a very clear picture of what seems to be a once powerful radio galaxy now on its last legs in our near neighbour Centaurus A, a mere 16 million light years away. This has very large radio components, stretching across two million light years, but is very weak as radio galaxies go, with just one-thousandth the power of Cygnus A. As well as the giant clouds of radio emission there are two smaller clouds just emerging from the edge of the galaxy, and the central radio source provides fully a fifth of the total radio noise, a high proportion by radio galaxy standards. So there is some evidence that after a huge outburst millions of years ago Centaurus A has had a second hiccup of activity on a lesser scale. This is just the kind of behaviour that would be associated with a once active central black hole, initially swamped by matter pouring onto it, which has now cleared most of

the space around itself and is kept modestly active by a last trickle of remaining mass being swept up as the hole moves into a quiet old age. In all probability, Centaurus A is the nearest massive black hole still showing the effects of this accretion. Once, the galaxy must have been as energetic as Cygnus A is today.

Many quasars are also powerful radio sources, and again they show a characteristic preference for the double-lobed structure, with trails of radio emitting material shot out and spreading across the light years. The relationship with radio galaxies seems clear, and there is no need to invoke any different processes to explain these superficially different phenomena after all. Most quasars must now be dead, their central black holes starved of matter to swallow up and turn into energy, so that they sit quietly at the centres of galaxies. Some, like Centaurus A, are dying; some holes never quite got enough mass together to become quasars, but have had an active life as the powerhouses of Seyferts, N-galaxies and the like. The Universe is a violent place, but less violent than it was, and getting less violent still as time goes by. Will it fade away into an ever expanding sea of ever quieter galaxies, with the lights of the stars and galaxies going out one by one until all that is left is a void filled with a scattering of black holes? Or is some more interesting fate in prospect?

From Big Bang to What?

We have already taken a look at the beginning of the Universe. What will be its end? And why is it constituted the way it is? The answers may bring us full circle and tell us what was there before the 'beginning', and what there may be after the 'end'.

The crucial factor turns out to be how much matter there is in the whole Universe. For, just as galaxies are combined by the attractive force of gravity, the total mass of the whole Universe provides an insistent tug on every galaxy and cluster of galaxies within the Universe. As the Universe expands outward from the initial explosion of the Big Bang, the gravitational tug gradually decreases. But it is always present, and if there is enough matter, eventually gravity will overcome the expansion, and the Universe will slow to a halt. The whole drama will then be played out in reverse. The Universe will collapse faster and faster under the overwhelming pull of gravity until it is squashed into another fireball, perhaps then to bounce back out again.

On the other hand, without enough mass the expansion can never be reversed, and the spread of matter in the Universe will indeed get thinner and thinner forever as the Universe ages.

By measuring the amount of matter – the number of galaxies – in the volume of space we can see, cosmologists get an indication of what the density of matter is throughout the Universe. Intriguingly, but annoyingly, the best estimates we have are that the Universe is balanced on a knife edge. As far as we can tell, there is either just enough matter to 'close' the Universe and make it, eventually, collapse; or there is not quite enough, so that it will expand forever. More as an act of faith than anything, most astronomers at present seem to prefer the continuously expanding model. They could well be wrong, for one thing is clear: we are not going to find any *less* matter in the Universe, while there

might well be material we don't yet know about – cold gas between the galaxies, for instance, or black holes at present undetected. So even now the two versions of the fate of the Universe should be given equally serious considerations.

This is just as well for the theorists, since there is really nothing more to be said of the continuously expanding – 'open' – Universe than that it will expand forever and die. If this is the ultimate destiny of the Universe, we are very lucky to be around at a time when it is relatively young and active, with so many interesting phenomena to observe.

But the closed universe model is much more interesting. Apart from anything else, it removes the puzzle of the 'beginning'. For now we can say that before the Big Bang there was another cycle of expansion and collapse, and that these oscillations have been going on for ever, and will go on for ever.

Once we start to ask these deeper questions about the nature of the Universe – philosophical questions, if you like – some fascinating puzzles emerge, not least being the way the Universe is just right for life as we know it. Our surroundings really are pretty unlikely, in terms of the standard rules of physics. We know from physical experiments that the most likely state of any system is one of uniformity, which would correspond to a random spread of matter across space, with no clumps of galaxies, stars and planets (let alone life) to provide order among the chaos. In physics terminology, the Universe contains a lot of information – it is a complicated place. And that is very unlikely, if we are dealing with a random state.

Does that mean that the Universe is *not* the result of chance? Have cosmologists rediscovered God? Maybe – if this Universe is a one-off. But if it is cyclic, it is possible to envisage not just one Universe arising out of the initial conditions of the Big Bang, but many alternative possible Universes, depending on just which way many, random (chance) processes turn out as the Universe expands. It might be that all of the infinite variety of possible Universes get a chance to develop into reality. In the infinite chain of cycles stretching off into the past and future, everything that can possibly happen does happen. 'Our' Universe is a very unlikely development, perhaps, but it must happen sometime. We are here to see it because life can develop in this kind of unlikely universe, with the kinds of stars and galaxies and planets that actually exist.

This must be the ultimate step in the long process of setting man in the context of Creation. Less than 500 years ago, he seemed its very centre. Over the centuries, we have discovered that the Earth orbits the Sun, not the other way around; that the Sun is just one insignificant star in the Milky Way Galaxy; that the Milky Way is but one modest galaxy with no special place in the Universe; and now that the Universe as we know it is, perhaps, just one Universe in an infinite number.

Two possible versions of the origin and fate of the Universe show the fine difference there seems to be at present between cyclical Big Bangs and infinite dispersion. The only way to find which (if either) is correct is to assess accurately the amount of matter in the Universe.

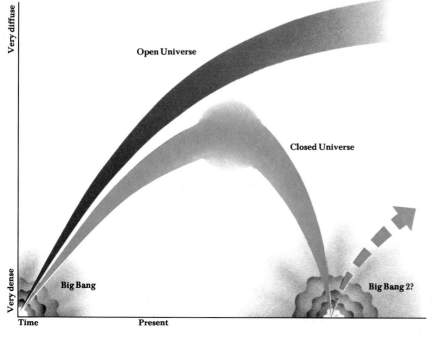

Very diffuse

Open Universe

Closed Universe

Very dense

Big Bang

Big Bang 2?

Time

Present

9/THE QUEST FOR KNOWLEDGE

HEATHER COUPER

As the history of astronomy itself shows, the difficulties of discovering anything firm about the Universe are immense. All we have to go on is the information conveyed by light. Until very recently, mankind could do nothing but observe – and that poorly, immersed as we are at the bottom of a sea of hazy atmosphere. Over the last few decades, however, scientists have been able to undertake research with astonishing new tools – radio-telescopes that analyse invisible regions of the spectrum (like the dish at left at the Mullard Observatory, Cambridge), devices that can count the basic constituents of light and information from satellites outside the atmosphere. These – and others – are the techniques that have sparked the current revolution in astronomy.

In many ways astronomy – unlike other sciences – has changed little with the centuries. There have been vast improvements in techniques; but these have been refinements rather than differences in actual approach. Until the last decade, astronomy has been almost exclusively a purely observational science. Experiment, the lynch-pin of most other sciences, has little place in astronomy. An astronomer cannot dissect a star: all he can do is observe its radiation, and then attempt to say something of its motions and structure. This has given astronomy an extraordinary historical continuity.

Astronomy's beginnings will always remain shrouded in the mists of antiquity. We know from the great megaliths that Man must have had an intimate and sophisticated knowledge of the workings of the cosmos at least 4,000 years in the past. Earlier perhaps than this, men living in the Near East and around the shores of the Mediterranean had begun to gaze heavenwards. Blessed with a good climate and clear skies, they soon became familiar with the starry vault. As we have learned in Chapter 2, these early astronomers joined up the stars into an imaginative array of groups and patterns which wheeled in procession across the sky from dusk to dawn, making up a huge celestial clock.

Their clock also doubled as a calendar, for it soon became obvious that changing star patterns

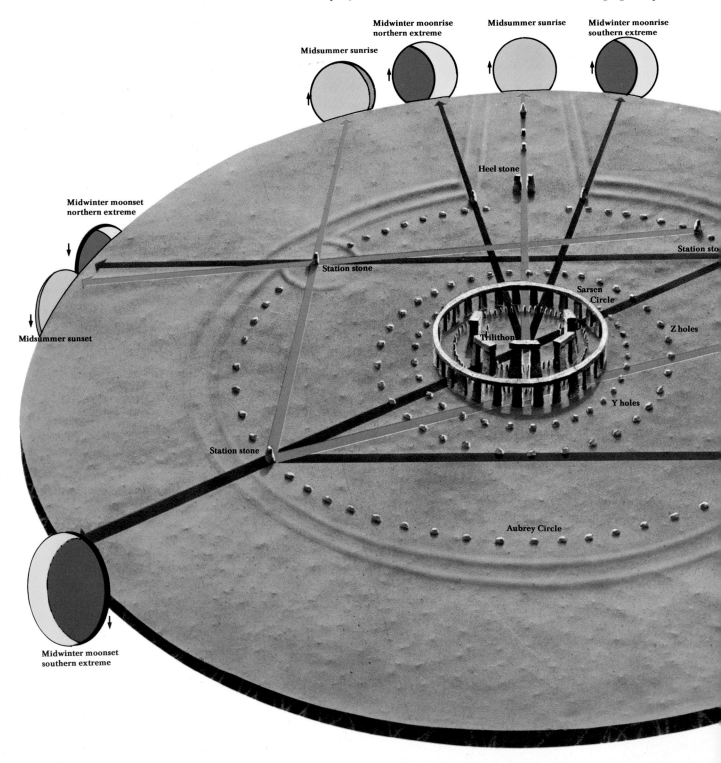

became visible as the days wore on; and the power of these first astronomers lay in their ability to predict regular earthly events from different configurations of this heavenly calendar. For example, the appearance of certain constellation patterns heralded the coming of spring and with it, the signal for the farmers to sow their crops; and the Egyptian astronomers knew that when Sirius, the brightest star in the sky, rose just after sunset, the mighty Nile would burst its banks and start its much-needed annual flood. To the uninitiated, it must have seemed almost miraculous that the stars could foretell – or even cause – events upon which Man's very survival depended. By extension, it was assumed the stars controlled *all* human life, and astrology was born.

The early watchers noticed, too, that other bodies besides the stars inhabited the heavens. Most obvious of these were the Sun and Moon, whose extreme brilliance made them very important. There were other objects, too, which behaved in a way which set them apart. Whereas the stars stayed fixed in their constellation patterns, the five bright 'wandering stars' (called *planetes* by the Greeks, hence our modern word 'planets') shifted slowly against the starry background in a manner which was independent of the motion of the sky.

Among the first peoples to leave written ac-

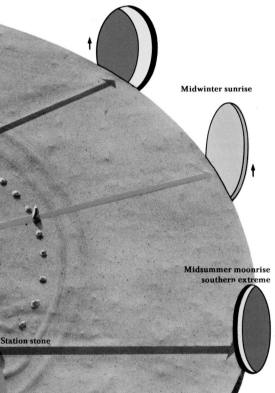

Midsummer moonrise
northern extreme

Midwinter sunrise

Midsummer moonrise
southern extreme

Station stone

Midwinter sunset

counts of their astronomical observations were the Babylonians, active around 1500 BC, who inscribed their records on clay tablets. Their interest in the heavens was not a purely objective one, as that of a modern astronomer would be: the observations they made, and their interpretation of them, were partly motivated by astrological concerns. They believed that the heavenly bodies had an influence not on individuals, but on the governing of a country and its affairs of state. In such times of political unrest, it would clearly be advantageous to see into the future. The sorts of observations which they needed for these divinations were of the movements of bodies in the sky, and of phenomena which were cyclic; but they showed little interest in recording transient events such as novae, supernovae and the appearances of bright comets. And so, from this observational bias, we see how earthly concerns were able to influence and even dictate the progress of a science.

Despite these selection effects, the Babylonian astronomers were careful and diligent observers. Watching the sky from the roofs of specially constructed towers ('towers of Babel'), they mapped the brighter stars and paid particular attention to the band of constellations along which the Sun, Moon and planets appeared to move, dividing it up into 12 sections (as we have

Stonehenge on Salisbury Plain, is an astronomical observatory of remarkable sophistication. Built in several stages from 2900 BC to 1400 BC, this neolithic computer records the sunrise, sunset, moonrise and moonset at mid-winter and mid-summer. The key alignment is that from the centre through the heel stone to mid-summer sunrise. Four 'station stones' – part of the so-called Aubrey Circle of 56 holes – mark a number of the observations. Because the Moon's orbit 'rocks' slightly, the Moon rises and sets at a slightly different place every day, in an 18.6-year cycle that is also recorded in the stones.

The Universe of Aristotle places the Earth at the centre with its four elements – earth, air, fire and water – surrounded by concentric spheres containing the heavenly bodies. The spheres revolve on their axes, driven by the primum mobile or 'prime mover' in the outer sphere, which medieval Christians could equate with God.

today) so that they could more easily chart the motions of the wandering bodies. They knew the length of the year, from the Sun's motion against the stars, to an astonishing accuracy of four and a half minutes, but preferred to use a calendar based on the motion of the Moon.

From these studies, the Babylonians drew up a model of the Earth and the Universe. The Earth was believed to be a flat, round disc, surrounded by water and ringed with mountains which supported the dome of the sky; the dome itself was pierced with doorways to enable the various celestial bodies to enter and leave. Completing the picture was a shell of water surrounding the dome, which was necessary to account for rain.

The Babylonians did not care to theorize as to what made the Sun, Moon and planets move, unlike the Egyptians, who flourished at about the same time. Their conception of the Universe was superficially similar to that of the Babylonians, with a flat Earth surmounted by a heavenly dome, but rather than being based on observation, it was founded on superstition. What were the heavens? The Egyptians replied that they were actually the star-spangled body of their goddess Nut, who arched in an ungainly fashion across the sky supported by the arms of the air god Shu. Why did the Sun move across the sky during the course of a day? Because it sat in a boat which made the daily voyage along Nut's body, they answered.

It is hardly surprising to learn that astronomy, as a science, did not flourish in Egypt. However, as an aid to timekeeping and calendar making it reigned supreme, for the Egyptians were first and foremost a practical race, and highly skilled in engineering. Like the Babylonians, their observations – made, at best, with crude wooden altitude-measuring instruments – were painstakingly accurate, as we know from the alignments of the pyramids and their predictions of the Nile floods.

The next race to inherit this tradition of ob-

servation and meticulous calculation were the Greeks, with whom the Egyptians and the Babylonians undoubtedly traded. But upon these foundations, they developed a discipline whose framework and conclusions were destined to guide subsequent civilizations in many fields far removed from science for the next 2,000 years.

Starting with Pythagoras, the great geometer, they replaced the concept of a flat Earth by one of a globe which remained stationary in the middle of a spherical Universe. Thus began their preoccupation with symmetry, which, with a few exceptions, coloured the interpretation of their observations for the whole time the Greek civilization flourished. Indeed, so powerful did this obsession with symmetry and perfection become under the great philosopher Plato, that those who aspired to wisdom were advised not to make any observations (of anything) at all; for the objects under observation would be but crude and misleading approximations to the real and perfect truth.

Fortunately, a number of Plato's pupils had other ideas, and sought to incorporate astronomical observations into their own teaching of philosophy. One was Eudoxus, who tried to account for the motions of the planets as the rotation of a nest of crystalline spheres centred on the Earth – the first of countless attempts to explain these complicated movements by variations on a circular theme.

Another of Plato's pupils – Aristotle – was destined to become the greatest and most influential philosopher of all time. His first appointment on leaving Plato's Academy was as tutor to the young Alexander of Macedon (later to become Alexander the Great), but he returned to Athens on the assassination of Alexander's father to set up his own school (the Lyceum). Here students flocked to hear Aristotle's teachings on philosophy, biology, medicine and astronomy. His work was startlingly original and ahead of its time, and his astronomical teachings in particular were to determine the course of Western science during a very crucial stage in its development.

On the surface, Aristotle's view of the cosmos differed little from that put forward by Pythagoras 200 years before: he too believed that the

In this representation of the Egyptian cosmos, the Earth God, Qeb, lies spanned by Shu (kneeling) representing the air and the Goddess Nut, representing the sky, arched above. The Moon (left) and the Sun (right) sail over Nut's body.

globe of the Earth remained stationary at the centre of a spherical Universe. The difference lay in Aristotle's approach to the problem. He argued his case from logical, scientific principles, relying heavily on observation. The Earth was a globe, he taught, because different constellation patterns appeared in the sky if one ventured substantially north or south of Athens; and there was the extra evidence that the Earth's shadow on the Moon looked curved during lunar eclipses. However, the Earth did not move in space, because if it did, we would expect to see the stars rising and setting in different places as time passed. All the heavenly bodies were perfect – because no imperfections had been observed. The Universe itself was symmetrical and the planets indeed travelled around the Earth, but not in as simple a manner as Eudoxus had proposed. Aristotle had to bring in some 55 homocentric spheres to clear up that problem.

Aristotle's picture of the Universe was logical, symmetrical, consistent, beautiful and, above all, scientifically reasoned. It claimed to be accurate, and therefore stimulated further observation and precise measurement. Eratosthenes, one of the first directors of the great museum and library in Alexandria (a true forerunner of a modern research institution) carefully measured the circumference of the Earth by observing the Sun's position at noon at different latitudes, getting much the same value as we do today; while Aristarchus of Samos, his contemporary, was busily measuring the distances and sizes of the Sun and Moon. Aristarchus' results were wildly inaccurate, but there was nothing wrong with the scientific principles behind his methods: it was simply that extremely precise angular measurements were required, and the requisite instrumentation did not then exist.

Aristarchus is widely remembered for his far-sighted proposals that the Earth is just one of the planets in orbit about the Sun, and that the rising and setting of the stars is simply a consequence of the daily rotation of the Earth. But his theories could not be tested by observation: they were therefore not accepted as valid.

Shaken by political upheaval, and threatened by the rise of the Roman Empire, Greek culture died around the first century AD. During the brief settled period which followed, scholarship flourished for a while, particularly in research centres like the library in Alexandria. But the mood was nostalgic, and efforts were made to incorporate the Greek teachings of the past whenever possible. One of the Alexandrian researchers was Ptolemy (Claudius Ptolemaeus), who compiled the magnificent *Almagest* – a synthesis of the astronomical and geographical knowledge of the Greek's, which incorporated many of his own ideas and observations.

He strongly supported Aristotle's picture of the Universe, believing that he had obtained additional evidence against the movement of the Earth from the compilation of his vast star catalogue. Ptolemy's was not the first. Two centuries before him, the great observer Hipparchus had drawn up a catalogue of 850 stars, arranged in the six different brightness classes (magnitudes) which we use today. By comparing his catalogue with the incomplete charts of his predecessors Hipparchus, in 130 BC, was able to establish the existence of precession. In a similar, and equally scientific fashion, Ptolemy compared his even larger catalogue (listing more than a thousand stars) with that of Hipparchus, and maintained that their similarity proved the Earth to be stationary in space.

Another of Ptolemy's innovations – and the one for which he is chiefly remembered – was in his explanation of planetary motions by a system of epicycles and deferents. On this theory, planets orbit the Earth in combination of two types of circle – a wide circle (the deferent) and individual loops (epicycles) around this mean path – thereby accounting for the strange (retrograde) motion sometimes observed. Apollonius of Alexandria had actually suggested this mechanism over 400 years earlier, but Ptolemy developed and refined it to such a degree that the *Almagest* contained tables predicting the positions of the planets for many years in the future.

Ptolemy was, in a sense, right; the planets *do* move in ways that can very nearly be described in terms of circles, except that one of the circles is the Earth's own orbit. His system was thus very accurate, but not perfect; it was also horrendously complicated. The *Almagest* was a fitting memorial to seven hundred years of Greek astronomy; and more than any other work it was responsible for laying the foundations of modern science. But such was not to be for another thousand years. Political and religious chaos soon plunged most of the civilized Western world into cultural darkness, and the flame of science was extinguished.

The Copernican Revolution

Our story now leaps some 1,500 years, to a chilly spring in Poland where an old man lay on his deathbed. Clutched in his hands was a book destined to sow the seeds of a revolution. The year was 1543; the man, Nicolas Copernicus.

In Copernicus' day, Aristotle once again reigned supreme, but certainly not in a way that Aristotle himself would have liked. The great man's teachings had been distorted into a form which would aid and abet the power of the Church. This had come to be by a roundabout route. In the 7th century AD, the Arabs had swept across the Near East in the name of Mohammed, penetrating as far west as Egypt and Spain. Stumbling across the Greek texts which remained in the devastated Alexandrian library, they realized that some could be put to practical use, and began the task of translating them. Among the works they discovered were the *Almagest* and the teachings of Aristotle. Both helped the Arabs in constructing lunar and solar calendars for religious and civil use in formulating a picture of the Universe.

This medieval drawing of Ptolemy shows him holding an astrolabe, a device used to measure the height of stars above the horizon. His view of the Universe resembled Aristotle's in that the Earth was placed at the centre, but the motions of the planets were described in theoretical and geometrical terms as combinations of various sorts of circle. His scheme, which was complex but surprisingly accurate, was taught alongside that of Aristotle in the Middle Ages.

Copernicus, a Polish cleric, for whom astronomy was little more than a hobby, began the scientific revolution in 1543, when he suggested that to place the Sun at the centre of the Solar System would provide a simpler and more accurate description of planetary motions than Ptolemy's system.

By the start of the mediaeval era, news of the Greek texts, as interpreted by the Arabs, began to filter through Spain and thence to the great monasteries and centres of learning in Northern Europe. The Crusades enabled scholars to get their hands on some Greek originals, and translation soon began. The Church – at the pinnacle of its wealth and power – was not slow to realize that the teachings of Aristotle could be re-interpreted in a Christian way. Here was a tautly argued, but non-partisan view of an Earth surrounded by a perfect, unchanging Universe. Aristotle's scientific conception was twisted into dogma to perpetuate religious power, and such an attitude stultified objective enquiry. Into this climate Copernicus was born.

Copernicus made his living as a priest, but had studied astronomy and mathematics at university and was well acquainted with Ptolemy's explanation of the planetary motions. Others, too, had been looking at these motions afresh, with a view to drawing up navigational tables for the long sea voyages of trade and exploration which had just commenced. Copernicus was dismayed at how unwieldy Ptolemy's system had become, now needing many extra epicycles to achieve reasonable predictions. How much simpler, he thought, if the Sun were to replace the Earth in the centre of the Universe; and that the Earth, along with the planets, moved around the Sun. Such a view – whatever its predictive appeal – was tantamount to heresy in the eyes of the Church.

It is still unknown whether Copernicus genuinely believed that the Earth moved, or merely regarded his suggestion as a computational device; whatever the truth, it remains that Copernicus did not publish his ideas until the year of his death. Tradition has it that the first copy of *De Revolutionibus* – the book which set forth his heliocentric theory – arrived at the dying man's bedside just in time.

Copernicus had no observational evidence to support his theory. Yet its effects were far-reaching: not only did it contradict the contem-porary dogma, but it prompted scholars to generate their own ideas at last, rather than continuing to look to the great masters of a by-gone civilization.

The one hundred years which followed were a watershed in astronomy: Copernicus' thinking was brought to fruition by the efforts of three great contemporaries – Tycho Brahe, Johannes

In this medieval painting, an astronomer takes navigational measurements – vital information with which sailors could check the date, time of day and position.

Right: Galileo's work with his own telescope enabled him to see the Moons of Jupiter – a Solar System in miniature – which convinced him that the Copernican theory was correct. His penetrating arguments on its behalf – combined with his own fiery temper and cutting wit – struck the Church as heretical, and he was forced to recant by the Inquisition.

Kepler, and Galileo Galilei.

Tycho was born in Denmark only three and a half years after Copernicus died, growing up to become a colourful, if rather hot-headed figure; but his astronomical observations were quite without equal. He began his research, generously sponsored by the King of Denmark, at a time when scientific instrument manufacture was increasing in leaps and bounds. The main emphasis was on surveying instruments – mapping and town construction being of paramount importance in those financially buoyant times – but the principles were easily adapted to angular measurements in the heavens. Tycho started to record the positions of Sun, stars, Moon and planets, at first with mural quadrants, and later with portable sextants of his own design. His instruments were unprecedentedly precise, partly because their great size allowed large, easily-read scales, and also because Tycho had reasoned that each observation he made was subject to various sources of error, which he sought to track down and eliminate. In this way, Tycho made positional observations which were five times more accurate than those of Hipparchus. But because of his religious beliefs, and also because he could see no evidence for the motion of the Earth in his observations, Tycho did not support Copernicus' theory.

It remained to his assistant, Johannes Kepler, to demonstrate how overwhelmingly these supported it. Kepler, 25 years Tycho's junior, was a brilliant mathematician who believed strongly in Copernicus' ideas. He had outlined these thoughts in his popular *Mysterium Cosmographicum* where he made it clear that his motivation was not purely scientific: he believed that the distances of the planets from the Sun bore a geometrical relationship to one another as a result of their 'spheres' (orbits) being separated by one of the regular geometrical solids (cube, tetrahedron, etc).

This search for heavenly harmony led him to analyse Tycho's comprehensive planetary observations after the latter's death in 1601. Mars

was particularly puzzling, and traced a path which could be accurately predicted by neither Ptolemy nor Copernicus. Kepler hit on the answer, making the decisive step which wrenched astronomy away from all considerations of perfection: Mars did not travel around the Sun in a circle, but in an ellipse. From this discovery stemmed Kepler's laws of planetary motion, which he considered to be the ultimate confirmation of the grand design. It is somewhat ironic that we no longer remember Kepler for his harmonies, but for their bare skeleton: a powerful and elegant set of laws which allow us to determine the dynamics and scale of our Solar System.

Our final member of the 'watershed' trio – Galileo – had been appointed professor of mathematics at Padua in 1592. Italy was then in the grip of rigid pro-Aristotelian feeling, the Church authorities equating non-orthodoxy in science with unauthorized religious beliefs. Galileo made himself unpopular with the authorities because of his open, and somewhat tactless support for the Copernican theory, and he was dismissed from Pisa university as a result.

In 1609 came the news that was to hammer the final nail into the coffin of Aristotelian dogma, and the first into Galileo's own. An optician in Holland – Hans Lippershey – had patented a tube containing two lenses which made distant objects appear closer. Already, Lippershey was aware of their military application and had sold a number of these 'telescopes' to the Dutch army. Galileo was familiar with lenses and the principles of optics, as well he might be, for Venice was a centre of glass manufacture; and it was not long before he constructed his own telescope and turned it to the skies.

He was astonished at what he saw. The Moon, instead of being a perfect body as the Church preached, looked like the Earth, with seas and mountains; the inviolate Sun had spots; Venus displayed phases, showing conclusively that she orbited the Sun and not the Earth; and Jupiter was accompanied by four tiny bodies which circled the planet perpetually. All these dis-

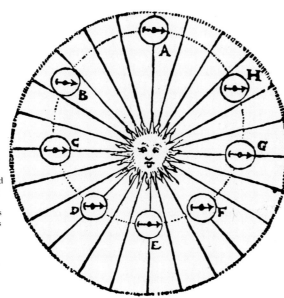

The Danish astronomer Tycho Brahe sits in the centre of this drawing, celebrating his life and works (*left*). He is bracketed by the great quadrant with which he made planetary observations of unprecedented accuracy. His observations enabled his assistant, Kepler, to show that planets do not move in circles, but in ellipses (*right*) sweeping out equal areas of space in equal units of time.

coveries were made with miniscule telescopes which would be regarded as toys today – none magnified more than twenty times.

Fired with enthusiasm for his findings, and by their support for the Copernican theory, Galileo rushed into print with a series of dangerously explicit books which succeeded in infuriating the Church authorities. Eventually, he was forced to recant under the Inquisition, but even this did not prevent him from pursuing science. For the last ten years of his life he was under virtual house arrest, and despite his growing blindness he turned his mind to the problems of moving bodies and their acceleration. Fittingly, it was the kind of work which foreshadowed that of Isaac Newton, who was born in the year that Galileo died.

England of 1642 was a very different place from Italy. Life was no longer dictated by religious observance; reading and learning flourished; exploration was the norm in all endeavours and a healthy spirit of scientific enquiry had emerged. Experimental science in particular prospered, as scientific instrument makers improved their craftsmanship dramatically; and on the theoretical side, reports of Copernicus's and Kepler's work came swiftly over from the continent and met with little resistance. News of the telescope was sweeping across Europe and astronomers and craftsmen alike tried to improve the instruments. The first telescope workshops began to be set up in London.

This heady atmosphere would have permeated the centres of learning at the time, and even quiet, introspective students like Isaac Newton – who went up to Trinity College, Cambridge, in 1661 – must have been affected. He showed little early promise, but his Professor of Mathematics, Isaac Barrow, realized that the young man had

great ability in optics. But Barrow was to see little of his student for a couple of years, for the University closed in 1665 as a precaution against the advancing plague, and Newton returned home to Lincolnshire.

Always a solitary person, Newton was glad of this sojourn. He spent the time thinking; and in two short years laid the foundations which were to dictate much of the course of science right up to the present day. He pondered on the behaviour of moving bodies, like Galileo had before him, and wondered if the stars and planets were subject to the same laws as bodies on Earth. He thought about the nature of light and colour, and came to the conclusion – after experimenting with a prism purchased from a travelling fair – that 'white' light was really made up of a mixture of colours. Had his prism been of higher quality, he would undoubtedly have discovered the principle of the spectroscope.

On returning to Cambridge – and quickly assuming the Chair of Mathematics – Newton translated his thoughts into actualities. His researches on light led him to design a telescope which collected light with a mirror, avoiding the (until then insuperable) problems of false colour and lens distortion. This 'Newtonian' reflecting telescope – which he exhibited at the newly-formed Royal Society in 1672 – was the forerunner of all the world's greatest telescopes today.

But Newton is renowned for his theory of Universal Gravitation, which tells us that the motion of a star around a galaxy can be described in exactly the same way as that of an apple falling to the ground. En route to a complete theory, Newton was forced to devise a powerful new mathematical tool, the calculus; and in 1687 he was finally persuaded to publish his results in a book entitled *Principia Mathematica Philosophiae*

Newton and the Apple

Every schoolchild has heard the story of Newton and the apple. It sounds apochryphal. But it was a real incident, as William Stukeley in his *Memoirs of Sir Isaac Newton*, of 1752 records:

'On 15 April 1726 I paid a visit to Sir Isaac at his lodgings in Orbels buildings in Kensington, dined with him and spent the whole day with him, alone . . .

'After dinner, the weather being warm, we went into the garden and drank tea, under the shade of the apple trees, only he and myself. Amidst other discourse, he told me, he was just in the same situation, as when formerly, the notion of gravitation came into his mind. It was occasion'd by the fall of an apple, as he sat in a contemplative mood. Why should that apple always descend perpendicularly to the ground, thought he to himself. Why should it not go sideways or upwards, but constantly to the earth's centre? Assuredly, the reason is, that the earth draws it. There must be a drawing power in the matter of the earth; and it

Sir Isaac Newton.

must be in the earth's centre, not in any side of the earth. Therefore does this apple fall perpendicularly, or towards the centre. If matter thus draws matter, it must be in proportion of its quantity. Therefore the apple draws the earth, as well as the earth draws the apple. That there is a power, like that we here call gravity, which extends its self thro' the Universe.

And thus by degrees he began to apply this property of gravitation to the motion of the earth and of the heavenly bodys, to consider their distances, their magnitudes and their periodical revolutions; to find out, that this property conjointly with a progressive motion impressed on them at the beginning, perfectly solv'd their circular courses; kept the planets from falling upon one another, or dropping all together into one centre; and thus he unfolded the Universe. This was the birth of those amazing discoverys, whereby he built philosophy on a solid foundation, to the astonishment of all Europe.'

'The Leviathan of Parsonstown'.

Throughout the early 19th century, the world's largest telescope was Herschel's 48-inch reflector, built in 1789. But in 1845, an Irish nobleman, William Parsons, the third Earl of Rosse (*right*), built a 72-inch (183 cm) telescope (*below*), which remained the world's largest until the opening of the Mount Wilson 100-inch (254 cm) in 1917.

Lord Rosse, born in 1800, was a wealthy landowner who entered Parliament when still an Oxford undergraduate. But his chief interest was science. At the age of 27, he began to design large reflecting telescopes – those that use mirrors to gather light – which were still in an experimental stage (Herschel had never published his methods).

Rosse set about developing a new alloy of copper and tin that would take the maximum amount of polish. In 1839, he successfully made a three-foot speculum (mirror), but found that it responded too rapidly to changes in air temperature, and he moved on to more solid structures. By 1843, he had made two massive mirrors – 72 inches across

and weighing four tons apiece – which he planned to use in rotation. In 1845, the complete telescope – called the Leviathan of Parsonstown – was ready for use. Rosse mounted it at his house, Birr Castle, between two walls, 56 feet high, which acted as windbreaks for the 58-foot tube.

Although the instrument was clumsy to use, its light-gathering power was tremendous. With it, Rosse became the first to see that some of the cloudy objects known as nebulae – actually other galaxies – were spirals. He also noticed the structure of the so-called planetary nebulae (stars that have cast off their outer atmosphere) and sketched the Crab Nebula (see Chapter 3).

The telescope was used for a decade after Rosse's death in 1867, but was later dismantled.

Naturalis. In rigorous mathematical form, Newton proposed in this mighty work that the laws which operate on Earth also apply to the whole Universe. It was a master stroke: and many believe *Principia* to be the greatest scientific treatise of all time.

How could any scientist follow Newton? None tried: there was enough work to do in the century after his death in following through what he had begun. Some astronomers made precise measurements of the positions of bodies to see how they behaved under gravity; this kind of research led to the discovery of the planet Neptune in 1846. Others concentrated on mathematics and theory, establishing the discipline of celestial mechanics. A few worked on the explanation of gravity itself, taking a course which led to Einstein's General Theory of Relativity.

Newton's optical researches inspired larger and better telescopes, now the principles were fully understood. Astronomers such as William Herschel built huge reflectors through which the very structure of our Galaxy could be discerned.

From *De Revolutionibus* to *Principia* is a span of only one hundred and fifty years. By the end of that period, Newton had synthesised the work of his predecessors into a more magnificent vision than even Aristotle's: and once again, astronomy was a science looking forward to the future.

The Coming of the Camera

By 1840, astronomy was in a buoyant and healthy state. The advances begun by Newton and sustained by his successors meant that the Solar System had been well-charted, and immense star catalogues had been drawn up, noting both star positions and apparent brightnesses. Astronomers had even started to extend into the third dimension, for two star distances had recently been measured. But what were the stars? Could astronomers ever find out? The French philosopher Auguste Comte obviously thought not: for in 1835 he made a pronouncement to the effect that science was quite incapable of giving certain information, and as an example he cited the constitution of the stars.

Two advances in 19th-century science were to prove Comte devastatingly wrong. One – photography – was in its development stages as Comte made his declaration, and by 1839 the first examples had appeared. At last there was a way of permanently recording light-emitting objects, and astronomers were exceedingly quick to take advantage. It must be remembered that all observations up until then had been made visually, by cold, tired and frequently error-prone astronomers. Despite the messy and uncertain nature of early photographic processing, John Draper (in the USA) succeeded in photographing the Moon in 1840, but it was not for a further 10 years that the first picture of a star (Vega) was taken. The early pioneers battled away, however, improving both their chemistry and their technique, so that by the end of the century astrophotography had become relatively commonplace and stars far fainter than were visible to the naked eye could be recorded.

In the meantime, the desire to build improved optical instruments had led Joseph Fraunhofer in Germany to test prisms of superior glass with a view to making colour-free (achromatic) lenses. Using sunlight as a source, he was surprised to find dozens of dark, straight lines crossing the familiar rainbow spread of colours. He realized at once that these did not arise from lens defects, but were something intrinsic to sunlight itself. By 1817, he had mapped the positions of hundreds of these lines, naming the brightest after letters of the alphabet.

Then, at the University of Heidelberg, the physics professor, Gustav Kirchoff, teamed up with a chemist, Robert Bunsen, to investigate chemical reactions which absorbed or produced light. Kirchoff, a great follower of Newton, suggested spreading out light into its individual colours by means of a prism, and from this, in 1859, the two developed the first spectroscope, in which they passed light first through a narrow slit before it was dispersed. The power in this method lay in its far greater precision: the scientists were able to measure accurately the positions in which the sharp images of the slit fell after dispersal.

Before going further, it is worth explaining why light should be spread out by a prism. Light can be thought of as a series of waves (although, as we shall see later, modern physicists also regard

Top: Gustav Kirchhoff, the Professor of Physics at Heidelberg University, explained the dark absorption lines in stellar spectra. He stated that all lines indicated the absence of light at particular wavelengths, indicating the presence of particular chemicals between the source and the observer.

Above: Robert Bunsen, one of Kirchhoff's colleagues, is remembered as the inventor of the burner, named after him. It admits air as well as gas to produce a hot, non-luminous flame, which proved useful in spectral research (in fact he did not invent the burner, but popularized its use).

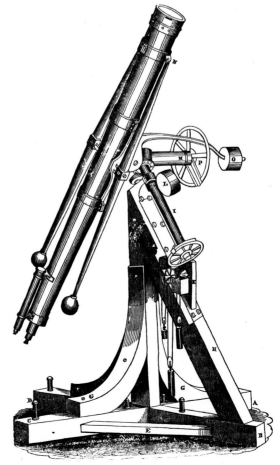

Left: Though Fraunhofer died when he was only 39, he was renowned for his instruments as well as for his research. This is a small, 231 mm. refracting telescope he built for the Dorpat Observatory one year before his death.

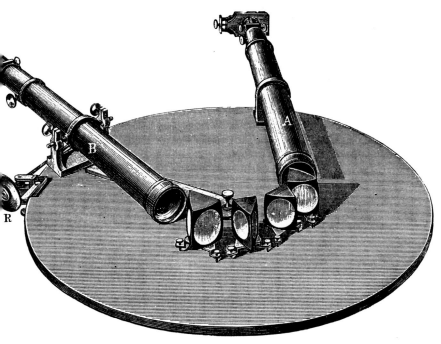

Top: Kirchhoff used this instrument to research the solar spectrum. Light enters at B, is focused, and refracted through a series of prisms to the observer's instrument (A), which has a moveable slit at the eyepiece for scanning the spectrum.

Below: The spectrum breaks up when passed through a prism and the various wavelengths are bent by different amounts.

Bottom: In this early photograph of a spectrum by William Huggins, a nebula (in the centre) is compared with the spectrum of incandescent iron above and below. Four chemicals in the nebula can be identified from their spectral lines.

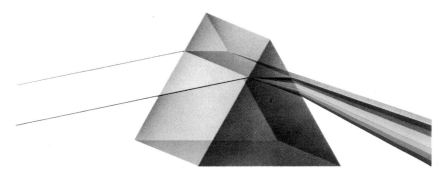

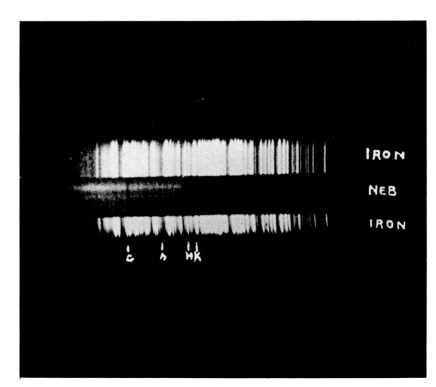

it as a stream of particles) and ordinary 'white' light comprises waves whose wavelengths – the distance from the peak or trough of one wave to the next – range from some 400 nm. (nanometres: thousand millionths of a metre) to 700 nm. Our eyes perceive the shortest wavelengths as blue light, and the longest as red. In everyday situations, all these colours blend together to give white light. But if we pass light through a transparent substance – a lens, a prism or even water – it is bent, or refracted, and short wavelengths are affected more than long ones. The rainbow spread of colours we see when sunlight passes through a prism (whether it is rain or glass) is called a spectrum, and it tells the scientist which particular wavelengths of light are present.

Of course, many other objects besides the Sun also give out light, but not usually in a continuous rainbow of colours. Hot gases, in particular, give out light at certain sharply-defined wavelengths – this is the reason why sodium vapour appears yellow, while mercury looks blue – and a spectroscopist can instantly identify a gas from its characteristic pattern of 'emission lines'. Kirchoff and Bunsen were the first to investigate these spectroscopic 'fingerprints', using Bunsen's newly-invented burner which heated gases to incandescence without producing much extraneous light.

Kirchoff noticed that the wavelength of the sodium line he had measured in the laboratory coincided exactly with the position of the dark line in the Sun's spectrum which Fraunhofer had labelled 'D'. It meant that sunlight must pass through some sodium on its way to Earth: what better place for it to originate than in the atmosphere of the Sun itself? Kirchoff and subsequent spectroscopists went on to find dozens of elements present in the Sun; and one element, helium, was actually discovered in the Sun's outer atmosphere before it was isolated on Earth.

Spectroscopy was the key that astronomers needed to unlock the secrets of the stars and prove Comte wrong. It was not that easy at first, though, for even the brightest star is 10,000 million times fainter than the Sun and its spectral lines are correspondingly dim. But astronomers like William Huggins, working in his own private observatory near London, persevered, and showed that other stars had compositions similar to that of the Sun. Father Pietro Secchi, in Italy, visually observed the spectra of some 4,000 stars – a mammoth task indeed – in an attempt to classify them. And as well as revealing the nature of the stars, spectroscopy turned up an unexpected bonus. The motion of a star directly towards or away from us produces a tiny shift in the position of its spectral lines because of the Doppler Effect (see Chapter 2). This enabled Huggins to derive the first radial velocity of a star – Sirius – in 1868, establishing the basis upon which distances to the most remote galaxies are now measured.

After the 1870s, the rise in the new discipline of astrophysics was rapid. Photographic emulsions became gelatin-based and far easier to use, and so astronomers were able to record spectra photographically and reach much fainter limits. But the greatest boost to the understanding of the nature of stars was in the development of the objective prism – a large, shallow-angled sheet of glass which fitted over the full aperture of a telescope and enabled coarse spectra of many hundreds of stars to be obtained at the same time. In this way, astronomers at the Harvard College Observatory published the details of almost a quarter of a million stars in the Henry Draper Catalogue (1918–1924), which they grouped into spectral classes on the now-standard Harvard Classification System.

By the start of the 20th century, astronomers could measure the composition, temperature, surface gravity, luminosity and velocity of any number of stars. With such a vast amount of data to hand, they were at last in a position to work out what made stars shine, and how they were born, lived and died. From that moment on, astronomy and physics became inextricably intertwined. Astronomers needed to know more about the behaviour of gases in the laboratory before extrapolating to the Universe; while physicists wanted to plumb the near-perfect vacuum of space. These links have been forged even closer as time has passed. The modern astronomer is a true physicist in every sense of the word – and his laboratory is the entire Universe.

To gain more information still, astronomers needed more light. At the beginning of this century, this need led to the building of successively larger telescopes which could reach to fainter and fainter limits. Because a mirror is more efficient at collecting light than a lens, all these great instruments had their origins in Newton's tiny reflecting telescope, adapted to the needs of 20th-century observers. Situated high on mountains away from the glare and pollution of cities, these great eyes on the Universe were, and are, giant cameras.

First of the modern giants was the 1.5 m. (60 inch) reflector on Mount Wilson, California – a telescope with a mirror five feet across – followed in 1917 by a 2.5 m. (100 inch). Both telescopes were inaugurated by the eminent astrophysicist George Ellery Hale (who founded observatories at Yerkes, Mount Wilson and Mount Palomar), and began the American specialization in extremely distant, faint objects. Largest of all the American reflectors is the five-metre (200 inch) belonging to the Hale Observatories on Mount Palomar, and only recently (1976) exceeded in size by the Russian six metre (236 inch) reflector at Zelenchukskaya in the Caucasus Mountains.

Strangely enough, the latest generation of telescopes do not follow in this Cyclopean tradition. They have moderate-sized mirrors of around 3.5 m. (140 inch) aperture, and are designed specifically to be used with the widest range of recording and measuring equipment. As we see later, the mirror does only part of the job in collecting the light from a faint object: the astronomer's task today is to record and analyse

it as efficiently as possible. Earlier astronomers had only one means of recording – the photographic plate – but there is now a veritable battery of hardware which can be attached to a telescope. These detectors are many times more efficient than the photographic plate, and so the light-grasp problem is no longer so acute. The new telescopes are a compromise between information acquisition and economics.

Several of the new observatories have found homes in the Southern Hemisphere – Chile has the European Southern Observatory at La Silla and a four metre reflector at Cerro Tololo, while the 3.9 m. (153 inch) Anglo-Australian Telescope is based at Siding Spring in Australia. Contemporary astronomy has trodden the route of the past in this respect, with the northern skies being plumbed first, but these new telescopes will more than redress the balance. As this is being written, the first comprehensive photographic survey of the southern heavens has just been completed, using the one metre Schmidt telescope at Siding Spring. Like its twin at the Hale Observatories, which surveyed the northern skies a quarter of a century ago, this telescope works only as a fast camera, giving undistorted photographs of very large areas of sky. Its result will keep astronomers busy for several decades.

On the whole, most modern telescopes bear quite striking resemblances to one another, as visitors to observatories will have noticed. This is partly related to the limited number of telescope manufacturers and designers, but has even more to do with the nature of a telescope's work. Basically, the instrument must collect light from a faint source and focus it on to a detector. However, its job is complicated by the Earth's rotation, and the telescope has to accurately track the object under surveillance as it appears to move across the sky. Most telescopes manage this by being equatorially mounted: part of the mounting is precisely aligned with the celestial pole and the instrument pivots around this 'polar axis' to follow the stars. As objects often need to be monitored by equipment for periods of several hours, the telescope needs to be very smoothly and precisely driven, without any mechanical flexures or vibration. Until recently, this was always a taxing engineering feat, but the widespread application of computers to astronomy has meant that it is becoming a problem of the past. In both the Anglo-Australian Telescope and the Russian six metre reflector, a computer takes care of the tracking, correcting continuously for changes in the angle of the telescope, positioning of equipment, atmospheric refraction and many other factors.

For all its complexity and expense, a modern telescope's main function is simply to gather light. At its focus there is a greatly brightened image of some distant celestial object, and the astronomer must decide how to analyse it. Traditionally he viewed it through an eyepiece, to obtain a highly magnified image; but the human eye is not a very sensitive detector – it cannot store up the light

Newtonian

Cassegrain

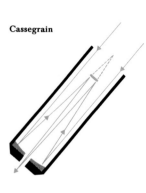

Coudé

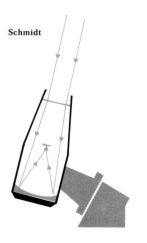

Schmidt

In a Newtonian telescope light is bounced from a parabolic mirror to a focal point near the top of the tube, where a small flat mirror deflects it out through a hole in the tube's side.

In a Cassegrain telescope light is collected on a parabolic mirror at the bottom of the tube. The light is then reflected to a smaller curved mirror near the top of the tube, where it is aimed back through a hole in the centre of the big bottom mirror and brought to focus at the point below the telescope itself.

The coudé focus is like the Cassegrain except that it has a flat mirror to deflect the light sideways. This mirror moves, compensating for telescope motion so that the resulting light beam can be kept in a fixed position.

The Schmidt telescope combines features of both reflector and refractor instruments. Used solely for photographic work, the Schmidt first passes light through a lens which corrects spherical aberration – the loss of sharpness produced by the different focal lengths of different parts of the same lens then bounces the light off a large mirror to a point of focus on a curved photographic plate.

The 48-inch (1.22 m.) Schmidt telescope at Siding Spring, New South Wales, Australia, was installed at the Anglo-Australian Observatory in 1973. It is carrying out a survey of the southern skies to match that of the northern hemisphere made by the Mount Palomar Schmidt of the same size in the early 1950's.

A worker, standing by the open framework of the Mount Palomar 200-inch (508 cm.) reflector, shows the scale of the instrument. The mount behind him is to hold the mirror which will be set at the end of the 110-ton tube. The telescope, which opened in 1948, is so delicately balanced that one hand could turn it on its oil-pad bearings.

A radio-telescope is a huge antenna for collecting weak radio signals from space. This type uses a large dish (Jodrell Bank is an example) to concentrate the waves on the antenna proper, mounted above the centre of the dish. The signals are first amplified by a receiver, then sent to a simple computer which sorts out static. A recorder finally transcribes the signal on a graph.

Karl Jansky, an American communications engineer and founder of radio-astronomy, pushes round the radio receiver, which, in the early 1930's led him to be the first to identify a radio source beyond the atmosphere.

falling on it – and the analysis of the image cannot help but be subjective. Moreover, it is difficult to measure positions accurately at the telescope.

Photographic plates are now universally used to record star images in permanent and easily measured form. Modern photographic emulsions are 50,000 times faster than those of the last century, latest techniques including the baking of plates under dry nitrogen for several hours before use to increase their sensitivity. Cooling them with liquid air during the exposure also helps when faint objects are under study, for it alleviates 'reciprocity failure', the unfortunate tendency of photographic emulsions to need disproportionately longer exposure times at low light levels.

Even the best photographic emulsions are fairly inefficient light detectors, however. To get some measure of efficiency, astronomers utilize the fact that light is not just a wave-motion, but these waves come in separate packets of energy, called *photons*. Normal astronomical plates record only one in a hundred of the photons which fall on them: 99 per cent of the light goes to waste.

In recent decades, astronomers have been experimenting with electronic devices which will alleviate this chronic wastage, and this has led to the trend away from large telescopes: a perfect detector of light mounted on a 50 cm. (20 inch) telescope would be as sensitive as the 200 inch telescope equipped with standard photographic plates. The first highly successful device was the

photomultiplier tube, still an indispensible tool, which can detect one photon in six. Photomultipliers complement, rather than replace, photographs because they cannot form an image: they simply measure (very accurately) the brightness of stars or galaxies. Astronomers use these devices to maximum advantage by confining their measurements to one colour at a time, cutting out the others with a filter.

Since photomultipliers can only measure star brightnesses, the past decade has seen a sustained effort in adapting photo-electric brightness-measuring devices – photometers – to record two-dimensional images. This drive has been not so much to obtain more accurate pictures of extended objects like galaxies and nebulae, but more to record spectra precisely.

Astronomical spectrographs no longer use glass prisms, which absorb valuable light. The wavelengths are separated by reflection off a *diffraction grating*, a very finely ruled sheet of glass. Due to the wave-nature of light, each wavelength can only reflect off in a few specific directions, so the light is split into several spectra, lying in different directions from the grating. By shaping the fine grooves suitably ('blazing') astronomers can ensure that almost all the light falls into just one of these spectra. The task is then to record this spectrum in the most efficient and accurate way.

Although most spectrographs still use photographic plates, electronic devices are very much on the way in. One of the most successful is the

Spectracon, in which a cathode emits electrons independently according to the light falling on each part of it. The electrons are then focused to form an exact image of the astronomical object or its spectrum.

Since the Spectracon tube is evacuated, while the photographic film must be outside – for technical reasons as well as ease of handling – the tube ends in small mica window, against which the film is pressed. The window must be very thin so it absorbs few electrons, but this makes it weak – the untimely demise of a Spectracon is usually caused by someone accidently puncturing the instrument's window and so letting air into the tube.

A highly sophisticated system – arguably the best in the world – is the University College London's Image Photon Counting System. Here the screen of the initial image tube is scanned by a television camera, and a computer automatically measures each spot of light on the screen to ensure it is actually caused by an electron from the photocathode, and not by a stray electron or charged atom in the tube.

The latest devices use just arrays of very tiny silicon diodes – like the 'chips' which are revolutionizing the whole of electronics – at the focus of the telescope or its spectrograph. These robust and highly efficient devices, which simply 'scan' themselves electronically, are probably the image detectors of the future.

Whispers from Space

Our preoccupation with the ingenious devices used by modern optical astronomers has tended to obscure the fact that light gives us only one set of information. Glancing through this encyclopedia, the reader will notice that objects in space emit all sorts of other radiations as well – radio waves, X-rays and infrared radiation to name but a few. Light is just one very narrow band of wavelengths belonging to the whole electromagnetic spectrum – and the recent revolution in astronomy has been mainly centred on exciting discoveries made at other wavelengths. The entire electromagnetic spectrum ranges from the ultra-short gamma-rays, whose wavelengths are measured in terms of a million-millionth of a metre, up to radio waves hundreds of metres long. (There are no actual physical boundaries between these radiations, apart from their range of wavelengths or frequency. The various names have arisen because the waves are generated by different processes, and we need different kinds of detectors to receive them.)

Although light waves cover a band of the electromagnetic spectrum which amounts to only a thousand-million-million-millionth (10^{-21}) of its total spread, they assume such importance to us because they are one of the few radiations to penetrate our atmosphere. Shorter-wavelength ultraviolet rays (with wavelengths less than 300 nm.) are blocked by the atmospheric ozone layer, whereas the longer infrared radiation (wavelengths greater than 1,000 nm.) is mainly

absorbed by water vapour. The atmosphere is once again transparent to radiation between 1 cm. and 30 m., the so-called 'radio window'. Longer wavelength radio waves are prevented from reaching the ground because they bounce off the upper layers of the ionosphere – a belt of charged particles at the top of the atmosphere whose reflecting properties (from the lower layers this time) make worldwide broadcasting possible.

Because of the atmosphere's transparency to radio waves, it is not surprising that they were the first of what were once called the 'invisible radiations' to be exploited by the astronomer. Yet their discovery came about by accident in 1931, when Karl Jansky of the Bell Telephone Laboratories, looking for static from thunderstorms, detected a faint background of radio 'noise' which rose and set with the stars. But it was not generally until after World War II when radar – the bouncing of radio waves off objects to determine their position – was developed that sufficient practical knowledge existed to support radio astronomy.

Modern radio telescopes work essentially like giant radio sets, the only difference being that the radio waves they receive come from natural transmitters in space. But a radio telescope needs to be millions of times more sensitive than even the most sophisticated hi-fi receiver – the total amount of energy picked up by all the radio telescopes in the world, over the entire history

Top: The Arecibo radio-telescope in Puerto Rico is the world's largest radio-astronomy dish, measuring 1,000 feet (305 m.) across. It is slung hammock-like in a natural dip. The receiver, mounted 870 feet above the wire-mesh surface and held in position by the side trusses, can be moved to give a 50° north–south coverage.

Above: These five dishes are part of the Mullard Radio Observatory, Cambridge. They, and others like them, are used to provide very long base lines to obtain accurate fixes on distant objects.

The Pic du Midi Observatory, set at 9,390 feet (2,862 m.) in the French Pyrenees, shows the extremes to which astronomers must go if they are to avoid the turbulence and pollution of the lower atmosphere. To use the new 78-inch (200 cm.) reflector, astronomers have to brave arctic conditions (*inset*).

of radio astronomy, is less than the energy you would expend in lifting up a feather.

Much of the ingenuity in this field of astronomy lies in receiving and amplifying these whispers from space. Most radio telescopes work rather like huge optical reflecting telescopes, collecting the radio waves in a curved dish and focusing them on to an aerial in the centre. From the aerial, the waves are fed to a receiver, which often needs to be designed with the characteristics of particular types of radio source in mind. Every attempt is made to cut down spurious 'noise' at this stage, by methods ranging from cooling the receiver with liquid air to incorporating a maser (the microwave equivalent of a laser) into the circuitry. After being amplified, the signals are generally recorded directly on magnetic tape and analysed by computer, which can process the information at a rate far surpassing any human researcher. Such automation has always been the norm in radio astronomy: and

its fearless exploitation of computers and modern electronic techniques has had an invigorating effect on optical astronomy, as we saw earlier.

In some ways, though, the radio astronomers' lot is not a happy one, compared to that of the optical astronomer. The latter builds large telescopes not only to capture more light, but also to see finer details. Eventually, however, the optical astronomer gets no return for a further increase in size, because very small images become blurred out by irregularities in the atmosphere. However, the resolution also depends on the wavelength, and so radio astronomers, using waves a million times longer than light, need to build radio telescopes to an appropriate scale if they are to see the same degree of detail as an optical astronomer.

Although this low resolution was a great problem in the early days, radio astronomers have become adept at getting round it. The most elegant solution was pioneered by Britain's

To analyse the light that they gather in their instruments, astronomers are increasingly turning to electronic data processing and computer techniques. A prime example is the Image Photon Counting System (*far right*), developed by Alec Boksenberg at University College, London, which records individual photons. The picture at *right*, showing photon 'impacts', represents less than 1 per cent of the IPCS's field. With it, any light source can be surveyed at any frequency over any time period – a vast advance over photographic methods.

Astronomer Royal, Sir Martin Ryle, who electronically linked the output from a line of small radio telescopes, and allowed the Earth's rotation to swing them round as they tracked a source. By stacking the results in a computer, and altering the spacing of the dishes each day, Ryle was able to achieve the resolving power of a single radio dish with a diameter equal to the length of the array. This technique of Aperture Synthesis allows Ryle's Five Kilometre Telescope in Cambridge to resolve detail as fine as would be seen in a good optical telescope: And by linking up radio telescopes thousands of miles apart – perhaps on different sides of the world – radio astronomers to pin down extremely small-scale structures even if they cannot get a proper picture. This technique, called Very Long Baseline Interferometry, is telling astronomers much about the sizes of emitting regions in quasars and radio galaxies, and even about the centre of our own Galaxy.

From remote galaxies we return to the Solar System, and to an illustration of astronomy in an active, rather than a passive role. The technique of radar is now being used to probe the planets, and has already met with considerable success. Early work was concerned with measuring planetary distances – a relatively simple matter of 'timing the echo' – but soon extended to timing the spinning of planets from the Doppler Effect this causes in the reflected radio pulse. In this way, the rotation periods of Venus and Mercury were discovered (in 1962 and 1965 respectively). Now radar is a valuable tool in planetary mapping, particularly in the case of cloud-covered Venus, where radar echoes have located craters and mountain ranges.

The region between light and radio waves in the electromagnetic spectrum – from 0.001 to 1 mm. – is the domain of infrared radiation. This is familiar to us all as 'heat', and there are several types of natural infrared sources in space, such as the clouds which surround forming stars. The

infrared astronomer is beset by the problem that, although some radiation does get through our atmosphere, most is absorbed by water vapour. So infrared observatories are situated on bitterly cold mountain tops where as much water vapour as possible is frozen out. However, the astronomers who man the Kuiper Airborne Observatory – which flies above most of the Earth's atmosphere – have a more comfortable time. An additional problem to be faced by the infrared astronomer is that he himself, his equipment and the observatory all emit relatively vast amounts of heat, creating brighter infrared sources than anything in the heavens. He is rather in the position of an optical astronomer attempting to observe a faint star under floodlights. All he can do in the circumstances is to resort to a number of sophisticated electronic techniques, and use cooled detectors especially sensitive to infrared radiation.

All wavelengths shorter than light – ultraviolet radiation, X-rays and gamma-rays – are absorbed by the atmosphere, and so the analysis of these radiations has had to await the advent of the Space Age. In many respects, these branches of astronomy share a similar approach. Their detectors need to be of a compact and robust design to be sent into space; and they may well have to vie for accommodation aboard a satellite which, on economic grounds, may be crowded with more than a dozen competing experiments.

First of the trio was X-ray astronomy, which began in 1962 with the accidental detection from a rocket of an X-ray emitting source in Scorpius (Sco X-1). Research groups in the UK and USA flew many more rockets with X-ray detectors on board to learn more about these sources, but until 1970, when the first X-ray satellite (UHURU) was launched, the total flight time only amounted to minutes. UHURU, and the British satellite Ariel V, launched four years later, have made comprehensive surveys and have detected many new – and often transient – X-ray sources.

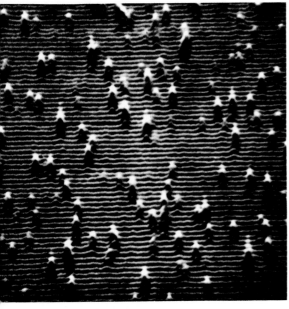

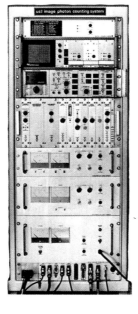

What of the future? It seems certain that the destiny of astronomy lies in space; not simply to detect radiations otherwise unattainable, but to branch out in a hundred other directions. Who can tell what revolutions the Space Telescope – the first optical telescope scheduled to soar above Earth's grey, churning atmosphere – will bring when it is launched early in the 1980s? And then there is 'active' astronomy too; the physical exploration of planets by probes, and soon, Man. Next will come the stars.

Our story has taken us through 5,000 years. We have traced the course of an objective science, and encountered superstition, fear, mistrust and inflexibility on one hand; and the glorious rewards of intellectual achievement on the other. And now we are at another watershed in astronomy. For thousands of years it has been a purely Earth-based science. But space is beckoning: a new era has begun.

10/PROBES AND SATELLITES

LEO ENRIGHT

The launching of the first Earth satellite by the Soviet Union in 1957 was one of the greatest shocks sustained by the West since the war. The US responded with a massive investment in rocketry and assorted space hardware. Beginning with the Redstone launcher – now enshrined in a 'historic Redstone test site' (left) – the most dramatic manifestation of this investment was the lunar landing in 1969 (see following chapter). But the developments in unmanned spacecraft – in particular Russia's Lunar rover and the US's Mars Lander – have shown that Man can explore the Solar System indirectly, with a corresponding saving of money and, almost certainly, lives.

Early on the morning of October 4, 1957, a pinprick of light rose in the western sky, crossed the heavens within minutes and set again in the east. It was the rocket that had just released the first artificial satellite. Those who saw the rise of that tiny star, and later heard and read of the satellite's existence, knew that a new age had dawned for mankind.

For many in the west, excitement was tinged with dismay, for the new star, named Sputnik, was Russian. Circling the Earth every 96 minutes, the beach-ball sized sphere of polished steel represented a supreme technological achievement. American experts listened to the faintly derisory 'bleep-bleep' transmitted from its elliptical 228 km. (142 mile) by 947 km. (588 mile) orbit, uncomfortably aware that the 83 kg. (184 lb) satellite had not only ushered in the Space Age – it had given the Soviet Union a head start in the space race that would inevitably follow. In the dark days of the Cold War, that head start was alarming. If the Soviet Union could launch a satellite, it had rockets powerful enough to rain H-bombs on the US. The days of America's strategic, technological – and by implication, political – primacy were over.

The U.S. satellite programme had actually begun in 1955, but no one had seen the need to give it much support. Now the few space scientists – part of the US Navy's Project Vanguard – suddenly found themselves catapulted into the political arena, and were forced into an attempt to match the Russians' achievement. Their attempt was disastrous: the first Vanguard never left the pad. On December 6, 1957, it blew up in front of a television audience of millions.

Meanwhile, the Russians had reasserted their supremacy. The second Sputnik, weighing an impressive half-a-ton (508 kg.), was placed in orbit on November 3, 1957, scoring another historic triumph. Its cargo was the world's first living creature in space, a black and white fox-terrier called Laika (she died in orbit a week later). The Space Age was barely a month old, America was still Earth-bound, and already the race to put a man in orbit had begun.

In January 1958, a second Vanguard attempt was called off 22 seconds before launch. As people turned to go home, someone shouted and pointed into the darkness. There above them, drifting across the Cape Canaveral sky, was another speck of light: Sputnik 2, already more than a month in orbit and still unchallenged.

America lost patience with the luckless Vanguard project. After years of pleading, the US Army was given the go-ahead to launch its rival Explorer satellite. The rocket was to be a modified Intercontinental Ballistic Missile (ICBM), the 17.7 m. (58 ft.) Jupiter C.

Jupiter C was the brainchild of Dr. Wernher von Braun, the German V-2 rocket engineer who surrendered to the Americans at the end of World War Two. His four-stage rocket, the Redstone, was ready as early as 1956, but incredibly he was ordered not to attempt a satellite launch. (He once tried to defy that order, but the Department of Defense discovered the plan in time to stop it.)

Now, however, Jupiter C and the Explorer programme had official blessing, and on January 31, 1958, Explorer 1 was rocketed over the Atlantic from Cape Canaveral and into orbit. Its weight, including the last rocket stage, was only 14 kg. (31 lbs.) – one sixth the weight of Sputnik 1 – but it packed a powerful scientific punch. In 84 days of operation, the tiny probe discovered the Van Allen Belts of intense radiation surrounding the Earth. But a month later, on March 17, 1958, Vanguard 1 finally made it into orbit. Soviet Premier Nikita Khrushchev nicknamed it 'The Grapefruit' because of its size, but it gave valuable service for six years.

In the two years that followed Explorer 1, the United States dominated Earth orbit. In May 1958, the Soviets launched Sputnik 3, a 1326.8 kg. (1.3 ton) heavyweight that clearly demonstrated the true power of their Sputnik launcher. But in 1958-9, the US launched 19 satellites, including a military satellite called Score – the first prototype communications satellite. And the civilian programme acquired a new impetus when on October 1, 1958, the National Aeronautics and Space Administration (NASA) officially took over responsibility for civilian projects. Eleven days later (at 3.42 a.m. on October 11), NASA launched its first space probe: Pioneer 1, a project inherited from the Air Force. Its destination was the Moon.

Moonfall

Without the Moon to practice on, it is hard to see how man could have progressed so fast in his exploration of the Universe. After 20 years of space travel, the Moon seems a close neighbour, and we are inclined to forget that a quarter of a

Wernher von Braun was the true founder of the space age. Born a German, his Ph. D thesis in the 1930's laid the foundation for the V-2, the world's first successful large missile. Always more interested in space flight than military matters, he surrendered to the Americans in 1945. In the US, he developed in turn the Redstone rocket, the basis for early American satellite launchers, and eventually, the Saturn rockets that took men to the Moon.

Three Pioneers of Rocketry

A century ago, men were still trying to conquer the air. But in the little town of Kaluga, south of Moscow, a partially deaf schoolteacher called Konstantin Tsiolkovsky was already dreaming of the stars. He and just two others – Robert Goddard in the US and Hermann Oberth in Germany – independently invented the theory and technology that would take man into space.

Tsiolkovsky's deafness early forced upon him on intellectual response to life. By 16, the bright youngster was already convinced man could escape from the Earth. But how? In searching for the answer, Tsiolkovsky worked out the fundamental laws of rocket propulsion that would land the next generation on the Moon. In 1883, his first article on space flight, *Free Space*, accurately described weightlessness in space. In his *Dreams of Earth and Sky* (1895), he wrote of an artificial Earth satellite. His pioneering calculations, proving that space travel by rocket was possible, were published in 1903, the year the Wright brothers first flew. Tsiolkovsky predicted and solved theoretically most of the basic engineering problems of astronautics – his proposed fuel, liquid hydrogen and liquid oxygen, propel many modern rockets.

Robert H. Goddard had not even heard of Tsiolkovsky when he began studying rocket-

ry before the First World War. By 1919, the Massachusetts physics teacher had written a now-famous 69-page pamphlet, *A Method of Reaching Extreme Altitudes*, and sent it to the Smithsonian Institution in Washington together with a begging letter. The Smithsonian gave him a $5,000 grant and accidental notoriety. A press release of January 11, 1920, uncovered the central theme of Goddard's pamphlet. The 'extreme altitude' of his work was the Moon, a notion that earned him a good deal of ridicule.

On March 16, 1926, Goddard launched the first liquid-fuelled rocket in history. It burned for $2\frac{1}{2}$ seconds and travelled 55m (184 ft.), landing in his aunt's cabbage patch.

Tsiolkovsky, Russia's 'Father of Astronautics'.

Such limited success didn't silence the detractors. One sneering headline read: 'Moon Rocket Misses Target by $238,799\frac{1}{2}$ miles.' Goddard, who never forgave the cynics, worked on in seclusion backed by grants arranged by the aviator Charles Lindbergh from the Guggenheim Foundation.

Across the Atlantic, however, another theoretician and visionary was being taken very seriously indeed. In Germany, Hermann Oberth, another teacher, came to the same conclusions as his two predecessors. His theoretical work, summarized in the classic *The Rocket into Interplanetary Space* (1923), sparked off a great deal of research. It culminated in the V-2 rocket. After the first attack on London, one of the team (the top US rocket scientist Wernher von Braun) is said to have remarked: 'A perfect mission, but it landed on the wrong planet.'

Oberth, Germany's rocket pioneer.

One of Goddard's early rockets being prepared for launching, 1935.

million miles was a very long way indeed in terms of the 1950's technology. Pioneer's mission was to orbit the Moon. It never got near its target. Its booster cut out early, but it travelled far enough to map the limits of the Van Allen Belts before falling back into the atmosphere.

The Russians were less ambitious, and correspondingly more successful. Their first lunar probe (Luna 1) was designed to do no more than fly past the Moon. They called it Mechta ('Dream'), a 361 kg. (797 lb.) battery-powered sphere that reached the Moon's vicinity in just 34 hours – less than half the travel time of the fastest Apollo Moonshot. It whipped past the Moon only 7,500 km. (4,660 miles) from the surface and swept into a 450-day solar orbit – the first artificial planet. The next Luna probe was a direct hit. Luna 2 smashed into the Moon near the crater Archimedes on September 13, 1959 – the first man-made object to reach another world. On October 4, the Soviet Union celebrated the second anniversary of Sputnik 1 by launching a third Luna probe, this one carrying a camera. For 40 minutes, Luna 3 photographed the Moon's hidden face and beamed back the pictures when it reappeared from its enforced radio silence behind the Moon.

The mission caused a sensation, but it was to be Russia's last major lunar success for six years. In that time, at least ten moonshots were attempted, of which only one had even partial success.

The Americans, meanwhile, had rallied to the challenge of the space race. President Kennedy had made his historic commitment to place Americans on the Moon by the end of the decade, and it was vital to create a bank of information and experience on which to build the Apollo project. The first efforts – with the Ranger missions to photograph the Moon on close-up – were disastrous. The first six Ranger flights failed. Ranger 7 was the 'now-or-never' mission. It got off to a perfect launch on July 28, 1964, heading for its target in the Sea of Clouds. Sixty-three hours later, and 17 minutes before impact, Ranger's six television cameras were switched on. Pictures flowed in at the rate of one every two and a half seconds, bringing the first detailed pictures of the lunar surface. For the first time, it was seen that lunar craters come in all sizes, down to tiny pock-marks a few feet across (though there was still dispute as to whether the surrounding areas were solid or dust into which a soft-lander might vanish).

The next Ranger (launched on February 17, 1965) performed even better, sending back 7,137 pictures of the Sea of Tranquility (a possible Apollo landing site). A month later, Ranger 9 ended the series with a spectacular television show watched by millions. On March 24 it relayed 5,814 pictures of its dramatic plunge into the crater Alphonsus. Television screens flashed the unforgettable words: 'LIVE FROM THE MOON.'

The next dramatic first was again a Russian

Russia's Luna 9 – shown below both with its shields unfolded and in position atop its lander – was the first probe to make a soft landing on the Moon, in January 1966.

achievement. After several soft-landing failures, in January, 1966, Luna 9, a black egg-shaped object bounced on to the Ocean of Storms after a 5 m. (16 ft.) drop from its spent propulsion unit. For four minutes ten seconds, Luna 9 just sat there. Then, like a flower, four petals folded out to force the craft upright and allow its antennae to announce that the first soft landing on the Moon had been accomplished.

Eight hours later, Luna 9's television camera was turned on. It had only a third the sensitivity of the human eye, but its first historic picture showed a porous, pitted, crated-strewn surface, with small rocks that cast long, dark shadows. Even the beaten Americans breathed a sigh of relief: the 100 kg. (221 lb.) Luna had not sunk in fine Lunar dust. Man could walk on the Moon.

Although in March, 1966, the Soviet Union gave the Moon its first artificial satellite – Luna 10, which broadcast 'The Internationale', from the Moon – it was becoming increasingly clear that Russia's firsts were little more than propaganda victories. The USA had pioneered a new technology that would provide repeated, reliable, projects sound enough to offer long-term returns on the huge investments involved in their creation. The Moon's next visitor, Surveyor 1, was the US's first soft-lander, launched from Cape Canaveral on May 30, 1966. The experts gave survival odds no better than ten to one. But 4 m. (13 ft.) above the Ocean of Storms, Surveyor's retro-rockets shut down and the 283 kg. (624 lb.) robot fell to the surface (the equivalent of only a 2 ft. drop on Earth because of the Moon's weaker gravity.) In six weeks of operation – Luna 9 had worked for only three days – Surveyor sent back 11,240 pictures from its television camera. The only other payload was an American flag sneaked aboard by a technician.

Three months after landing, Surveyor 1 was joined by a companion orbiting 257 km. (160 miles) above: Lunar Orbiter 1. The £83.3 million ($200 m.) Lunar Orbiter series was the third generation of American lunar explorers. Their job was to map the Moon – pathfinders for the Apollo missions. Lunar Orbiter 1 was launched on August 10, 1966, and was followed by four more orbiters within a year. Each weighed 390 kg. (860 lb) and was an orbiting photographic laboratory capable of processing a 61 m. (200 ft.) roll of film.

By the summer of 1966, both space powers were flying concurrent orbiter and lander programmes. Luna 13 was the Soviet Union's second successful soft landing. During Christmas week 1966 it touched down in the Ocean of Storms to photograph its surroundings and test the strength of the soil by pushing a rocket-propelled probe into the ground.

The US's second Surveyor crashed, but Surveyor 3 made it safely to the Ocean of Storms on April 17, 1967. It took 6,315 pictures in a fortnight and dug four trenches with a tiny shovel. (Two and a half years later, the dead robot had visitors when two US Navy men dropped in during man's second Moon landing.) Surveyor 4 was a bit of a mystery. It was launched successfully in July 1967, but just 1.4 seconds before touchdown in Sinus Medii (Central Bay) all contact was lost. Perhaps some future Moonwalker will happen upon the little robot, still waiting for someone to tell it what to do next.

The last three Surveyors worked perfectly. They sent back a total of 69,000 pictures, and Surveyor 6 even re-ignited its retro-rockets to make a short hop across the surface. Between September 1967 and the following February they completed the pre-Apollo exploration. The next American lunar explorers would be men (see following chapter).

The Soviet Union's first phase of lunar exploration was also drawing to a close with the Luna 14 orbiter. Luna 15 left for the Moon on July 13, 1969 – only three and a half days before the Apollo 11 astronauts. The 5,800 kg. (5.7 ton) Luna could have only one mission: to beat Apollo to the Moon and get back first with a rock sample.

The space drama continued for over a week, then – two hours before the Apollo astronauts blasted off from Tranquility Base – the orbiting Luna fired its retro-rockets for four minutes to put it into a nose-dive towards the Sea of Crises. Half a mile above the surface, Luna's retros apparently failed and the craft hit the Moon at nearly 800 k.p.h. (500 m.p.h.). The Americans were the first to return samples of lunar rocks.

But the next Luna to reach the Moon was an historic trailblazer – a blueprint for the unmanned exploration of the planets. Luna 16 settled gently on the Sea of Fertility at 8.18 a.m. (Moscow time) on September 20, 1970, about a mile (1.6 km) from its target point. An hour later, with the help of a stereo camera, controllers had picked a spot on the surface and ordered an automatic drilling rig into action. Drilling silently in the lunar vacuum, it worked for about 30 minutes to extract a 35 cm. (14 inch) sample of soil and rock weighing just over 100 gr. ($3\frac{1}{2}$ oz.). Twenty-six hours 25 minutes after landing, Luna 16's 440 kg. (970 lb.) ascent stage blasted off the Moon with its precious cargo sealed inside a re-entry capsule. Using the landing stage as a launch pad, Luna aimed directly at Earth and landed the 0.5 m. (1.6 ft.) spherical capsule 129 km. (80 miles) south-east of the Tyuratam cosmodrome.

Scientists examining the returned sample found it quite unlike the Apollo 11 rocks collected 900 km. (560 miles) away. They also took the opportunity to make an historic East–West swap: three grams of Luna 16 soil for six grams of Apollo samples.

Meanwhile, Luna 17 was on its way to the Moon carrying one of the Soviet Union's greatest technological triumphs in space: the first Moon rover. Luna 17 landed inside a small crater in the north-west corner of the Sea of Rains on November 17, 1970. After checking for obstructions, controllers lowered a ramp and the three-quarter ton (756 kg.) Lunokod 1 rolled out. It was a

This lunar landscape shot by Luna Orbiter 2 became an instant classic when it was released in late 1966. It was the first time most people had seen the Moon close up, at the low angle familiar from landscape shots on Earth. It shows the crater Copernicus, one of the Moon's most salient features, 93 km. (58 miles) across, with mountain ranges 3,000 m. (10,000 feet) high.

The Lunokhod, Russia's automatic Moon-car, was delivered by the Luna 17 soft-lander in November 1970. Though it looks a little amateurish, the 222 cm. (7 ft. 3 in.) long vehicle was a triumph. It televised its surroundings, detected cosmic rays, measured the chemical composition of lunar rock and bounced laser beams back to Earth, allowing scientists to determine the precise distance of the Moon.

An Atlas D booster is hauled out at Cape Kennedy, to act as the launch vehicle for the Mariner spacecraft that flew past Mars in July, 1965.

peculiar eight-wheeled device, looking like a 2 m. (7 ft.) long casserole dish. It was powered by solar cells tucked inside its clam-like lid and a radioactive source kept its insides warm during the chilly lunar night.

Its amateurish look notwithstanding, in its ten-month lifetime (more than three times its design life), the Lunokod sent back over 20,000 individual pictures, 200 panoramas and countless surface and astronomical measurements.

The rover travelled a total of 10.5 km. (6.6 miles) under the guidance of a five-man team at the control centre (probably near Moscow). Navigation pictures poured in at the rate of 20 a minute, but the team – commander, driver, navigator, systems engineer and radio operator – had to remember all the time that Lunokod had moved several metres by the time their instructions reached it across the void.

A year later (after another failure and a second successful recovery of lunar samples), Luna 21 placed Lunokod 2 inside the Le Monnier crater on the Sea of Serenity, 112 miles (180 km.) from the Apollo 17 landing site. This second Lunokod failed in early May, but it still achieved more than Lunokod 1 in half the time. It drove to the top of a 400 m. (1,300 ft.) hill to photograph distant mountain peaks, and moved gingerly along the edge of a 50 m. (164 ft.) deep precipice similar to the Hadley Rille explored by the Apollo 15 astronauts two years earlier. At high noon on the first lunar day, Lunokod 2 nearly had the first traffic accident on the Moon. Without shadows to help them judge distances, Lunokod's masters nearly crashed it head-on into the Luna lander. It is thought that a similar misjudgement may have prematurely ended the robot's life at the bottom of a ravine or crater, but the Russians have never explained its demise. Whatever its fate, Lunokod 2 travelled 37 km. (23 miles), sometimes up to its hubcaps in dust, and sent back 80,000 television pictures and 86 panoramas. It analysed the soil, detected cosmic rays and stomped on the ground to test its rigidity. A French-built laser-reflector was used to measure the distance to the Moon, but it may also have tested navigation techniques for future marathon rovers.

The next successful sample return mission was Luna 24, which landed in the Sea of Crises on August 18, 1976. It was the first Luna mission in two years (a repeat of the failed Luna 23) and the fourteenth Soviet spacecraft to land on the Moon. The programme now seems to have been halted, but perhaps only temporarily.

Luna's creators claim robots are ten times cheaper than manned lunar flights, and we could readily believe this were it not for the consistently high failure rate in the lunar programme. The latest moratorium on lunar flights is probably being used to look again at the Luna design in preparation for a number of new missions. If, for instance, the Russians can put a communications satellite in lunar orbit, the way will be open for landings on the far side. A sample return mission from there would be a major 'first' and valuable experience for planetary missions. Equally, they could send Lunokods to the far side, and it is certain they are examining ways of combining the talents of the rover and sample return craft to allow them to explore large areas and return the most interesting samples to Earth. The Moon still looms large in Soviet plans.

The Planets

Early flights to the planets were a bitter disappointment to the romantics who dreamed of exciting new worlds beyond our own.

The first blow came on June 19, 1963, when Mariner 2 passed within 35,400 km. (22,000 miles) of Venus and found a world as frightening as anything conceived by Edgar Rice Burroughs – a seething mass of thick cloud above a boiling surface crushed under nearly half a ton of atmosphere per square inch.

Then Mariner 4 went to Mars and the dreams of a planet if not criss-crossed with ancient canals, then at least rich in plant life, were shattered within minutes. Passing 9,800 km. (6,000 miles)

from the planet on July 14, 1965, Mariner sent back 21 pictures of Moon-like craters gouged into an arid landscape. Mars was, apparently, a dead world, and like the Moon, a geological backwater.

Four years later, the next Mars probes, Mariners 6 and 7, sent back 201 pictures from within 3,540 km. (2,200 miles) of Mars. The conclusion was the same, but it was too late to think of abandoning the next missions. That turned out to be a remarkable stroke of luck.

Mariner 9 became the first manmade object to orbit another planet on November 14, 1970. (Mariner 8 had crashed into the Atlantic after its rocket broke up). Mariner 9 immediately switched on its two television cameras and three sensors slung beneath the craft on a moveable platform. But the cameras saw virtually nothing, because the probe had arrived during a terrible dust storm. At least Mars was not geologically dead.

Weeks later, when the dust settled on Mars, US scientists got the surprise of their lives. The early Mariners had simply not shown enough in sufficient detail. Every day, Mariner 9 made new and important discoveries. The first sensation came on January 7, 1971, when the cameras focused on a highland region identified by Earth-based radar. The view was staggering. One of the mountains – Olympus Mons – was a huge volcano broader than any on Earth. To find out its height, scientists used Mariner's ultra-violet sensor to measure atmospheric pressure at the summit. The result was almost unbelievable: Olympus Mons is about 34 km. (15 miles) high – almost three times the height of Everest on a planet half the size of Earth.

Mariner 9 had been expected to work in orbit for 90 days, mapping 70 per cent of the surface. It continued instead for 349 days, took 7,329 photographs covering *all* the planet and sent back 54 billion items of scientific information – at a cost of a mere 14 cents per item.

But for all its success, scientists conceded that Mariner 9 could not have spotted a herd of 10 million elephants on Mars, let alone smaller life forms. The search for life would have to be made on the spot. The first attempt to do just that was made only days after Mariner went into orbit, but it took a year to achieve success. After two failures, the Soviet Union landed Mars 3 on the Martian surface on December 2, 1971. Ninety seconds later, the 350 kg. (772 lb.) probe's camera started transmitting. But the signals lasted only 20 seconds and the picture, to use the official description 'did not reveal any noticeable difference in the contrast of details' (it was blank!). Circling above, Mars 3 (like Mars 2 before it) wasted all its film on a planet hidden beneath five miles of dust because it could not be reprogrammed to wait out a storm. The rest of the Mars series were all failures. Mars 4 missed the planet completely; Mars 5 added little to Mariner 9's findings; the Mars 6 lander crashed and Mars 7 fired its lander into

solar orbit. That cosmic convoy ended the Soviet Union's latest phase of Mars exploration.

But in America space scientists were preparing their own search for life on the Red Planet with the Viking probes. The first two contained astonishing automatic laboratories designed to test the Martian soil for the existence of life.

On June 9, 1976, Viking 1 made the longest deep-space engine burn in history (38 minutes) to place itself in orbit around Mars. The 3,399 kg. (3.3 ton) Viking was three times heavier than Mariner 9 and took the best part of a year to complete its 708 million km. (440 million mile) journey. The first two weeks in orbit were spent searching for a smooth landing site amid the jumble of boulders, mountains and ravines. Controllers found it, with the help of Viking's cameras and Earth-based radar, on the western edge of the 'Gold Plain', Chryse Planitia.

Three hours before landing the lander separated from the Viking orbiter and fired its retro-rockets for 20 minutes. Slowed first by the thin atmosphere, then by giant parachutes and finally,

Atop a Titan-Centaur rocket, Viking 1, containing the first Mars lander, is blasted off from Kennedy Space Center on August 20, 1975.

The Viking Mars lander
(*below*) proved a triumph for the
US. Its sophisticated scientific
package not only took classic
pictures of the Martian
landscape (*right*) and sunset
(*far right*), but also tested for the
existence of life (see Chapter 6).
For the sunset shot, the camera
scanned from the left as the Sun
dipped below the horizon and
took ten minutes to complete
the 120 degree coverage from
left to right; hence the better lit
section on the left and the
stroboscopic effect in the Sun's
glow. Parts of the lander can be
seen in the centre and right.

after dropping its heat shield, by three main descent engines, the three-legged robot – a six-foot (2 m.) tall, hexagonal aluminium box measuring 3 m. (10 ft.) across and weighing over half a ton (575 kg.) – settled on the Martian surface at 4.53 a.m. (California time) on July 20, 1976 – the seventh anniversary of the first manned Moon landing. Twenty-five seconds after touchdown, one of two cameras was switched on to record the first black and white picture from the surface of Mars. Six minutes later it began a 300-degree panorama of the Mars-scape.

The pictures – received over that immense distance 19 minutes later – were astonishingly good. It's worth comparing them with early Moon-lander pictures to see how space photography has been revolutionized by new computer processing techniques. The later colour pictures were even more remarkable, proving at last that the surface of Mars really is red.

We knew so little about the Martian surface that no one dared rule out the possibility of seeing a landscape teeming with plant life. There were no plants. The next step was to look for microscopic life forms in the soil (see Chapter 6). To do this Viking had three tiny biological laboratories powered, like everything else aboard the lander, by a pair of small nuclear generators. Electricity was at such a premium that engineers designed

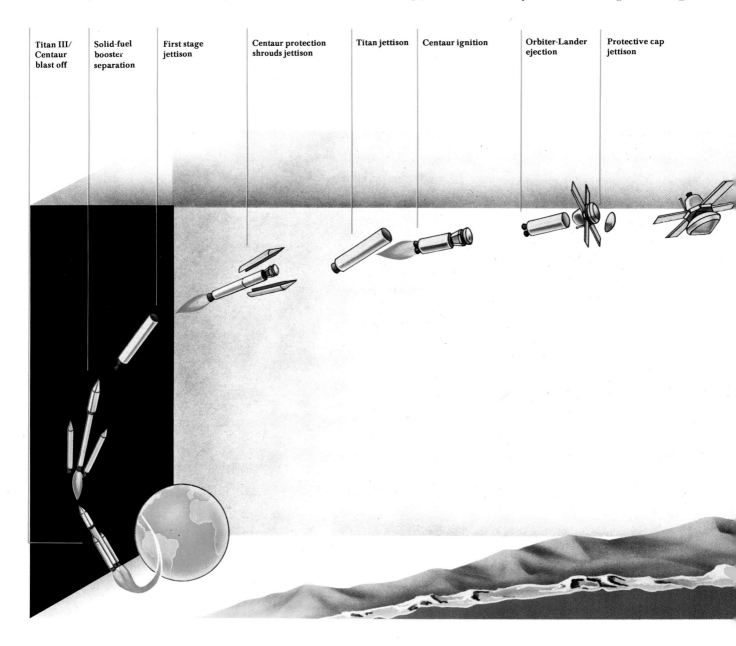

| Titan III/ Centaur blast off | Solid-fuel booster separation | First stage jettison | Centaur protection shrouds jettison | Titan jettison | Centaur ignition | Orbiter-Lander ejection | Protective cap jettison |

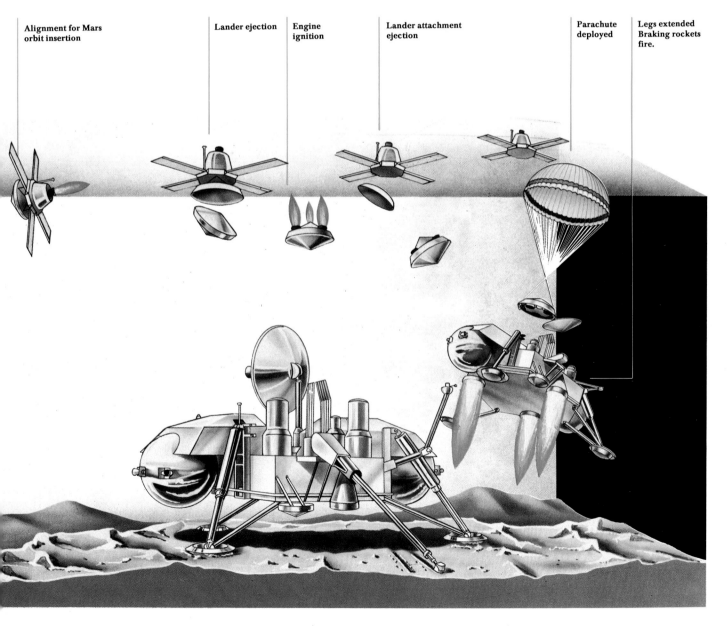

Alignment for Mars orbit insertion

Lander ejection

Engine ignition

Lander attachment ejection

Parachute deployed

Legs extended Braking rockets fire.

the laboratories to operate on little more power than a domestic lightbulb. Yet they could detect life of any kind, even in soil many times more barren than Earth's most barren desert.

All three experiments worked on the assumption that Martian life would be similar to organisms on Earth, simply because it was impossible to fit more ambitious experiments into less than a cubic foot (0.28 m.^{3}) inside the lander's body. Soil samples for each experiment were lifted from the surface during Viking's ninth day on Mars by a shovel at the end of a 3 m. (10 ft.) extendable boom. The results were dramatic (see Chapter 6 for details) but inconclusive.

The perplexed scientists – and the world – would have to wait and see what happened with Viking 2, now streaking for Mars. The second Viking lander touched down on Utopia Planitia (Utopia Plain) on September 3. It was dumb for several hours before landing with a radio failure aboard the orbiter, but it descended automatically and communications were restored eight hours after touchdown. Viking 2 repeated its companion's search for life, but the results were equally uncertain. Something, animal or mineral, was giving unexpected readings – and scientists concluded sadly that it was probably not a life form. The Viking 2 orbiter was switched off on July 25, 1978, after orbiting the planet 706 times, but the other orbiter and the two landers continued operating well into their third year in space. Viking 2 even sent pictures of ground frost

on a chilly ($-80°$C.; $-114°$F.) winter morning. The two landers were shut down in the spring of 1979, although Viking 1's radio continued to be used to map distances on the surface.

The Viking search for life ended officially on June 1, 1977. They hadn't proved there is life on Mars, but nor had they proved there isn't. Viking's ten-foot reach meant that between them the landers could only search 190.6 square feet (17.7 m.^{3}) on a planet of many millions of square miles.

It remains for future Mars probes to resolve the life question, helped immeasurably by the Viking experience. Those new explorers are already on the drawingboards: A Viking on

1. German V-2 Ballistic Missile (1942-5)
2. US Redstone Medium Range
   Ballistic Missile (1953-65)
3. US Atlas ICBM (1959-65)
4. US Titan ICBM (1963-   )
5. US Saturn 1-B Space Launcher (1966-75)
6. Soviet G-1 Space Launcher (1971-   )
7. US Saturn V Space Launcher 1967-73)

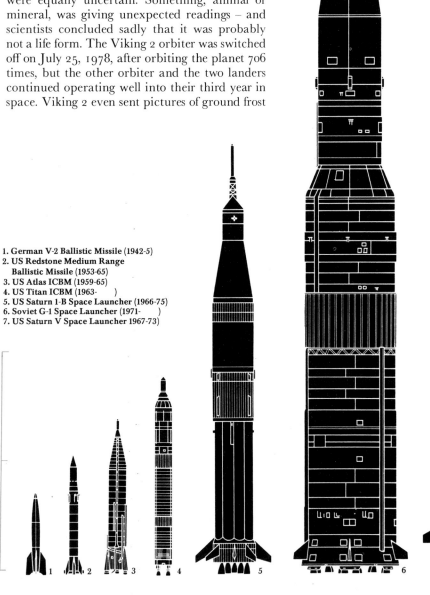

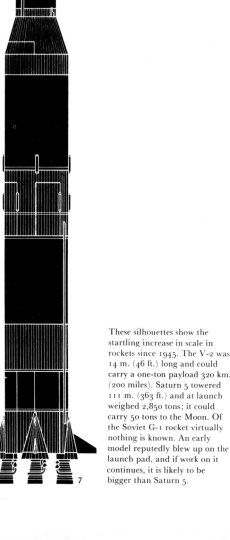

These silhouettes show the startling increase in scale in rockets since 1945. The V-2 was 14 m. (46 ft.) long and could carry a one-ton payload 320 km. (200 miles). Saturn 5 towered 111 m. (363 ft.) and at launch weighed 2,850 tons; it could carry 50 tons to the Moon. Of the Soviet G-1 rocket virtually nothing is known. An early model reputedly blew up on the launch pad, and if work on it continues, it is likely to be bigger than Saturn 5.

Venus's rock-strewn surface was recorded for the first time in June 1975 by Russia's Venera 9 soft-lander. Only one other picture (by Venera 10) has been taken of the surface, for the conditions are truly frightful. In heat that will melt lead and atmospheric pressure 90 times that of Earth, for the probe to operate at all, even for a few minutes, was a triumph.

wheels with a range of over 500 km. (300 miles) and a two-year lifetime, for launch possibly in 1984; a low-altitude pilotless aircraft with a 19.8 m. (65 ft.) wingspan and a range of 4,023 km. (2,500 miles) which would carry up to 100 kg. (220 lb.) of instruments; a 'tumble-weed' probe blown by the Martian wind with a 30 kg. (66 lb.) instrument package. The Soviet Union has made it clear that the Lunokod programme is a forerunner of 'Planetokhods', but they must first iron out the problems that have crippled their Mars missions.

If they can do that, we need only look at their spectacular Venus successes to appreciate the great potential. Mariner 2 had shown Venus to be a most forbidding place, but Soviet scientists decided to take up the challenge of a world where a man would be at once poisoned in a lethal atmosphere, fried in the oven heat and squashed by atmospheric pressure many times that of Earth's.

The early Venus flights were notable technological failures. Between February 1961 and February 1966, nine Venus probes failed abysmally – three flew silently past the planet with broken radios and six never made it out of Earth orbit. On March 1, 1966, Venera 3 crashed onto Venus, the first manmade object to reach another planet. But this craft too fell silent before reaching its target, making it a scientific failure.

In October, 1967, Venera 4 gave an indication that success was possible by transmitting data from inside the Venusian cloud cover. It penetrated 27 km. (17 miles) into the atmosphere and worked for 94 minutes before being crushed by the terrible pressure. (A day later, America's Mariner 5 passed by 4,000 km. (2,500 miles) from the surface.) In 1969, Venera 5 and 6 both penetrated to about 25 km. (16 miles) on the night side of the planet before being crushed, and on December 15, 1970, a tougher and more sophisticated Venera 7 transmitted the first signals from the surface of another planet. Venera 8 repeated that success on July 22, 1972.

The next Venus probe was Mariner 10, an American probe which killed two birds with one stone by flying past Mercury too. The first two-planet explorer passed within 5,760 km. (3,580 miles) of Venus on February 5, 1974, transmitting the first pictures of the planet's lethal cloud cover. Venusian gravity then bent the probe's trajectory for its rendezvous with Mercury, a trick known as a 'slingshot manoeuvre'.

The innermost planet has always been difficult to study because it is so close to the blinding Sun. But on March 29, Mariner 10 gave us our first view of the Moon-like surface, from a distance of only 271 km. (168 miles). Its orbit was so impeccably calculated that the probe, now in solar orbit, passed Mercury two more times (September 1974 and March 1975), sending back more than 10,000 pictures, together with magnetic and radiation measurements – all for a mission cost of £41.6 million ($100 million).

By this time, Soviet scientists felt certain they had mastered the problems of Venus's hellish

In this time exposure, a Titan Centaur rocket blasts off into the night from Kennedy Space Center on December 10, 1974, carrying a Helios 1 space-craft, which approached to within 45 million km. (28 million miles) of the Sun. The Helios programme consists of joint German–US projects.

Clarke's Orbit

When communications satellites first replaced hilltop transmitters in the 1960s, they had one big disadvantage. They moved. Keeping track of a transmitter travelling at 28,000 k.p.h. (17,500 m.p.h.) would be expensive, and even then the satellites were not always where they were needed (trans-Atlantic communications via Telstar were impossible for 14 weeks each year). Customers naturally demanded a more reliable service.

Fortunately the answer already existed. As far back as October 1945, Arthur C. Clarke, the space prophet and science fiction writer, pointed out in *Wireless World* that a satellite orbiting above the equator once every 24 hours would appear to be stationary in the sky, with important consequences for global communications. Here is his suggestion in extracts from his classic paper *Extra-Terrestrial Relays*:

'Although it is possible, by a suitable choice of frequencies and routes, to provide telephone circuits between any two points or regions of the earth for a large part of the time, long-distance communication is greatly hampered by the peculiarities of the ionosphere, and there are even occasions when it may be impossible. A true broadcast service, giving constant field strength at all times over the whole globe, would be invaluable, not to say indispensable. . . .

'Many may consider the solution proposed in this discussion too far-fetched to be taken very seriously. Such an attitude is unreasonable, as everything envisaged here is a logical extension of developments in the last ten years – in particular the perfection of the long-range rocket of which V2 was the prototype. While this article was being written, it was announced that the Germans were considering a similar project, which they believed possible within fifty to a hundred years [!] . . .

'It will be possible in a few more years to build radio controlled rockets which can be steered into orbits . . . and left to broadcast scientific information back to the earth. A little later, manned rockets will be able to make similar flights with sufficient excess power to break the orbit and return to earth.

'There are an infinite number of possible stable orbits, circular and elliptical, in which a rocket would remain if the initial conditions were correct . . .

'It will be observed that one orbit with a radius of 42,000 km. (26,000 miles), has a period of exactly 24 hours. A body in such an orbit, if its plane coincided with that of the earth's equator, would revolve with the Earth and would thus be stationary above the same spot on the planet. It would remain fixed in the sky of a whole hemisphere and, unlike all other heavenly bodies, would neither rise nor set . . .

'Using material ferried up by rockets, it would be possible to construct a 'space-station' in such an orbit. The station could be provided with living quarters, laboratories and everything needed for the comfort of its crew, who would be relieved and provisioned by a regular rocket service . . .

'Let us now suppose that such a station were built in this orbit. It could be provided with receiving and transmitting equipment . . . and could act as a repeater to relay transmissions between any two points on the hemisphere beneath, using any frequency which will penetrate the ionosphere . . .

'A single station could only provide coverage to half the globe, and for a world service three would be required, though more could be readily utilized. The stations would be arranged approximately equidistantly around the earth.'

The orbit Clarke envisaged is said to be 'geostationary' or 'geosynchronous'. There are now many scores of satellites in this orbit; and, not surprisingly, there have been moves in recent years to have this important orbit officially named the Clarke Orbit.

---

environment. On June 18, 1975, they launched a giant Proton rocket from Tyuratam carrying a Venera probe four times heavier than earlier craft – a 5,000 kg. (5 ton) orbiter and lander called Venera 9, complete with searchlights to counter the pall of darkness that the planets clouds presumably cast.

On October 22, 1975, the orbiter became the first artificial Venusian satellite while the 1,560 kg. (1½ ton) lander began a descent unique in the history of lunar or planetary exploration, using a combination of six parachutes and its own shape to pass quickly through the hottest parts of the atmosphere. The hair-raising drop ended at 2.13 p.m. (Moscow time), and 15 minutes later Venera 9 gave man his first view of another world: a landscape of large and small rocks, but without much dust and rocky debris. One big surprise was that Venera's powerful searchlights were not necessary – Venus was 'as bright as a cloudy day in Moscow'.

Venera 10 arrived at Venus three days later and touched down 2,200 km. (1,370 miles) away. Venera 9 had landed on a high plateau (possibly the slopes of a volcano) with 'young' angular rocks scattered about, while Venera 10's pictures suggested a stony desert, with old weathered rocks like pancakes. The 53-minute lifetime of Venera 9 and 65 minute life of Venera 10 have revolutionized our understanding of Venus. Man can now work from first-hand knowledge, not speculation.

The next Venus mission had a more specialized role: its purpose was to use the 'Evening Star' to tell us something about our Earth. American scientists reasoned that a study of Venusian weather patterns would yield useful insights into our own complex weather system. That was the mission of Pioneer Venus 1978, a dual-launch project begun in the summer of 1978. An orbiter began circling Venus in early December while another spacecraft released four atmospheric probes – one large and three small – into the Venusian cloud cover. It was one of the cheapest interplanetary missions ever attempted, because everything had been tested on earlier flights and no prototypes were built.

Looking away from the Sun and beyond Mars, the giant planets – Jupiter, Saturn, Uranus and Neptune – present space scientists with even greater challenges. Already, the Americans have started to meet those challenges. On March 2, 1972, Pioneer 10 left Earth faster than any man-made object had travelled before, covering nine miles every second as it headed for its appointment with Jupiter. The probe weighed 270 kg. (595 lb.) and carried eleven experiments powered by four small nuclear generators. Its 2.75 m. (9 ft.) dish antenna could transmit to Earth from beyond 9,000 million miles.

After passing the orbit of Mars, Pioneer 10 entered the unknown, hurtling towards the asteroid belt and – said the pessimists – probable destruction in a cosmic collision with one of the mini-planets. But that didn't happen, and Pioneer emerged unscathed seven months later. The biggest theoretical hurdle to flights to the outer planets was not after all a major problem.

At 2.25 GMT on December 2, 1973, Pioneer passed 130,300 km. (81,000 miles) above the

A Junk-Yard in Space

Sputnik 1 was too small to be seen in orbit. The tiny speck of light that for Earth-bound onlookers ushered in the Space Age was actually nothing but a piece of junk, Sputnik's huge final rocket spage. Since then the amount of man-made litter in space has reached alarming proportions. There are over 900 satellites and 4,000 bits of debris orbiting the Earth. Sooner or later most of them will fall – and many hundreds more over the next decades as the range of satellites (*right*) increases.

Friction with the outer traces of the atmosphere has already forced nearly 6,000 space objects to re-enter the atmosphere. Most of them burn up in the heat of friction. But parts of some of the sturdier ones have survived.

On September 5, 1962, two policemen in the town of Manitowoc, on the shores of Lake Michigan, found a 20-lb chunk of red-hot metal embedded in the street. It was the remains of Sputnik 4.

In January 1978, the people of Yellowknife, on the shores of Great Slave Lake in Canada, were at the centre of a nuclear alert. At 6 o'clock on the morning of January 24, the Northwest Tettitories were showered with debris from Cosmos 954, including 45 kg (100 lb) of radio-active uranium. Cosmos

should have been boosted into a higher orbit out of harm's way, but it went out of control on January 6. As it happened, after an initial panic, it was found that the satellite had broken up over a remote area and no damage was done; but the implications were ominous.

The next large satellite to go out of control was America's Skylab space station, which in

early 1979 was due to fall to Earth later in the year or early 1980. No-one knew exactly when or where it would fall. Experts stressed that it would probably burn up in thousands of harmless incandescent particles, but it is surely only a matter of time before one of these man-made meteorites strikes a major population centre.

Jovean cloud tops, snapping pictures and measuring the environment of Jupiter and its moons. Scarred by the intense Jovean radiation, and with some of its instruments damaged, it continued on its journey – one that will shortly become interstellar. In 1987, Pioneer will become the first craft to leave the Solar System; it should reach the vicinity of Aldebaran (brightest star of the constellation Taurus) about the year 8,002,000 AD!

Pioneer 11 repeated Pioneer 10's success a year later – a 30-mile-a-second blur passing 43,000 km. (26,500 miles) from the planet on December 3, 1974. But then it went one better. Jupiter's gravity catapulted the probe back across the Solar System for a rendezvous with mighty Saturn. The interplanetary billiard ball was aimed to pass *between* Saturn and its rings in September 1979.

By then, Voyagers 1 and 2 were in pursuit. Voyager 2 was launched first, on August 20, 1977, but it took a slower path than its partner, launched on September 5. The two are identical, using the basic Mariner design that has stood the test of 15 years exploration. But nuclear power plants were added (three each) because – as with the Pioneers – solar panels would not work properly so far from the Sun.

After a 805 million km. (500 million mile) journey, Voyager 1 arrived at Jupiter in early March 1979, followed four months later by Voyager 2. A slingshot trajectory sent both probes towards Saturn, with expected arrival dates in November 1980 and August 1981. Then, if all goes well, Voyager 2 will be aimed for Uranus (to

arrive in January 1986) and possibly even Neptune (September 1989).

The American Space Shuttle will dispatch Jupiter's next visitors (unless the Soviet Union makes a major change of policy). Due to be launched in 1982 as the first planetary mission aboard the Shuttle, JOP (Jupiter Orbiter Probe) should reach Jupiter in late 1984. A probe will parachute into the atmosphere while the orbiter circles overhead.

As man gradually extends his reach in the universe, he may well forget the first trailblazers. But it is possible that, millions of years hence, an intelligent being from another star system will come across one of our wandering probes. If it is Pioneer 10, the inquisitive traveller will see an engraved plaque depicting a man and woman of our species as they looked in the days when man first left the cradle Earth (see Chapter 6). If it is a Voyager, the finder will have picked up a 'message bottle' cast into the ocean of the cosmos: a 12-inch disc (needle and cartridge supplied) with pictures and messages, and the sounds and music of Earth.

Tools in Orbit

The skills developed to explore our Solar System are rapidly being turned towards shaping the destiny of our own planet. Today's satellites give us the weather reports we take for granted, the flood warnings that can save lives and the knowledge that could help us preserve our fragile environment. The spectacular development of communications satellites illustrates best the way man has harnessed his new-found power. They have made the 'Global Village' a reality.

On July 10, 1962, the American flag appeared on European television screens with the now familiar words: 'Live via satellite'. The satellite was Telstar 1, the world's first commercial communications satellite, and the pictures were being beamed from 5,600 km. (3,500 miles) out in space.

Two years after Telstar, a public company called Comsat put its first $200 million worth of shares on the market. Four minutes later they were sold out, and within six months they were worth $700 million. The company had no factory, no management structure, no product – just a licence to put communications satellites in orbit.

Comsat teamed up with 19 governments in 1964 to form Intelsat, a global communications system that today carries more than two-thirds of all international communications. Its first satellite was the famous Early Bird, launched on April 6, 1965. The first transatlantic cable was laid in 1956, and provided 36 static-free lines at a cost of $133 million. A decade later, Early Bird was providing 240 lines, or one television channel, for only $14 million. The ensuing years saw the circuits multiply as more complex Intelsats were placed in orbit. In 1978 there were 13 Intelsats in geostationary orbits around the globe, handling between them over 20,000 hours of television annually and relaying up to 14,000 telephone calls at once, at a fraction of the cost 20 years ago.

The Soviet Union's first comsat was launched 17 days after Early Bird and had about ten times the power output of its American counterpart. Molnya 1 (the word means 'Lightning') was the foundation of the world's first domestic space communications system – vital for a country stretching across 12 time zones. Now, dozens of Molnyas link Vladivostok with Moscow, transmitting newspapers to the Eskimos and reading Marx to desert herdsmen.

But now that comsats have created the 'Global Village', salesmen are appearing on its streets to sell the new technology in a rapidly escalating trade war. The Russians – snubbed by Intelsat when they proposed co-operation in the 1960's – are now planning to compete directly with Intelsat in supplying global communications. In 1978 they announced that eight 'Statsionar' comsats will be placed in geostationary orbits in the next few years, some so close to rival Intelsats that there are fears of radio interference and a major international row.

In the equally lucrative field of domestic comsats, the main competitors are currently the United States and Europe. The American technology has already been tried and tested in Canada, Japan, and Indonesia and by the three US domestic systems: Western Union's Westar (launched April and October 1974), RCA's Satcom (December 1975 and March 1976) and Comsat Genreal's Comstar (May and July 1976). But the biggest coup was the Applied Technology Satellite (ATS) programme, culminating on May 30, 1974, with the launch of ATS-6 – the world's first educational satellite.

ATS-6 was a remarkable satellite, combining all American comsat know-how into one super-satellite 30 times more powerful than the most advanced Intelsat. It was, moreover, manoeuvrable. The two-story high satellite weighed 1,400 kg. (1.4 tons), but its business end was a 9 m. (30 ft.) reflector antenna capable of transmitting to simple Earth-based receivers made of chicken-wire. It was first used to teach schoolteachers in the Appalachian Mountains of the US and then to provide televised conferences between clinic doctors in Alaska and specialists throughout the US. Then the £83 million ($200 million) ATS was moved, making the longest rocket burn in history (5 hrs. 37 mins. 17 secs.) to ease itself half way round its orbit to a new fixed position above India. Beginning on August 2, 1975, ATS transmitted for 12 months to 5,000 villages in seven states. 500 million people took basic lessons in agriculture, health and family planning.

The Indian experiment was taken over by America's market competitor – Europe. The Franco-German Symphonie 2, which undertook the job, was launched by the US on August 27, 1975, but only after the two competitors had agreed strict limits on its use. (France had foreseen such conflicts of interest in 1972 when it pressed for an independent European launcher. The European Space Agency's three-stage Ariene is due to become operational in the early 1980's). Like Symphonie 1 (December 19, 1974), the second Symphonie weighed 237 kg. (523 lb.) and carried two television channels or 300 telephone circuits.

Europe's other market showpiece is the 444 kg. (980 lb.) Orbital Test Satellite (OTS), orbited by the US – after one embarrassing failure – on May 11, 1978. OTS is a project of the 11-nation European Space Agency (ESA) and the prototype for the European Communications Satellite (ECS) system, planned as a four-satellite network with launches (aboard Ariene) beginning in 1981.

Comsats are by no means the only revolutionary satellites. Our knowledge of the weather has undergone extraordinary changes with the launch of weather satellites. The first weather satellite was Tiros 1, launched on April 1, 1960. The hatbox-shaped Tiros (Television and Infra-red Observation Satellite) was followed by nine more weather watchers in the next few years. Between them they sent back about 650,000 weather pictures. In 1963, Tiros 8 carried the first Automatic Picture Taking (APT) equipment for beaming pictures to hundreds of cheap receivers worldwide. Anyone with a few hundred pounds and a little electronic know-how can today set up their own satellite ground station to receive APT pictures.

Tiros was replaced in 1966 by the more advanced TOS (Tiros Operational System). The series ended in 1969 after nine launches, and in that year the seventh TOS made history by

These are just five of the many hundreds of satellites that now link the communication systems of the world, keep a watch on its weather and gather data on its resources.

ATS (Application Technology Satellite)

TIROS (Television and Infra-Red Operational System)

ITOS (Improved TIROS Operational System)

Nimbus (Weather satellite)

ERTS (Earth Resources Technology Satellite)

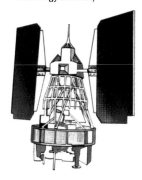

spotting huge snowdrifts in America's mid-west. Flood control experts were able to declare a disaster *before* it happened.

The next weather watchers were the ITOS (Improved TOS) satellites flown between 1970 and 1976. For the first time, the system was paid for by the US Weather Service, not NASA. Carrying more advanced equipment, each of the six ITOS satellites could 'see' clouds in the dark and each sent back four times more information in a day than the earlier TOS. But even ITOS has now been replaced.

The first Tiros N – twice as heavy as ITOS – was launched on October 13, 1978. Eight of these satellites will be launched before 1985, and the last three (called ITOS-E) will double as the first spaceborne marine rescue co-ordination system. They will be tuned to frequencies used by ship, boat and aircraft distress transmitters. Tiros N, meanwhile, is backed up by the world's first geosynchronous weather service, operated by three GOES (Geostationary Operational Environmental Satellite) launched between 1975

and 1978. These in turn will be replaced by GOES D, E and F, to be launched by Space Shuttle in 1980–83.

Much of this rapid development has been made possible by the Nimbus programme, begun in August 1964. Nimbus looks rather like a cosmic butterfly with huge solar panel wings and an instrument platform at the base of a conical spaceframe. Seven Nimbus satellites have served as test-beds for new technology – in particular, Earth sensing devices – as well as being operational satellites in their own right. They have monitored water pollution, tracked migrating animals, kept a benevolent eye on solo aviators and shown map-makers their mistakes.

The Soviets have their own weather eyes aloft, set there as part of the all-embracing Cosmos programme. They inaugurated their first weather satellite in March 1969. The two-ton Meteor 1 could cover a 620 mile wide strip beneath its path with two television cameras, and it is said the Meteor programme has cut 10 per cent off sailing times by telling Soviet shipping what weather to expect. Twenty-eight Meteor 1-type satellites have been launched and the Meteor 2 satellites are now combining Nimbus-type experiments with their weather-watch duties.

Successor to the Nimbus series is the Landsat earth resources programme, begun in 1972. The Landsats, originally called Earth Resources Technology Satellites (ERTS), have extended our view beneath the clouds.

Three Landsats have been launched at a total cost of £142 million ($251 million). Each was placed in a polar orbit circling the Earth 14 times a day and returning over the same point, at the same time of day, every 18 days. They look identical to Nimbus, but they carry more advanced remote sensors and are backed up by 300 scientists in 50 countries.

Landsat operates on the principle that everything on Earth – animal, vegetable and mineral – transmits or reflects electro-magnetic radiation, both at visible and invisible wavelengths. Our eyes register only a small part of the spectrum, whereas Landsat's 'eyes', recording with special filters, see far more. They can show, in 'false colour' images, the darker colour of diseased crops and polluted water.

With more than half of Asia, Africa and South America still inadequately mapped, Landsat has demonstrated the power of orbiting cartographers. Egypt, for instance, has cancelled a £1.5 million, 10-year mapping project because Landsat can do it cheaper and better in 5 years.

For the poverty-stricken two-thirds of mankind Landsat is probably the most important space project to date. Landsat's successors will predict crop yields and available grazing with 90 per cent accuracy, calculate the water yield of melting snows, spot watering holes and locate new mineral deposits. If these skills are developed and sensibly applied, perhaps it may not be too much to hope that the benefits of space technology will be spread more uniformly.

Landsat's false colour photography identifies ground cover, surveying areas and their day-to-day changes with vivid accuracy.

Below: The Alpine glaciers of the Saint Elias Mountains on the borders of British Columbia and Alaska lick down from the Highlands (white) forming Moraines (yellow) in the forested valleys (green). Silt shows up with a reddish tinge in some of the glacial lakes.

Bottom: Blue, red and green filters combine to emphasize the lush vegetation that clothes the Shan Plateau, in East Burma. Bare rock can be seen (*left*), as can the savannah lands that indicate forest clearances.

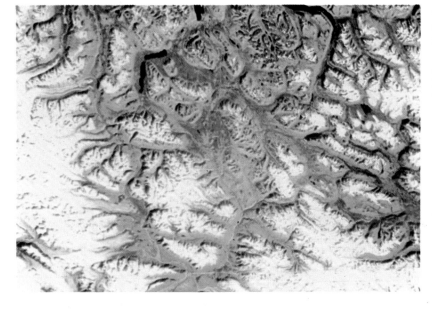

11/MAN IN SPACE

LEO ENRIGHT

The Soviet Union's success in placing the first man in space sparked a 'space race' that made the 1960's a decade of heady achievement, culminating in the US Apollo programme, which placed men on the Moon (at left, Jim Irwin of Apollo 15 salutes the Stars and Stripes beside the Lunar Module and Lunar Rover). Over a decade later, that achievement still seems astounding in its complexity. Almost certainly, men will never again be blasted so deep into space from the Earth to return again to its surface. The future of Man in space lies with the Space Shuttle and with giant space stations, of which space laboratories, like those tested by both the US and the Soviet Union, are the forerunners.

Yuri Gagarin, the first man to fly in space, on April 12, 1961, died in 1968 in a plane crash while training for another space mission.

Alan Shepard (*inset*) was the first American in space. He went into sub-orbital flight in a Mercury capsule, atop a Redstone rocket, shown *below* in a test with a 'cherry picker' which would allow Shepard to escape if anything went wrong on the launch pad.

Below, right: John Glenn, the first American to orbit the Earth, in February 1962, later entered politics and became a Senator.

On the morning of April 12, 1961, by the banks of the Volga, Mrs Anya Takhtarova saw an extraordinary sight: a space-suited man unhitching himself from a parachute. She stepped up to him and stammered: 'Have you come from outer space?' 'Just imagine it?' came the reply, 'I certainly have!' It was Yuri Gagarin, the first astronaut, who had ejected from his capsule at 23,000 feet.

Gagarin, the son of a carpenter, went into orbit inside a 2½-ton steel ball called Vostok 1 (the word means 'East'). The capsule, a modification of an unmanned satellite design, was packed with 241 electric lamps, 6,000 transistors, 56 electric motors, 800 relays and switches, a parachute and a rocket-propelled ejection seat, leaving barely enough room for Gagarin himself. The 27-year-old Air Force lieutenant was launched from Tyuratam, Soviet Central Asia, at 9.07 a.m. aboard a Sputnik booster with a third stage added. His 2.3 m. (7½ ft.) wide capsule nestled like a scoop of ice-cream atop a cone-like instrument compartment which included the retro-rocket. If that rocket failed, Gagarin had enough food and air to last the ten days needed for Vostok's orbit to decay naturally and return him to Earth.

The young pilot – little more than a passenger aboard an automatic craft – gazed awestruck at the view unfolding beneath him as controllers below watched his temperature, pulse, blood pressure and breathing for signs of trouble. There were none. The readings showed that man had a future in the cosmos.

One hour and 18 minutes after lift-off, the retro-rocket fired above Kilimanjaro in Africa and Gagarin's capsule scorched into the atmosphere at 28,000 km./hr. (17,500 m.p.h.). De-

celeration, the force of which took his weight from zero to almost a ton in a few seconds, slowed him to just 20 m.p.h. in 30 minutes. Landing by parachute 108 minutes after launch, Gagarin immediately inspected the charred capsule nearby. 'It was good enough to be used again', he reported. The cosmonaut was examined by an army of doctors who came to much the same conclusion about him. Man could survive spaceflight, for a couple of hours at least.

America's response to this Russian 'first', less than a month later, was not a strong one. Some said disparagingly it was a mere 'flea-hop' in in comparison. On May 5, Alan Shepard flew a 15 minute 22 second sub-orbital 'lob', 485 km. (300 miles) out into the Atlantic from Cape Canaveral. But the significance of the flight should not be underrated. This, the first flight of Project Mercury to place a man in orbit, used a new conical cabin design purpose built for the job.

Virgil ('Gus') Grissom admitted he was 'a bit scared' when he repeated Shepard's flight atop a Redstone missile on July 21. He had a lucky escape when, on splashdown, his hatch ejected by accident and his capsule sank, but otherwise the flight went perfectly and NASA decided to speed up the Mercury programme. Their next astronaut would go into orbit.

But by then the Russians had widened their lead. On August 6, Gherman Titov (at 24, the youngest person ever in space) headed for orbit aboard Vostok 2. He called his craft 'Eagle', and it soared aloft for seventeen orbits lasting 25 hours 18 minutes. Titov was the first man to live in space, eating, working and sleeping in weightlessness. But a serious bout of nausea revived doubts about the feasibility of marathon flights.

The next man in orbit was John Glenn, America's first. Millions watched on television as the Atlas missile left Cape Canaveral at 9.47 a.m. on February 20, 1962. Glenn's bell-shaped Mercury capsule (called 'Friendship 7') was 2.7 m. (9 ft.) high, measured 1.8 m (6 ft.) across at the base and weighed 1.3 tons. Crammed inside a space about the size of a telephone box were 10,000 components, 7 miles of wiring and a man. It anything went wrong during launch, Glenn could push the 'chicken switch' to his left and a 5.2 m (17 ft.) escape tower would rocket him clear. He didn't need it.

But Glenn did face an apparent emergency up in orbit, where nothing could have saved him. During the 4 hour 50 minute flight, an instrument called Segment 51 at Mission Control in Cape Canaveral said Glenn's heat shield had come loose. If it was right, he would vaporize in the heat of re-entry. Controllers ordered Glenn not to dump his retro-rockets after they'd fired at the end of the third and last orbit, hoping their straps would keep the heat shield in place. Fortunately, Segment 51 had malfunctioned and Glenn returned safely.

Scott Carpenter flew three orbits – as Glenn had – in May but in August the Soviets scored another success, this time a 'double', in August. Andrian Nikolayev blasted off in Vostok 3 on August 11, and had been only a day in orbit when he was joined in space by Pavel Popovich aboard Vostok 4. 'Nick and Pop', as they were called in the West, passed within 5 km. (3 miles) of each other at one point – a milestone on the road to space rendezvous. They gave us the first T.V. show from space and landed within six minutes of each other on August 15. Nikolayev had become the first man to pass the 'One Million' milestone in space, a distance a jet would take 2½ months to cover.

The Americans completed the £163.6 million ($392.6 million) Mercury programme without further embarrassment. Walter Schirra made six orbits during a flight lasting 9 hours 13 minutes in October, 1962, and in May, 1963, Gordon Cooper splashed down after a 22-orbit flight which lasted 34 hours 20 minutes.

Almost immediately, the Russians seized the headlines again by placing the first woman in space, Valentina Tereshkova. She was reputedly so star-struck by the Gagarin flight that her fellow workers at an Upper Volga textile factory nicknamed her 'Gagarin in a skirt'. Although no more than a parachutist, she wrote to Moscow asking to be made a cosmonaut. She learned to fly during training, and on June 16, 1963, she was rocketed aloft aboard Vostok 6 to join her 'space brother', Valeri Bykovsky, already in orbit in Vostok 5. The 26-year-old Tereshkova circled the Earth 48 times in 70 hours 50 minutes while her partner went on to set a new record of 81 orbits in almost five days. (Five months after her flight, 'Valya' married the Vostok 3 pilot, Nikolayev. Within a year, she gave birth to a daughter, Yelena – the first 'space age' child.

The next Soviet project was the mysterious Voskhod. This, the shortest manned project in history, now seems to have been an audacious political coup masterminded by Soviet Premier Nikita Khrushchev. When he heard the Americans were planning a two-man spacecraft (Gemini), he ordered his team to launch a three-man capsule. This was made possible by modifying a Vostok capsule, replacing the ejector-seat with a couch. An emergency escape was therefore impossible. In addition, the need to save weight meant that the astronauts could not wear space suits. Voskhod 1 went into orbit on October 12, 1964, carrying Vladimir Komarov (pilot), Konstantin Feoktistov (scientist) and Boris Yegorov (physician). During their 24 hour 17 minute flight, the three conducted medical experiments and spoke on the telephone with Khrushchev.

When Komarov asked for a 24-hour extension, controllers enigmatically quoted *Hamlet*:
There are more things in heaven and Earth, Horatio,
Than are dreamt of in your philosophy.
There were indeed: by the time the three arrived back in Moscow, their sponsor had been deposed.

The new rulers, Brezhnev and Kosygin, were less enthusiastic about the Voskhod project, but they allowed one more flight. On March 18, 1965, as Voskhod 2 – with two men this time, the extra space being taken up with an air-lock – passed over the Soviet Union for the second time, Alexei Leonov squeezed through the air-lock to make history as the first man to 'walk' in space. 'I saw the Universe in all its grandeur', he reported. 'It was as though I was swimming over a vast colourful map of Earth.'

Voskhod 2 was in orbit for 26 hours 2 minutes, making an extra orbit when the automatic re-entry system failed and Leonov's commander, Pavel Belyayev had to take manual control. He landed safely, but they were far from home, on the snowbound Ural mountains 2,000 km. (1,250 miles) from the target point. They froze there for two days before being rescued. Thereafter, the Voskhod programme ended abruptly, presumably because the new rulers felt (like many Western observers) that it was more a stuntship than a spaceship.

Gemini

'By heavens, methinks it were an easy leap to pluck bright honour from the pale-fac'd Moon,' said Hotspur in *King Henry IV*. In practice, plucking anything from the pale-fac'd Moon would prove a monumental undertaking. No one had any illusions. When on May 25, 1961, President John F. Kennedy committed his country to making the leap within nine years, he admitted to Congress that no project 'will be more difficult or expensive to accomplish'.

For two years, NASA agonized over three possible routes to the Moon. Everyone liked the simplicity of the Direct Flight method, which placed one large craft on the Moon's surface but needed a blockbuster rocket to get there. The other favourite was Earth Orbit Rendezvous, a

This sketch of astronaut Gordon Cooper, who made the sixth and last Mercury flight in 1963, captures something of the glamour that surrounded the pioneering spacemen in the early days of space travel.

Edward White becomes America's first space-walker as he floats outside his Gemini 4 spacecraft on June 3, 1965 (the first space-walker had been Alexei Leonov two and a half months previously). Eighteen months later, White was to be one of three astronauts killed in an Apollo capsule during a simulated countdown.

Right: Both these shots were taken in December 1965, from Gemini 6, carrying astronauts Walter Schirra and Thomas Stafford, as it manoeuvres towards Gemini 7, piloted by Frank Borman and James Lovell. The astronauts were building up their experience of rendezvous techniques before undertaking the delicate docking manoeuvres necessary for the Moon mission.

method calling for two smaller rockets which would join forces in Earth orbit and head straight for the Moon. The third approach began as an outsider – Lunar Orbit Rendezvous (LOR), championed by a team at NASA's laboratory in Langley, Virginia. LOR needed only one rocket which would place two spaceships in lunar orbit. Explorers would then descend in one craft like sailors going ashore in a small boat. The Langley team won the day and LOR was chosen to take the first men to the Moon, a journey that still seemed a long way off at the beginning of Gemini to test rendezvous and docking techniques.

The Gemini astronauts were to test the techniques of LOR in a two-man capsule nearly twice the size of Mercury and twice the weight. The designers added an Adapter Module to the base of the capsule, allowing them to clear vital equipment out of the crew cabin and dump it in a separate compartment. The adapter also housed 16 control thrusters which gave Gemini the ability to raise and lower its orbit at will. Another new feature was the remarkable collection of fuel cells which produced electricity through the chemical reaction of oxygen and hydrogen; a by-product of this reaction was over half a litre of water an hour – 'like bringing your own well into the desert', enthused one expert.

All this exotic equipment meant Gemini weighed almost 4 tons and needed a powerful rocket to lift it into orbit. The one chosen was the 27.5 m. (90 ft.) Titan 2 booster, a missile with the power of 32,000 automobiles.

The first Gemini flight went almost unnoticed in the excitement of Leonov's space-walk only five days earlier, but Gemini 3 (the first two were unmanned tests) marked America's resurgence as a contestant in the Space Race. On March 23, 1965, astronauts John Young and Gus Grissom carried the first computer into space. This device could make 7,000 trajectory calculations a second. For the first time, the pilots could truly 'fly' their spacecraft. Without such an instrument, man could never go to the Moon.

Three months later, on Gemini 4, the computer failed and commander Jim McDivitt had to make

a manually-controlled re-entry. But his partner, Edward White, also captured the limelight by making America's first space-walk – 'strolling' from Polynesia to the Caribbean in 21 minutes and using a small gas gun to manoeuvre around.

The next step was to see if men could survive in space long enough to get to the Moon. Gordon Cooper and Charles ('Pete') Conrad spent eight days aloft in Gemini 5 during August 1965; on the third day they made a rendezvous with an imaginary target in space – a first and successful test of LOR theory.

Gemini 6 got off to a slow start. The plan was to make history's first space link-up, using a 7 m. (23 ft.) Agena rocket stage as the target. Walter Schirra and Tom Stafford were only 42 minutes from lift-off when the Agena fell into the Atlantic during launch. The two men were still earth-bound when Gemini 7 carried Frank Borman and Jim Lovell into orbit for a 14-day space marathon beginning on December 4. Gemini 7 manoeuvred alongside its orbiting second stage for 15 minutes on the first orbit in a rehearsal of a spectacular now being planned. If all went well, the crew would be joined in orbit by Gemini 6 for the first true rendezvous of men in space.

Back at Cape Canaveral, the Gemini 6 crew were demonstrating the difference between astronauts and ordinary mortals. A split second after their Titan rocket roared to life on December 12 something went wrong. The motors shut down and everyone waited for the crew to follow the rules and blast to safety in their ejector seats. But they stayed put and risked a fireball to save the mission. The gamble paid off; the fault was traced and Gemini 6 finally made it into orbit on December 15, streaking for a rendezvous with Gemini 7. Only six hours later they were manoeuvering within a foot of each other, having surmounted one of the biggest obstacles on the

hard road to the Moon.

On March 16, 1966, Neil Armstrong and David Scott became the first men to dock with another spacecraft in orbit, forging the last link in the LOR chain. But the jubilation turned to horror when a control thruster on the Agena target went mad 27 minutes after docking. The linked Gemini/Agena went into a sickening spin, facing Armstrong with the first real space emergency. With remarkable calm, he separated the tumbling craft and fired thrusters for several minutes to bring the Gemini under control.

The remaining Gemini flights developed the techniques proven on the earlier missions. Eugene Cernan made a 2-hour spacewalk on Gemini 9; Gemini 10 docked with an Agena which then boosted John Young and Michael Collins to a record height of 761 km. (473 miles); Richard Gordon tied a 30 m. (98 ft.) tether between Gemini 11 and an Agena, then Pete Conrad rotated the two craft to create artificial gravity for about three hours (although each man 'weighed' only about half an ounce); Edwin ('Buzz') Aldrin spent a total of $5\frac{1}{2}$ hours outside Gemini 12 during three spacewalks on the last Gemini flight in November 1966. The £534,750,000 ($1,283.4 million) Gemini project paved the way for Apollo and, for the first time, placed the United States firmly ahead of Soviet space technology.

Apollo

The foundations of the epochal Apollo voyage were laid in a Florida swamp. As space experts wrestled with the problems of getting to another world, Army engineers raised an eighth 'Wonder of the World' out of the alligator-infested mire called Merritt Island, north of the main Cape Canaveral rocket range. They called their creation the Kennedy Space Center (KSC).

The Space Center defies easy comprehension. The single room of the Vehicle Assembly Building (VAB), for instance, can hold two Buckingham Palaces inside its 129.5 million cubic feet working area. Towering 160 m. (526 ft.) into the sky, it was the biggest building in the world when completed and needs a 10,000 ton air conditioning system to stop clouds forming inside.

Each Moon-rocket was built stage-by-stage on

In an early docking attempt, Gemini 9 approaches what Thomas Stafford called the 'angry alligator' – their unmanned target vehicle. The shroud that had protected the docking mechanism during launch had not fully released and the mission had to be abandoned.

Jettison motor and launch escape system

Command module

Service module

Engine

Lunar module

Instrument unit

THIRD STAGE

Liquid oxygen tank

J-2 rocket engine

SECOND STAGE

Liquid hydrogen tank

Liquid oxygen tank

J-2 rocket engines (5)

FIRST STAGE

Liquid oxygen tank

Kerosene fuel tank

Oxygen feed lines

Thrust structure

F-1 rocket engines (5)

The cutaway *above* and the pictures *at right* dramatize the immense difference in scale between the Apollo Command Module, the cabin for the three-man crew, and the Saturn 5 moon rocket that launched it. The Command Module was a mere 3.2 m. (10 ft. 7 in.) high, compared with the 111 m. (363 ft.) rocket. Behind the Command Module was the Service Module which supplied oxygen, water and power to the Command Module. The Service Module had its own engine for major speed changes on the way to the Moon. Below were the three stages that took the men (Apollo 11 is shown *centre*) into orbit and set them on course for the Moon.

top of a 5,000 ton Mobile Launcher which had then to be moved three miles to one of two launch pads, Launch Complex 39A or B. To handle this mammoth task, they built the biggest truck on Earth – the Crawler, a football-field size platform weighing 3,000 tons. Each link of its caterpillar tracks weighs a ton and the lumbering giant has what must be the slowest speedometer range of any vehicle in the world: 2 m.p.h. maximum.

The 'brain' behind the Apollo launch complex was the Launch Control Centre (LCC), next door to the VAB. The communication 'nerves' linking this 4-storey firing room with the pad cost £5.5 million ($13 million) to install. The launch pads themselves cost £8.8 million ($21 million) apiece and began life as 80 ft. tall pyramids of gravel. This gesture to another great civilization was needed to compact the swampy soil before construction could begin.

All this – the world's first Moonport – was nothing without the awesome creation it was built to serve. Saturn 5 – the most powerful rocket ever built – was 111 m. (363 ft.), at 2,850 tons the weight and size of a destroyer.

Saturn's first stage (the S-IC) was fuelled by liquid hydrogen (LH). The oxidiser, needed because a rocket can't take in air like a jet engine, was kerosene (a high-grade paraffin called RP-1). Starting as a liquid at minus 217°C, the hydrogen was pumped with the power of 30 diesel locomotives through the hollow walls of the huge rocket nozzle to cool the nozzle and turn the LH to gas. The gas was then forced into a chamber where it ignited with kerosene, burning at the rate of 15 tons a second to produce a total first-stage thrust of 7.5 million pounds (3,350 tons).

The massive 138 foot first stage took the rocket just 61 km. (38 miles) above the Atlantic, to be discarded after only 2½ minutes work. The stage that took over (the S-II) was smaller, yet it was an even greater technological achievement. The kerosene oxidiser was replaced with liquid oxygen (LOX), an explosive combination with hydrogen.

The deadly liquids of oxygen and hydrogen were combined again in the 18 m. (59 ft.) third stage (the S-IVB) which completed the job of placing Apollo in Earth orbit. It would be re-ignited later to blast its cargo to the Moon and then sent into solar orbit. (Later third stages were crashed on the Moon to set the interior ringing with artificial moonquakes).

It took years to perfect the colossus that would perform for barely 12 minutes and travel a small fraction of the way to the Moon. But Saturn's power was needed to heave the 43-ton spaceship out of the pit of terrestrial gravity to a distance at which it could continue the journey freely.

Apollo itself came in three sections: the Command Module (CM) mothercraft; the Service Module (SM) containing vital systems and the Apollo main rocket engine; the Lunar Module (LM), composed of a four-legged octagon, the descent stage, attached to a stubby ascent stage.

The 5½-ton Command Module housed the three astronauts and was the control centre for

the mission. Built by North American Aviation under the biggest civilian contract in history, the 3.5 m. (11.4 ft.) cone measured 3.9 m. (12.8 ft.) across at its base and was twice the size of Gemini. It was the most sophisticated spacecraft ever built, comprising over two million parts. Life was still cramped, but at least there was room for one man to stand when the middle couch was stowed away. The space sextant at the astronauts' feet was linked to a remarkable computer that allowed the crew to navigate in deep space and return home unaided in an emergency; parts of that computer are now used routinely by long-distance jets. The cabin walls were lined with lockers containing food, water, clothing and shaving gear.

The CM fitted snugly into the front of the 4.7 m. (15.4 ft.) long cylindrical Service Module. Like Gemini's adapter module, the 23-ton SM was the 'public utility' section of Apollo, providing power and water from its three fuel cells and enough life-giving oxygen to last several weeks. It also housed the large Service Propulsion System (SPS) rocket engine to slow Apollo into lunar orbit and blast it home again. Sixteen smaller rocket jets gave the CSM the manoeuverability it needed.

The LM was the first true spaceship, built exclusively for use in the vacuum of space and therefore lacking the aerodynamic good-looks of earlier craft. If often looks deceptively small in lunar surface photographs, but in fact the LM was over two storeys high and weighed $14\frac{1}{2}$ tons. The two lunanauts lived in the pressurized ascent stage cabin, standing at windows on either side of the front door during the landing. Behind them, and sticking out of the floor, was the ascent stage engine that would blast them back into lunar orbit.

Explosive bolts linked the LM's ascent stage with the four-legged descent stage and the unique variable thrust rocket that would slow the craft gently onto the Moon. When the time came to leave the Moon-base, the bolts would fire to separate the two halves and leave the descent stage behind as a launch pad.

America now had a plan and the technology to execute it. What remained between man and Moon were the flight tests to knit all the equipment and experience into a fool-proof assault plan. Space veterans Gus Grissom and Edward White, and newcomer Roger Chaffee, would test the CSM on Apollo 1. They were to be hoisted aloft aboard the 68.3 m. (224 ft.) Saturn 1B, a smaller member of the Saturn family.

Days before their flight, tragedy struck. The three astronauts were taking part in a practice countdown atop the Saturn 1B at launch complex 37 in Cape Kennedy. At 6.31 p.m. on January 27, 1967, one of them shouted: 'Hey, we've got a fire in the cockpit!' Seconds later came another shout, then a final scream. The three were incinerated within half a minute, their spacesuits offering no protection against an inferno fuelled by the cabin's pure oxygen atmosphere. Not only was the tragedy intense in personal and national terms, but it pinpointed a major error in design that might ruin the assault on the Moon.

Apollo survived the disaster, but it took 18 months to recover from the blow. Then, on October 11, 1968, three men went into orbit aboard Apollo 7, with the same mission as their dead comrades and a much safer capsule. The crew were Walter Schirra – only man to fly Mercury, Gemini and Apollo – Donn Eisele and Walter Cunningham. They circled the Earth for nearly 11 days, testing every system aboard the CSM and firing the vital SPS booster four times.

Apollo 7 was still in orbit when NASA did a remarkable thing. Departing completely from their step-by-step programme, they announced that Apollo 8 would ride mighty Saturn to the Moon. It was a couragous – many said foolhardy – step, because the Saturn 5 had had only two test launches out of an original 13 planned, and the most recent launch had been far from flawless, with problems developing in all three stages.

But there was little time to spare, and the space spectacular at Cape Kennedy began on December 21, 1968. The Earth shook for miles around as Saturn mustered the awesome power for its thrust to the Moon. On top of the pillar of fire were three astronauts: Frank Borman, Jim Lovell and Bill Anders. The launch was flawless, as was the 5 minute 20 second Translunar Injection (TLI) burn of the S-IVB third stage 161 km. (100 miles) above Australia on the second orbit. Travelling faster than any men before them, the first humans to break totally with Earth's gravity streaked for their target at almost 11 km. (7 miles) a second.

On Christmas Eve, Apollo 8 fired its SPS motor for six minutes to slow it into orbit 111 km. (69 miles) above the Moon. During Christmas Day they sent back live pictures of the air-less moonscape: 'It makes you realize just what you have back there on Earth,' said Lovell. 'The Earth from here is a grand oasis in the great vastness of space.'

Apollo 8's triumphant six-day mission – including 20 hours orbiting the Moon – vindicated the Apollo plan. Apollo 9 then tested the LM for the first time in Earth orbit, during March 1969. The crew were James McDivitt, David Scott and Russell Schweickart. Only one final dress rehearsal now remained.

Apollo 10 must have been the most frustrating

In this capsule on January 27, 1967, Virgil Grissom, Edward White and Roger Chaffee died in just half a minute when a spark caused the oxygen-rich atmosphere to burst into flames. The disaster delayed the Apollo programme by 18 months as the craft was redesigned to eliminate the hazard and introduce a quick-opening hatch.

The method by which the Apollo programme carried men to the Moon and back (*below* and *overleaf*) was not self-evidently the best method when President Kennedy made his famous commitment in 1961 to reach the Moon by 1970. But by the end of 1962, NASA has decided on the lunar orbit rendezvous method, in which one rocket launched both the Apollo spacecraft and the Lunar Module. The operation will always remain an astounding one. 98 per cent of the 3,000 tons belonged to the Saturn launch vehicle. There were 87 motors all together. The rocket burned half a million gallons of fuel in the first three minutes. At launch, to avoid undue stress, the rocket was programmed not to fight cross-winds and as it began to topple, to compute a new trajectory with each gust.

mission ever flown by an astronaut. On May 22, Thomas Stafford and Eugene Cernan flew their LM within 14.5 km. (9 miles) of the Moon, testing every step of Apollo except the landing itself. With only a 12 minute rocket burn between themselves and a place in history they dutifully discarded their descent stage and ignited the ascent engine to rejoin John Young in the CSM 96.5 km. (60 miles) above. A perfect 2½ minute SPS burn set them on course for a dawn splashdown on May 26, when a huge sign went up in the Mission Control Center (MCC) at Houston, Texas. It read: '51 Days to Launch'.

It was an extraordinary time, for the whole gigantic project was a TV spectacular relayed around the world, from count-down to splashdown. For a few days, the launch pad, the banked TV monitors at Mission Control, the radio exchanges with their dividing whistle, the images of Earth and Moon, the arcane space terminology – all were almost as familiar to tens of millions of viewers as to those involved on the ground and in the Apollo craft.

The first crisis the lunar explorers faced came just short of moonfall. The Apollo 11 Lunar Module, code-named 'Eagle', was still 9.5 km. (6 miles) up when the vital guidance computer began flashing an alarm – it was overloading. Any second it could give up the ghost under the mounting pressure and nothing the two astronauts could do would save the mission. Emergencies were nothing new to commander Neil Armstrong (remember Gemini 8), but he and his co-pilot Buzz Aldrin hadn't even practiced for this one on the ground – no one believed it could happen. Sweeping feet first towards their target, they pressed ahead as controllers on Earth waited heart-in-mouth. Racing against the computer, Eagle slowed and then pitched upright to 'stand' on its rocket plume and give Armstrong his first view of the landing site. The wrong one! They had overshot by four miles into unfamiliar territory and were heading straight for a football field size crater filled with boulders 'the size of Volkswagens'.

With his fuel running out, and only a minute's

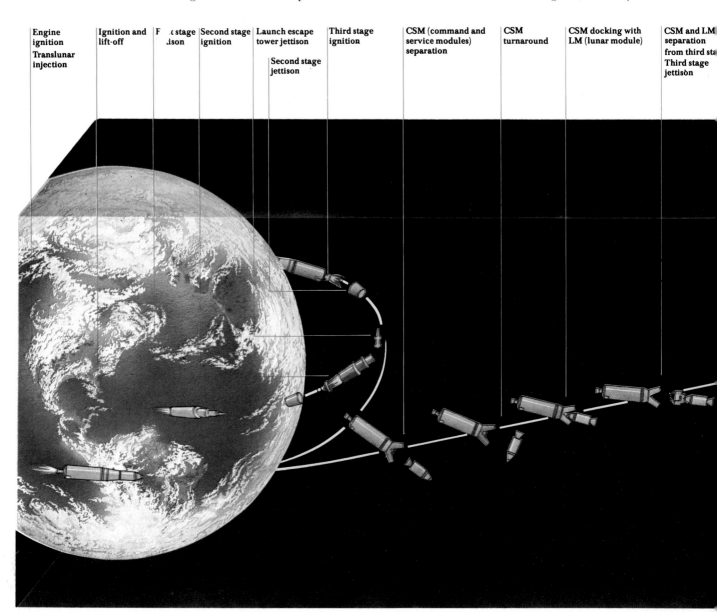

| Engine ignition Translunar injection | Ignition and lift-off | F t stage .ison | Second stage ignition | Launch escape tower jettison / Second stage jettison | Third stage ignition | CSM (command and service modules) separation | CSM turnaround | CSM docking with LM (lunar module) | CSM and LM separation from third sta Third stage jettison |

flying time left, Armstrong coolly accelerated the hovering Eagle beyond the crater, touching 88 km./hr. (55 m.p.h.). Controllers were puzzled and alarmed by the unplanned manoeuvres. Mission Director George Hage pleaded silently: 'Get it down, Neil. Get it down.'

The seconds ticked away.

'Forward, drifting right,' Aldrin said. And then, with less than 20 seconds left, came the magic word: 'Contact!'

Armstrong spoke first: 'Tranquility Base here, the Eagle has landed.' His words were heard by 600 million people – a fifth of humanity.

About 6½ hours later, Eagle's front door was opened and Armstrong backed out onto a small porch. He wore a £42,000 ($100,000) moonsuit, a sort of Thermos flask capable of stopping micro-meteoroids travelling 30 times faster than a rifle bullet. The 'service module' for this one-man spacecraft was a 49 kg. (108 lb.) backpack with enough oxygen for up to four hours. The complete ensemble weighed more than Armstrong, though of course it was much lighter in gravity

one-sixth that of Earth's. Heading down the ladder, Armstrong unveiled a £200,000 ($480,000) TV camera so the world could witness his first step: 'That's one small step for a man, one giant leap for mankind.' It was 3.56 a.m. BST, July 21, 1969.

Armstrong spent several minutes finding his Moon-feet and snapping pictures of his desolate surroundings. 'There's absolutely no difficulty in moving around,' he observed. 'It's like much of the high desert of the United States. It's different, but it's very pretty out here.'

'Magnificent desolation', gasped Aldrin as he stepped on the surface 18 minutes later. The two men set up a seismometer to measure moon-quakes and a reflector which scientists have used to measure the distance to the moon by aiming laser beams at the silicon prisms and timing the round trip.

Forty-eight lbs. (22 kg.) of lunar rock and soil were safely salted away in two special suitcases during the astronaut's 2¼-hour moonwalk. In return they left behind them a junk-yard of

In this, perhaps the most famous of all space photographs, 'Buzz' Aldrin faces Neil Armstrong and the Eagle lander in the Sea of Tranquility, July 1969, during the Apollo 10 first lunar landing.

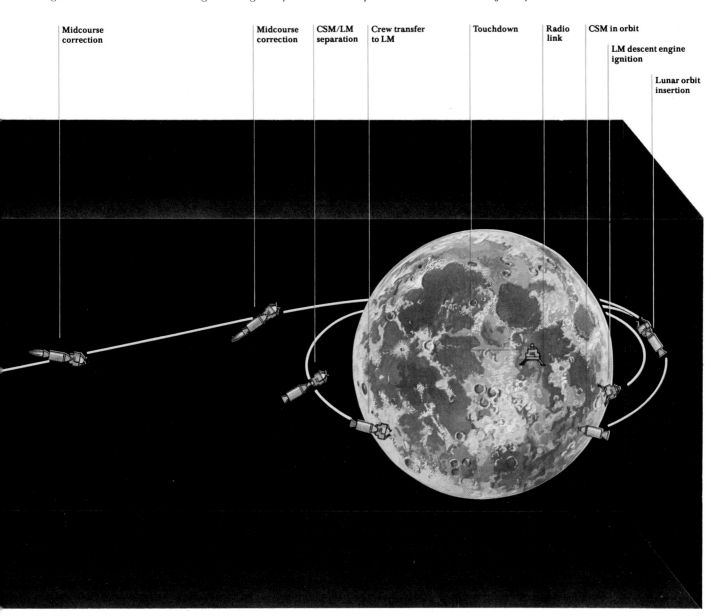

Midcourse correction

Midcourse correction

CSM/LM separation

Crew transfer to LM

Touchdown

Radio link

CSM in orbit

LM descent engine ignition

Lunar orbit insertion

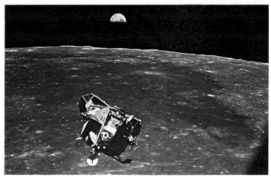

Right: Standing by the Eagle, Buzz Aldrin sets up a solar wind experiment.

Centre: The Lunar Module in orbit as seen by Michael Collins in the Command Module.

Far right: The Command Module which turned in a few minutes from spacecraft to meteorite to boat, floats safely in the Pacific.

unwanted equipment, ranging from bags-full of urine to the 3-ton descent stage of Eagle.

Armstrong and Aldrin left Tranquility Base at 6.53 p.m. BST, July 21, after a tense countdown. The ascent engine had to work, for there was no hope of rescue. A smooth lift-off and an eight minute burn placed Eagle's ascent stage back in orbit, ready for the linkup with the CSM (code-named Columbia) and the forgotten man of Apollo 11 – Michael Collins. After off-loading its precious cargo of men and moonrock, Eagle was cast adrift and Columbia's SPS gave the CSM a 8,530 km./hr. (5,300 m.p.h.) kick in the tail to fire it back to a waiting Earth.

The three heros had to delay celebrating their homecoming for 17 days after splashdown on July 24. Fearing Moon-bugs, NASA whisked them into isolation in a specially-built £12 million ($29 million) Lunar Receiving Laboratory at Houston. (No germs were found and .moonrock samples from later missions have been mailed to scientists around the world without such treatment.)

On November 14, 1969, Apollo 12's all-Navy crew headed for the most important beach-head of their careers. The Saturn 5 survived a major crisis when it was struck by lightning 50 seconds after a storm-tossed launch, but on November 19,

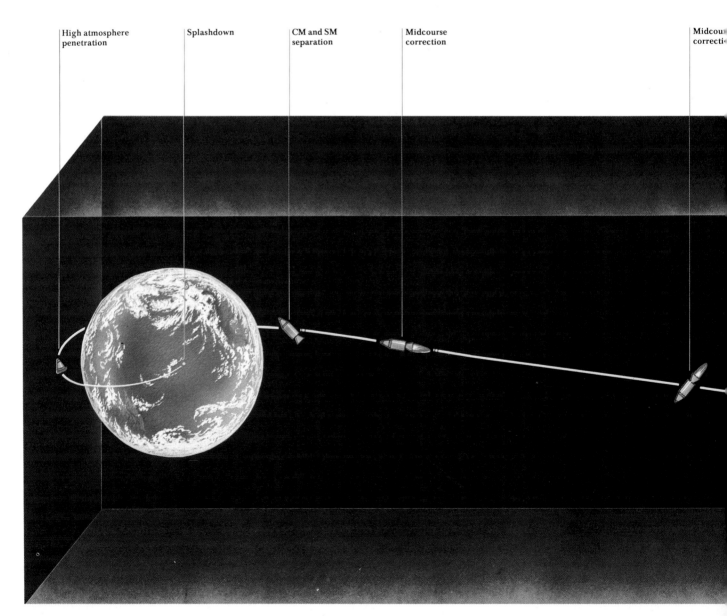

High atmosphere penetration

Splashdown

CM and SM separation

Midcourse correction

Midcou correcti

astronauts Pete Conrad and Alan Bean began their descent in the Lunar Module Intrepid. Their target, ironically, was the Ocean of Storms.

The landing was a bulls-eye, only 183 m. (600 ft.) from where Surveyor 3 came to rest 31 months before. Five hours after touchdown, Conrad (the second shortest astronaut) stepped onto the surface: 'Man, that may have been a small one for Neil, but that's a long one for me!'

Conrad and Bean spent an hour setting up five elaborate instruments to test the Moon's incredibly tenuous atmosphere (there is one, though it is many millions of times more rarefied than Earth's), measure magnetic fields and study solar particles. The £10 million ($25 million) nuclear-powered science station was NASA's answer to the growing scientific revolt against Apollo's engineering 'bias', but the scientists were angered again when the colour TV camera broke down after 20 minutes and the crew threw out 6 kg. (13 lbs.) of priceless rocks to make room for it on the return trip. Without television, disappointed millions could only listen as the explorers spent their second moonwalk treking to Surveyor and removing parts for analysis back on Earth.

Intrepid left the Moon on November 20 with 34 kg. (75 lbs.) of rock samples. Reunited now with their colleague Richard Gordon in the Command Module 'Yankee Clipper', the crew spent a day photographing future landing sites before returning to Earth and quarantine.

Apollo 13 lifted off, the superstitious may like to note, at 13 minutes after the 13th hour (local time) on April 11, 1970. Jim Lovell and Fred Haise were joined only hours before launch by Jack Swigert, because third-man Thomas Matt-

Of the original rockets, only the tiny Command Modules were ever recovered. The other parts remaining after the lunar ascent were all discarded en route for Earth.

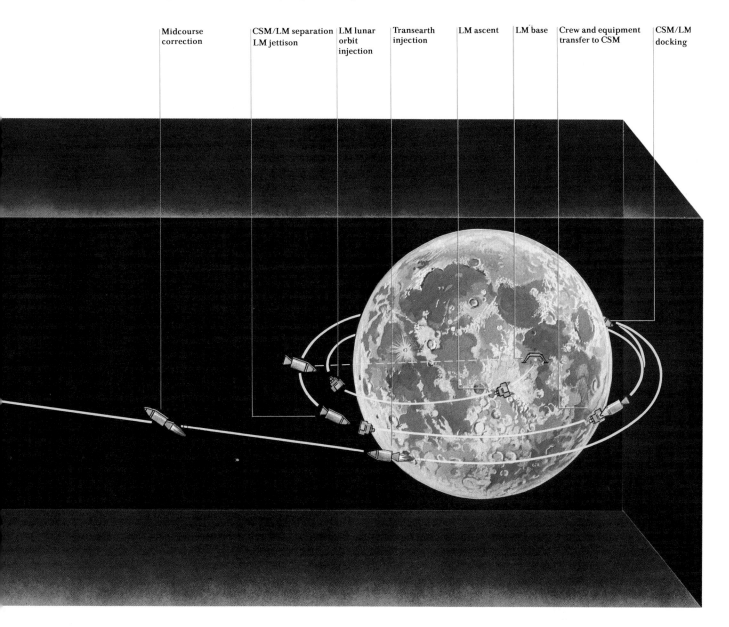

| Midcourse correction | CSM/LM separation LM jettison | LM lunar orbit injection | Transearth injection | LM ascent | LM base | Crew and equipment transfer to CSM | CSM/LM docking |

ingly was exposed to German measles. There were anxious moments when the second stage shut down two minutes early, but they made it into orbit and on course for the Moon when disaster struck. Lovell recalled later: 'We were flying with a bomb aboard. It exploded on April 13, 321,800 km. (200,000 miles) from home.'

The 'bomb' was one of two oxygen tanks in the Service Module, a life-source turned lethal because someone fitted it with the wrong thermostat, causing it to overheat. 'Hey,' said Lovell when he heard the banshee scream of the master alarm, 'we have a problem here' – a classic piece of understatement if ever there was one. Oxygen tank Number Two had blown up, filleting one side of the SM and destroying all the module's systems. Apollo 13 was fast running out of oxygen, water and power.

Months before, some pessimist had dreamed up a plan called 'LM Lifeboat'; suddenly three lives depended on that plan working. As the Command Module, 'Odyssey', slowly died, the crew scampered into 'Aquarius', the Lunar Module. Within hours it was powered up and providing oxygen, electricity, water and air conditioning for the long limp home. But everything was still scarce and Apollo 13 was still going the wrong way.

With the SPS booster wrecked, the crew had no choice but to use the LM's descent engine to change their course. They were so far from Earth that the quickest route was round the Moon and back, so they fired the rocket for 30 seconds and waited – waited nearly four days in a crippled spacecraft.

80,500 km. (50,000 miles) from Earth, the Service Module was finally discarded and the crew saw for the first time the full extent of the havoc caused by the explosion. Three-and-a-half hours later, Lovell spoke a heart-felt tribute to

the cast-off LM: 'Farewell, Aquarius, we thank you.' Man's most perilous week in space ended with splashdown on April 17.

On February 5, 1971, Alan Shepard stepped out onto the lunar highlands of Fra Mauro. 'Not bad for an old man,' quipped Mission Control, reminding him that he was the only Mercury astronaut to reach the Moon and (at 47) the oldest Moon-man. Shepard was quickly joined by fellow Apollo 14 crewman Edgar Mitchell.

The planned highlight of Apollo 14 was a 4½-hour trek to the towering rim of Cone Crater, a stadium-sized impact crater east of the LM. Pulling their equipment in a rickshaw-like cart, the two headed out across the undulating terrain, and promptly got lost. For two hours they trudged on, alarming controllers by their breathlessness, until finally – with only 8 hours to lift-off – they abandoned the attempt and headed home.

Apollo 15 crewmen David Scott and Jim Irwin would not tire or get lost – they brought a battery-powered taxi with them to the foot-hills of the Apennine Mountains on July 30, 1971. The 209 kg. (461 lb.) Lunar Rover was a manned spacecraft on wheels, built and delivered in only 17 months to carry twice its own weight and provide a television and radio 'outside broadcast unit' for the astronauts.

Scott and Irwin spent 67 hours on the Moon after a breathtaking descent over the 4,000 m. (13,000 ft.) Apennines onto a small basin bordered by a 366 m. deep (1,200 ft.) chasm called Hadley Rille. First time out, the two men drove 4 km. (2½ miles) to the rim of Hadley Rille and transmitted superb TV pictures back to Earth direct from the Rover. For the first time, scientists could control a camera on the Moon and draw the astronauts' attention to features of interest. The third lunar science station was set up when they got back to the Lunar Module.

Next day, Scott and Irwin went on a 12.5 km. (8 mile) tour of the mountain slopes. There Scott found 'what we came for' – a crystal rock quickly dubbed the 'Genesis Rock' because it was believed to date to the creation of the Moon (it didn't, but was still 150 million years older than anything found before).

By the end of a third moon-drive out into the Hadley-Apennine plain, the two men had collected 77 kg. (170 lbs.) of samples, travelled 28 km. (17.4 miles) and spent 18½ hours outside Falcon. On August 2, they rejoined Alfred Worden in Endeavor – a CSM turned orbiting laboratory. A Scientific Instrument Module stowed in the Service Module carried panoramic and mapping cameras, four spectrometers, a laser altimeter and a 35.4 kg. (78 lb.) subsatellite for ejection into lunar orbit.

John Young and Charles Duke of Apollo 16 touched down on the Cayley Plains near the crater Descartes on April 21, 1972, and spent a total of 20 hours 40 minutes outside the LM. They collected 95 kg. (210 lbs.) of samples and drove 27 km. (17 miles) in their Lunar Rover. Apollo was drawing to a close.

Night turned to day at Cape Kennedy when Apollo 17 rose on a pillar of golden flame into the star-studded sky. The night spectacular was needed to ensure that astronauts Eugene Cernan and Jack Schmitt arrived at the Taurus-Littrow landing site in daylight. Lift-off was 2 hours 40 minutes late because of a heart-stopping 'hold' in the countdown only 30 seconds before the planned launch.

Cernan flew the LM Challenger between two majestic mountain peaks to touchdown in the Taurus Mountains near Littrow Crater, on the edge of the Sea of Serenity. Four hours later, on December 11, Cernan dedicated the first step of the last Apollo mission to the half million space workers who made it possible. He was quickly joined on the surface by geologist Schmitt – the first scientist on the Moon.

During their 22 hours of surface science, Schmitt purposely toed the soil as he walked – a simple ploy that uncovered one of the most dramatic finds of Apollo during the second moon-drive. 'It's orange,' cried Schmitt. 'Orange soil!' There was pandemonium in Houston as fellow geologists jostled to see the TV pictures. There, on the edge of Shorty Crater, was an unmistakable orange tint later attributed to tiny glass beads formed by intense heat eons ago.

Apollo's scientific booty – including 385 kg. (850 lbs.) of rock and soil – will take generations to analyse fully. Its historic legacy may take even longer to assess.

Apollo, it must be remembered, always had powerful critics. Even President Kennedy seemed to have second thoughts when he broached the short-lived idea of a joint US/Soviet Moon

During the Apollo 15 mission of August 1971, James Irwin stands by the Lunar Rover with Mt. Hadley in the background. The Rover had a top speed of 16 k.p.h. (10 m.p.h.) and a range of up to 20 km. (13 miles).

Apollo 16 astronaut Charles Duke collects lunar samples at the rim of Plumb Crater, rounded by eons of micro-meteorite erosion, and heads back to the Rover after placing an experiment in the dusty soil.

During the last of the Apollo missions, Apollo 17, in December 1972, Schmitt rakes up some samples of lunar soil (*above*) and examines a huge split boulder, probably thrown up by a meteorite impact thousands of millions of years ago.

landing. Former President Dwight Eisenhower said: 'To spend $25 billion to go to the Moon is nuts.' A 'Poor People's March' picketed the Apollo 11 launch, and eminent scientists complained that Apollo was diverting star-struck talent from more pressing human problems. Some compared the programme with Egypt's pyramids: impressive, but of no earthly use.

Apollo's original justification lay way back in the early 1960's, when the US needed to recover a sense of national pride and security after the jolt of Russia's space successes. Such justifications were, indeed, outdated by the end of the decade. And it is true that Apollo will probably be seen as a horrendously complex and expensive programme the like of which will never be seen again. But the true returns have yet to be made – the technology it demanded will open a new era in space travel and astronomy during the 1980's.

Space Laboratories

The most spectacular disaster in the history of rocketry happened in early June, 1969, at the Tyuratam Cosmodrome. It struck, according to one account, during the closing stages of a practice countdown for the most powerful rocket ever built – the Soviet's top secret G-1 super-

booster, about 40 per cent more powerful than Saturn 5.

Only the Russians (and Western spies) know what the rocket looked like and what happened that fateful day, but leaked intelligence reports speak of a cataclysmic explosion during fuelling of the booster. The giant rocket blew up with the force of a megaton H-Bomb, destroying any Soviet hopes of beating America to the Moon. A decade later, the G-1 had still not flown.

Russia's Moon-plans are hotly disputed. Many Western observers remain convinced that a landing was planned, and they cite as proof the round-the-Moon missions of four unmanned Zond spacecraft between September 1968 and October 1970. Others believe the Soviet priority was the construction of space stations in Earth orbit. The project to achieve that was Soyuz ('Union'), begun disastrously two years before the G-1 eruption.

Vladimir Komarov rode Soyuz 1 into orbit on April 23, 1967. The expected second Soyuz launch, and historic first docking of manned craft in space, never came. Instead, Komarov began re-entry on April 24, after only 18 orbits. Reports that this followed flight problems were described by Yuri Gagarin himself as 'cock-and-

The Soyuz spacecraft weighed about 6 tons in orbit and comprised three sections. The cockpit, called the re-entry module, was shaped like a car headlamp and contained three couches, a parachute, soft-landing retro-rockets and controls for the mission. The cosmonauts, who wore pressure-suits only on spacewalks, had 'walking sticks' for prodding buttons beyond their reach during launch. Two wing-like solar panels powered the spaceship and recharged batteries. Attached to the nose of the re-entry module was a 2.7 m. (7.3 ft.) diameter sphere called the orbital module, used by the crew for work and rest during the flight. This brought Soyuz's total habitable volume to about nine cubic metres (318 cubic ft.) – 50 per cent more than in Apollo.

Soyuz 6, 7 and 8 were launched on successive days in October, 1969, to form the first orbiting space fleet. The mission did not include docking, but the Soyuz 6 crew performed the first welding of metals in the airless vacuum of space, a vital test of space station fabrication. The world's first space station, however, was sent up ready-made.

Salyut 1 was launched on April 19, 1971, by a Proton booster – Russia's most powerful. Soyuz 10 docked with the station five days later, but problems prevented the crew entering the Salyut. It was not until June 6 that Soyuz 11 carried aloft the first men to make their home in space. Georgi Dobrovolsky (commander), Victor Patsayev (test engineer) and Vladislav Volkov (flight engineer) spent a record 23 days aboard Salyut. Their waking hours were spent stargazing with three telescopes, photographing

The space suit or 'pressure garment assembly' left the astronauts, in Norman Mailer's words, 'about as much co-ordination as a two-year-old in three sets of diapers.' Designed to protect the astronauts in the vacuum of space, it consisted of 15 layers of plastic material. For a Moonwalk, the suit was turned into an Extra-Vehicular Mobility Unit consisting of three outer layers, a liquid cooling garment, a pressure garment, a thermal and micro-meteoroid garment, a portable life-support system, a radio, a waste-disposal system, a maintenance kit, a set of special visors for the helmets and a bio-medical belt. The extraordinary garment – a cross between a balloon and a suit of armour – had four gas connectors (two for oxygen-in, two for oxygen-out) and a urine transfer collector.

bull' stories, but Western experts think otherwise.

One guess is that Komarov's flight computer failed, forcing him to re-enter unaided straight through the atmosphere. Soyuz, like Gemini and Apollo, was built to skip computer-guided through the atmosphere, 'duck-and-drake'-style; to remain stable during a straight (ballistic) re-entry, Komarov would have had to put Soyuz into a dizzy spin (up to 30 r.p.m.). Failure to stop that spin before the parachute opened may explain the terrible tragedy that followed. Komarov's main parachute tangled at 7 km. (23,000 ft.) and he plunged to his death near Omsk in the Ural Mountains. It was the first fatal mission in six years of spaceflight.

Nearly two years later, the Soviets achieved their first big Soyuz success – what they called 'the world's first experimental space station', formed by docking Soyuz 4 and 5 on January 16, 1969. The technique had been tested 18 months earlier with the first automatic docking in space by Cosmos 186 and 188, both apparently unmanned Soyuz craft. Immediately after the 1969 link-up, Soyuz 5 crewmen Yevgeny Khrunov and Alexei Yeliseyev spacewalked to join Vladimir Shatalov in Soyuz 4, while Boris Volynov remained in the other craft.

Earth and conducting medical tests. Patsayev even tended the first 'space garden'.

Life was pleasant in the 21 m. (69 ft.) Soyuz/Salyut combination. The crew had eleven times Soyuz's work volume to move about in and 20 portholes through which to view the Earth and space. The Salyut had a forward docking tunnel, three habitable sections and an unpressurized service module with a booster to adjust the station's orbit. Power came from four solar panels, two at either end. Soyuz/Salyut weighed about 25 tons.

On June 30, Dobrovolsky announced: 'Our condition is excellent. We are ready for landing.' The Soyuz 11 retro-rockets fired for 2½ minutes at 1.35 a.m. (Moscow time) and the crew discarded the orbital and service modules in preparation for re-entry. At that moment, communications ceased abruptly, but controllers were not worried because the re-entry sequence went perfectly and Soyuz touched down on time in Kazakhstan. The jubilant recovery team scrambled to the hatch and flung it open. What they saw stunned the world.

'The recovery group found the crew of Soyuz 11 in their seats without any sign of life,' said the newsflash on Moscow Television at 6.30 a.m. The three heroes were dead, killed painlessly by a rapid pressure drop caused 30 minutes before landing when a valve was jarred open by the explosive bolts separating the orbital module.

The Soyuz disaster was followed by a series of Salyut fiascos, beginning in July, 1972, when a station launch ended in the Pacific Ocean. On April 3, 1973, Salyut 2 was placed in orbit, but it disintegrated a fortnight later. Several weeks after that, a third Salyut is thought to have been destroyed when its carrier rocket blew up. Yet another Salyut (called Cosmos 557) was launched on May 11, but this too was a failure. The reason for all this haste was clear: The Americans were catching up. Three days after Cosmos 557, they launched their titanic Skylab.

'Apollo gave us our Columbuses,' said one American official. 'Now we need our Pilgrims.

Skylab is the precursor of these.' The huge laboratory, heavy as the *Mayflower*, left Cape Canaveral at 1.30 p.m. on May 15, 1973. Sixty-three seconds later, disaster struck as atmospheric drag clawed at the fragile girdle shielding Skylab from the Sun's heat and deadly rays. The shield disintegrated, but even worse was to come.

Skylab in orbit let out plaintive cries to the ground. It was unprotected, and one of its wing-like solar panels had ripped off during the 1,500 m.p.h. rise to orbit. The other was jammed shut. America's £1,100 million ($2,600 million) programme was literally powerless and roasting alive. The first occupation was postponed while experts tried to find solutions, helped – they admitted later – by pictures of the orbiting Skylab from a top secret spy telescope.

The answer to the overheating problem was found in a Houston sports shop. With $63 worth of interlocking fishing rods, a space expert showed how a parasol might be extended over the scorched Skylab. Someone else came up with a solution to the jammed solar panel: a 25 ft. pair of cutting shears like the ones used by roadside tree surgeons. The unlikely space tools arrived by jet hours before launch. 'We fix anything,' cried commander Pete Conrad as he lifted off with Dr. Joseph Kerwin and Paul Weitz on May 25.

Television viewers watched as the 76-ton Skylab grew from pin-size dot to huge windmill on their screens. The 'windmill' vanes were four solar panels powering the eight telescopes clustered inside the Apollo Telescope Mount (ATM). The ATM girders were attached at right angles to the Multiple Docking Adapter (MDA), the two-part docking unit which also housed the ATM controls. Next came the Airlock Module (AM), 'foyer' for the spacewalks, and finally the cavernous Orbital Workshop – an S-IVB turned space station with the capacity of an average house. The liquid oxygen tank at the rear was now a huge dustbin.

The £23 million ($55 million) Apollo repair-van edged up to its giant companion and Paul Weitz tried to free the stuck solar panel. Leaning out of Apollo's hatch, a latter-day Don Quixote tilted at the panel with the 'tree cutters', but it refused to budge. Then came a major emergency. Conrad spent four hours trying to dock with Skylab's front port. But it took a space-suited repair job to fix the docking mechanism and save the mission. Inside the orbiting greenhouse, the crew quickly pushed the parasol through a solar telescope airlock on the side of the workshop. As the station cooled, Conrad and Kerwin spent three days rehearsing the most hazardous space-walk in history. Braving unexploded bolts on the solar panel, Conrad tied a rope to the tip and heaved. The panel swung free. 'We see amps,' cried Houston as the starved Skylab immediately began taking power from the panel. Incredibly, the three men still found time to perform more than 80 per cent of their planned science mission during the record 28-day flight.

A Soviet Soyuz spacecraft caps a Vostok launcher, with its strap-on boosters that can just be seen at its base.

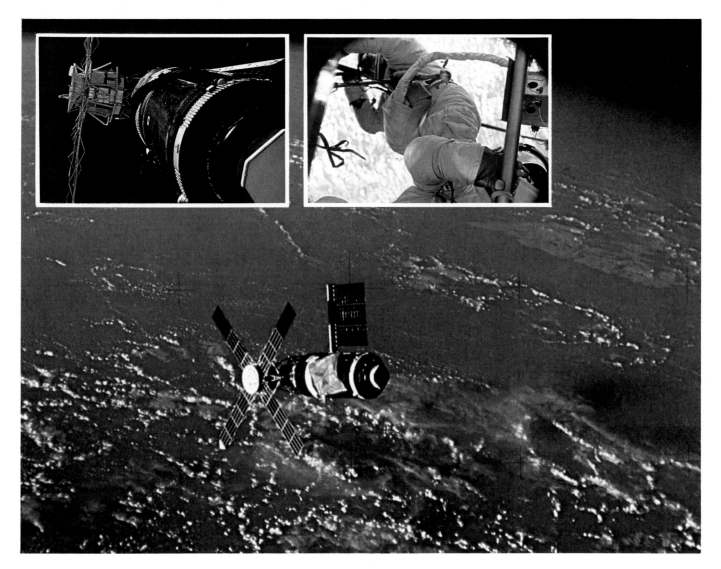

Conrad's Apollo 12 moon-mate, Alan Bean, led the second Skylab team, Dr. Owen Garriott and Jack Lousma, on what officials hoped would be 'a dull and boring flight'. It was not. Five days into the mission, on August 2, NASA sounded the first space 'scramble' and rescue teams raced to prepare the first CSM lifeboat ever launched. But the mercy flight proved unnecessary as major problems with the orbiting CSM ferry-craft's thrusters became better understood. The crew erected an improved parasol and settled down to a record 59 days of science, working up to 12 hours a day. They conducted 39 Earth surveys to assess the potential of future manned and unmanned resources satellites, took 71,700 pictures of the Sun as part of the most intense solar observation project in history, and subjected their bodies to gruelling tests of the effects of prolonged weightlessness. The medical results gave NASA the confidence to send up the third crew for nearly three months.

The Bean crew's prodigious work-mania made trouble for space rookies Gerald Carr, Dr. Edward Gibson and William Pogue. Tension between Skylab and ground control increased with the workload until finally the crew literally went on strike. However, a heart-to-heart with Houston quickly cleared the air and the crew returned to a less packed schedule. Carr and Pogue ventured forth on Christmas Day, 1973, to follow their own Christmas 'star' – the Comet Kohoutek.

The third crew splashed down on February 8, 1974, leaving behind a ghost ship that gave man almost a billion miles of taped scientific information, 175,000 pictures of the Sun, 46,000 photographs of Earth and the confidence for longer voyages. But Skylab is not a record. Salyut was still in business.

Beginning in September 1973, the Soviets started their slow recovery from the Soyuz 11 tragedy. Over the next four years they launched three Salyut stations and fourteen manned missions. There were some successes, but the programme was dogged by persistent failure. Then, on September 29, 1977, the tide started to turn dramatically with the launch of Salyut 6. Ten days later, Soyuz 25 left Tyuratam from the pad used 20 years before by Sputnik 1. It was to celebrate that birthday, and the 60th anniversary of the October Revolution, by docking with the orbiting Salyut. It did not succeed, and the crew had to return after only two days.

Two months later, however, Soyuz 26 opened a happier era of Soviet spaceflight. At two minutes past midnight (Moscow time) on December 11, Yuri Romanenko and Georgi Grechko docked successfully with Salyut 6. Their ferry-craft, like all others since Soyuz 11, had room for only two men because of the addition of spacesuits and extra life support equipment. Its solar panels had been replaced by two-day batteries in the service

After a Saturn 5 rocket took off with the 75-ton Skylab space station in May 1973, Skylab's micro-meteoroid shield was ripped away, destroying one solar panel and jamming another. The shield doubled as a heat deflector, and without it the space station began to over-heat. The mission was saved by the crew, who erected a sun-shield and cut free the jammed solar panel.

Top left: The remaining solar panel can be seen jammed shut, in the top right of the picture.

Top right: Charles Conrad (*top*) and Joseph Curwin (*bottom*) work outside Skylab to free the panel and rig up a make-shift sun-shield.

Bottom: Skylab, repaired, hangs in orbit, its one remaining solar panel successfully deployed and with the replacement sun-shield wrapped around the body of the craft. The cross-shaped solar panels generated power for the Apollo Telescope Mount.

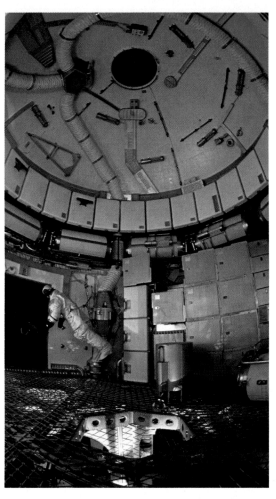

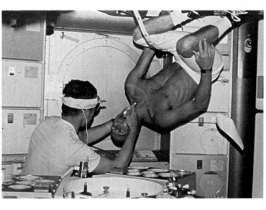

Inside, Skylab was remarkably roomy, as befitted a space station. It contained four sections. Largest was the workshop, consisting of the living quarters (*top left*) and a work section (*right*). In these relatively luxurious conditions, three Skylab crews – following the Skylab 1 unmanned mission that placed the station in orbit – worked during 1973.

Above: William Pogue and Gerald Carr of Skylab 4 demonstrate the effects of weightlessness.

Centre: In Skylab 2, Kerwin gives Conrad an oral examination.

module. Salyut had evolved into a sophisticated orbiting platform with docking ports fore and aft and a new stellar navigation system called 'Delta'. The station was powered by three rotating solar panels which could be pointed at the Sun without moving the entire complex.

On December 19, Grechko made the first Soviet spacewalk for nearly nine years, while Romanenko is said to have defied Mission Control and joined in the fun outside. On New Year's Eve the Salyut's swift passage over the Earth's time zones gave the crew 15 opportunities to celebrate the new year (they brought a real fir tree into orbit for the purpose), but they chose 11.40 a.m., December 31 (Moscow time), to mark their first passage into 1978 – one of the most productive and exciting years in the history of manned spaceflight.

The first sensation of the year came on January 11 when two cosmic 'postmen' knocked on Salyut's door. Vladimir Dzhanibekov and Oleg Makarov docked Soyuz 27 with the Salyut/ Soyuz 26 to become history's first 'visiting crew' aboard a space station. After unloading letters, gifts and new scientific equipment, the four men began five days of joint experiments. The most important was code-named 'Resonance' and it simply required the crew to cavort about the station while instruments measured the stability and strength of the first three-craft space complex.

The next step opened a new era of spaceflight on January 20. That morning, an unmanned Soyuz variation called Progress 1 was launched from Tyuratam for an historic link-up with

Salyut two days later. It carried a ton of rocket fuel and oxygen for the station and 1.3 tons of scientific equipment, film, linen, food and water. As the first space dockers manhandled the cargo into Salyut and prepared for the hazardous fuel transfer, the Soviets reminded the world that, within the space of a month they had shown that a space station could be kept continuously manned and supplied with vital stores.

On February 6, Romanenko and Grechko packed Progress with two months' garbage and cast it adrift to burn up in the atmosphere, leaving Salyut's forward docking port free to welcome its next guests.

History's first international manned spaceshot (US and Soviet astronauts met in space, but were not launched together) left Tyuratam on March 2. Soyuz 28 carried Alexei Gubarev and Czech cosmonaut Vladimir Remek, the first man in space who was not an American or Russian. His seat was the first of eight reserved under an 'Intercosmos' agreement signed by Socialist states in 1976.

Millions of viewers witnessed the unbridled jubilation as the Czech cosmonaut entered Salyut; the revelry in space continued late into the night, to the annoyance of Mission Control. By contrast, 11.36 p.m. (Moscow time) on March 3 passed almost without comment – that was the moment Romanenko and Grechko broke Skylab's space endurance record of 84 days 1 hour and 16 minutes.

On March 9, the Soyuz 26 crew began stepping up physical exercises in preparation for their return to Earth. They wore 'Chibis' pressure suits for 10 to 12 hours a day, forcing blood down into their legs to put the normal strains of gravity on their hearts. On March 16, six days after the return of Soyuz 28, Romanenko and Grechko touched down after a record 126-day flight during which they completed almost twice the planned number of experiments. Three months later, on June 17, the next boarding party arrived at Salyut 6 aboard Soyuz 29. Vladimir Kovalyonok and Alexander Ivanchenkov began a mission that was to beat Soyuz 26's record by 13 days.

The new tenants had barely settled in when the second Intercosmos flight left Tyuratam with Pyotr Klimuk and Polish cosmonaut Miroslaw Hermaszewski. Soyuz 30 docked with Salyut on June 28 for a week-long joint flight which included work on an experiment called 'Relaks', to study ways of 'putting your feet up' in orbit.

On July 9, the second Progress space tanker brought enough food and water to last the two men 50 days. It remained docked for three weeks while the crew manufactured metal alloys and crystals in two furnaces and conducted the first experiments in the promising field of lens and mirror production in weightlessness. A week after Progress 2 burned up in the atmosphere, Progress 3 docked with Salyut and the crew transferred a two-month stock-pile of food, water and equipment. Progress also carried the first guitar into space for Ivanchenkov.

On August 27, the third Intercosmos crew, Valery Bykovsky and East German Air Force pilot Sigmund Jähn, docked Soyuz 31 with Salyut/Soyuz 29. The joint crew worked together for eight days before Soyuz 31 headed home, leaving the long-stay crew alone to celebrate a new endurance record on September 20. Progress 4 brought more fuel and supplies to the men on October 6 to allow them complete their record $4\frac{1}{2}$ months in space. They returned to Earth on November 2.

1978 marked the emergence of the Soviet Union from years of uncertainty in its man-in-space programme. Almost unnoticed in the West, Soviet, Czech, East German and Polish cosmonauts spent about 13,500 man-hours in space.

Commuting to Space

Spaceflight in the first two decades made about as much economic sense as sinking the Q.E. 2 after its maiden voyage and building a new one for each voyage thereafter. Each time men left for the Moon, they littered the Atlantic floor with £77 million ($185 million) worth of Saturn wreckage.

Both the United States and the Soviet Union are perfecting revolutionary new spaceships to end this wastefulness and open up the cosmos. The key is reusability, a spacecraft that can be used time and again – a Space Shuttle. It will be to space what the railroad was to the Wild West.

Barring further delays, the first American Space Shuttle will be launched early in 1980. There is evidence, however, that the Soviet Union is at least as advanced and might well beat the Americans into a new era of spaceflight. The Soviet entrant in this new Space Race is

The Apollo-Soyuz Test Project of July 1975 tested a way of linking American and Russian spacecraft. Besides its political significance, the project could be vital in a space rescue.

Left: The Soviet Soyuz craft, seen here from Apollo, consists of three major components – the spherical Orbital Module, with its specially adapted docking system, the bell-shaped Descent Vehicle and the cylindrical Instrument Assembly Module.

Below: The two craft stand linked in the Smithsonian Space Museum, Washington.

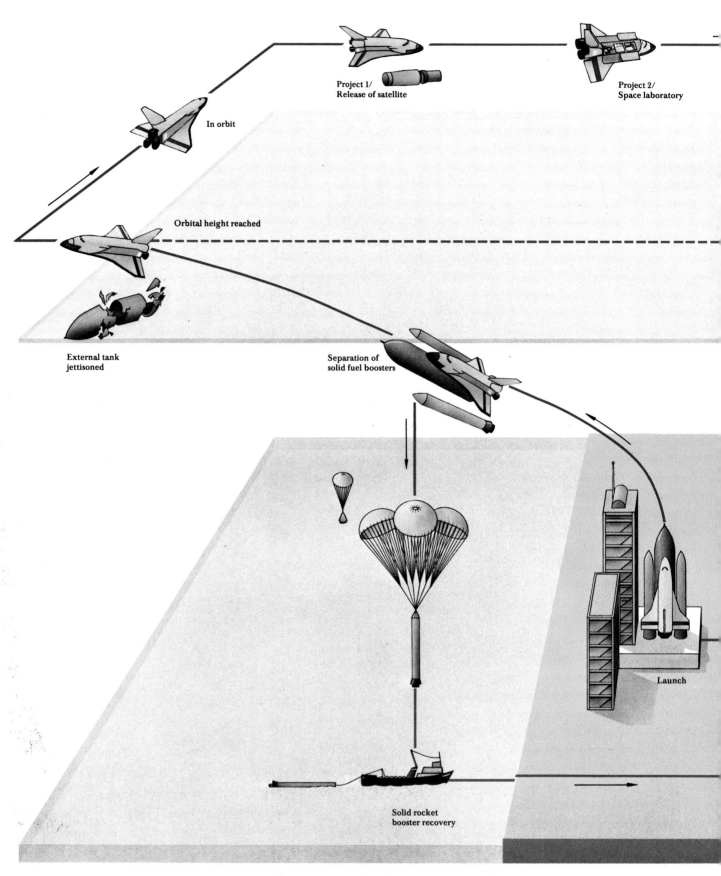

Project 1/
Release of satellite

Project 2/
Space laboratory

In orbit

Orbital height reached

External tank
jettisoned

Separation of
solid fuel boosters

Solid rocket
booster recovery

Launch

The Space Shuttle is designed as a versatile, commercially viable workhorse. In this diagram, just three of many possible projects are shown. All components bar the large external fuel tank are reusable. Its economic impact in the 1980's – on technology, science and commerce – will be immense.

top secret, but it must resemble the United States' ambitious project.

The Space Shuttle took ten years to develop and cost £3,000 million ($6,900 million). The result is an aircraft-like Orbiter, about the size of a DC-10, which is launched vertically like a rocket, flies in orbit like a spacecraft and lands like an aeroplane. It opens a new era of cut-price space travel: $100 per pound orbited, compared with $100,000 per pound for Explorer 1.

Each £104 million ($250 million) Orbiter is 37 m. (122 ft.) long and measures 24 m. (78 ft.) from wing-tip to wing-tip. It weighs about 67 tons when empty. The two-storey crew compartment – flight deck and lower deck – has seating for seven people (commander, pilot, mission specialist and up to four passengers), but it can hold 10 on a rescue mission. Stretching for 18 m. (60 ft.) behind the crew section is the cavernous cargo bay, 4.6 m. (15 ft.) in diameter and capable

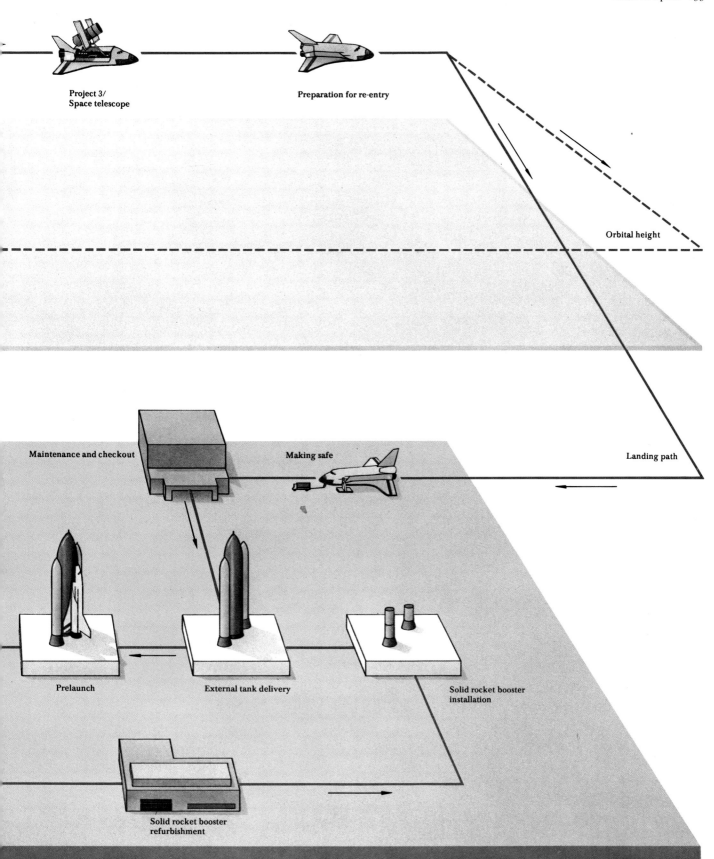

**Project 3/
Space telescope**

Preparation for re-entry

Orbital height

Maintenance and checkout

Making safe

Landing path

Prelaunch

External tank delivery

**Solid rocket booster
installation**

**Solid rocket booster
refurbishment**

of carrying payloads weighing up to 29 tons. The
three main rocket engines used during launch are
contained in the aft fuselage, together with two
smaller motors for making orbit corrections and
the de-orbit retro-fire. Thirty-six small thrusters
are used to adjust the giant craft's attitude in
orbit.

Launching a 112 ton fully-laden Shuttle requires an immense amount of power. Most of it
comes from the LH/LOX propellants fed to the
main Shuttle boosters from a 47 m. (154 ft.) high
disposable fuel tank called the External Tank
(ET). The Orbiter is attached to this 8.7 m (28.6
ft.) diameter tank during launch, and its three
boosters are supplemented by two solid propellant rockets, called Solid Rocket Boosters
(SRB's), attached to either side of the ET. The
45.5 m. (150 ft.) SRB's drop off 43 km. (27 miles)
above the Earth and parachute into the Atlantic
for recovery, refurbishment and re-use. The ET –

Top: The Shuttle's 18 m. (60 ft.) cargo bay will be able to place loads of up to 30 tons in orbit.

Above: The Shuttle is hung vertically before the fitting of its fuel tanks.

the only non-recoverable part of the Shuttle system – falls into the Indian Ocean.

After six orbital test flights in 1980, NASA plans a total of 560 Shuttle flights in the first twelve years of operation. Peak traffic is expected in 1988, when 65 missions are planned – 46 launched from Cape Canaveral and 19 from Vandenberg Air Force Base (VAFB) near Los Angeles. NASA expects to fly half of the 560 missions itself, while about 110 will be military missions funded by the Department of Defence, 30 will be flown by other U.S. Government agencies, 70 by private companies and 30 by foreign customers. One of the early VAFB launches will be the first manned mission into polar orbit, a long-awaited venture of considerable scientific interest.

On these flights, the Shuttle will be able to do things no spacecraft has done before. It can place several satellites in orbit and wait around while ground controllers make sure their multi-million dollar investment is working correctly. If it is not, the Shuttle crew can spacewalk to try on-the-spot repairs or return the damaged craft to Earth – a system that will knock millions off the cost of satellites currently equipped with back-up systems made obsolete by the Shuttle 'maintenance truck'. Satellites which have 'died' in orbit, either prematurely or of old age, can now be retrieved and given a new lease of life.

The Shuttle, too, will be rejuvenated when it returns to Earth. Everything has been built for easy replacement or repair if it gives trouble, and a production-line cycle is used to install new experiments. A Shuttle can be launched again after only two weeks on the ground, and rescue missions can be sent up within a matter of hours. Each Shuttle is designed to last for about a hundred flights, by which time it will have repaid its creators many times over.

The complicated manoeuvering needed to cast new satellites adrift or pluck old ones from orbit is controlled from the payload station at the back of the flight deck, where two small windows overlook the cargo bay. The commander stands at the right-hand window to manoeuvre the Shuttle while the pilot, to his right, opens the bay doors and operates a 15 m. (50 ft.) Canadian-built manipulator arm which grasps satellites

and extends them beyond the Shuttle or pulls them into the bay.

About 43 per cent of satellites launched from the Shuttle will need to be placed in higher orbits (many of them geostationary orbits) or inter-planetary trajectories. A re-usable Space Tug is under development to do this, but in the meantime the Department of Defense has developed an Interim Upper Stage (IUS) solid propellant booster to do the job.

The most ambitious project planned for the first decade of Shuttle flights is the European Spacelab, a £215 million ($516 million) enterprise undertaken by 10 member-states of the European Space Agency (ESA).

Spacelab is an orbiting laboratory nestled in the Shuttle's hold and drawing power and life support from its onboard systems.

The first Spacelab flight is due in 1980 to test the Shuttle/Spacelab combination and conduct about 40 experiments in space manufacturing, pure vaccine production, Earth resources and astronomy. Between 1980 and 1984, Spacelab missions (with European scientists aboard most) are planned at a rate of almost one a month, with some lasting up to 30 days. The scientists – men and women – will live in five-star accommodation compared with the early pioneers. The Shuttle has about as much living space as a double-decker bus, and is fitted with a compact kitchen, bunks, lockers and a toilet (first turn right at the side hatch). The maximum acceleration forces during launch will be only 3 g's (1 g. is Earth gravity), and less than half that during the deceleration of re-entry.

Private companies who want to use the Shuttle to launch their communications satellites, foreign governments, industrialists testing space manufacturing, and some research establishments must all pay for the service. It costs $21 million to hire a complete Shuttle, with a deposit of up to five per cent 'earnest money' payable in advance. That includes basic service charges (design and safety reviews, three-man crew, flight planning and data transmission), but special crew training, interim upper stages, revisits and retrievals, and specialized in-flight servicing all cost extra.

'Getaway specials' for as little as $3,000 are available on a standby-flight basis for experiments smaller than 5 cubic feet (.15 m³) and weighing less than 90 kg. (200 lbs.). In addition, there are 'special offer' prices for experiments of exceptional merit and 'ABC' fares for payloads booked well in advance. Only payloads of real scientific or industrial value will be accepted for the Shuttle.

The Shuttle flights in 1980 and 1981 are almost booked out, and NASA is already looking further to the future and has begun designing liquid fuelled boosters to replace the SRB's and give the Shuttle power to lift nearly 50 tons into orbit. Initial studies have even begun on a project to build a 74-seat 'passenger module' to ferry the 'cosmic labourers' who will build America's first giant space station in the 21st century.

PIGGY-BACK TO A NEW ERA

Man's immediate future in space lies with the Space Shuttle, due to become fully operational in 1980. By the late 1980's, at least one Shuttle flight a week will be launched into orbit around the Earth.

The heart of the Shuttle is the Orbiter which will deliver payloads into orbit, manoeuvre and glide back to Earth to be prepared for another flight. By the late 1970's, the Orbiter's gliding capabilities had been well tested in a peculiar piggy-back operation with a Boeing 747–100. The Boeing was specially modified for the test, and for its longer-term role as a transporter of Orbiters to and from launch and landing sites. The Shuttle – named *Enterprise* after the star-ship in the T.V. series 'Star Trek' – first flew in August 1977, as part of a series of flights from NASA's Dryden Flight Research Centre at Edward's Airforce Base, California. In front of 70,000 spectators, the *Enterprise* separated, banked in a series of turns over California's Mojave Desert at some 7,350 m (24,000 ft) and, $5\frac{1}{2}$ minutes later, landed safely in a cloud of sand and dust at over 320 k.p.h. (200 m.p.h.). The principle of the project had been established.

The Boeing Jumbo jet roars aloft bearing the 'Enterprise' on its back. Here – as in several tests – the 'Enterprise' has a tail-cone which acted as an aerodynamic faring to reduce buffeting on the 747's tail.

Above: The Boeing throttles back to 500 k.p.h. (310 m.p.h.) as the 'Enterprise' crew prepare to fire the explosive bolts on the attachment points.

Right: Immediately after separation, the Shuttle rolls to the right and the Boeing to the left to give each other good clearance.

Centre: The 'Enterprise' banks sharply over the Mojave Desert to test performance under stress.

Far right: Accompanied by a chase-plane, the 'Enterprise' lands on a dry lake bed in a cloud of sand and dust.

12/FRONTIERS OF THE FUTURE

IAN RIDPATH

Man is about to enter a new era of space travel.
The US Space Shuttle will make journeys into orbit as
routine as commercial airline flights. Released from the
constraints of Earth's gravity, space stations will be able to
evolve into city-sized orbiting communities, with power
drawn from the Sun and industries supplied from the
Moon and asteroids. Astronomy will be revolutionized
with a new generation of huge, orbiting telescopes.
Detailed plans for an unmanned starship already exist and
within a generation such a probe could be on its way.
We may mark the second millennium AD by peopling
interplanetary space, with its inexhaustible supplies of
clean solar energy and mineral resources. Beyond this, our
dreams of journeying to the stars ourselves may be made
reality, as suggested by the reconstruction at left, which
shows a spaceship arriving at a Jupiter-like planet
circling our dull, red neighbour, Barnard's Star.

Space will play an increasingly important role in mankind's future. The launch of America's Space Shuttle in 1980 marks the opening of a new era in space exploration – one that will not only be extraordinarily dramatic, but will also show great practical benefits.

Since the huge expenditures on space exploration in the 1960's drew much criticism for their supposed impracticability, it is as well to look first at some of the possible benefits we can derive from living and working in space. It is a startling but not, as yet, widely appreciated fact that a new industrial revolution is being planned – one that will be established in orbit. Preliminary experiments, being conducted via space stations such as the Soviet Salyut and the Euro-American Spacelab, have already utilized the weightless conditions of space to produce new materials – such as alloys and special types of glass that are unobtainable on Earth.

Alloys can be made in space from metals that do not mix successfully under the pull of Earth's gravity, which tends to separate the lighter metal from the heavier. In the same way, crystals of new semiconductor materials for use in electronics can be developed in space. And the absence of gravity also helps biologists in extracting super-pure vaccines. The number of products and processes that will be of benefit to Earth-based industry is vast. Eventually, whole industries will be transferred off the Earth, housed not in cramped space stations but in spacious colonies capable of supporting a workforce of thousands.

Among the first products of space industrialization may be giant satellites beaming power to Earth, and processing centres in which metals are extracted from the rocks of the Moon and asteroids. Next century could see the construction of starships and, later still, the possible modification of the climates of Mars and Venus to make those planets habitable by mankind. Present-day space activities are the first steps in mankind's colonization of the Galaxy – which could culminate in our first meeting with extra-terrestrial life.

To the Moon and Planets
Let's take a more detailed look at some short-term space plans. Soviet space experts have also spoken of large space stations, holding a dozen men or more. Possibly such structures could be made from several Salyut space stations docked together, although it is more likely that some new launch system would be required. There were unconfirmed reports in the late 1960's and early 1970's that the Soviet Union was developing a super-booster larger even than the Saturn 5. Apparently the project was scrapped after several launch failures, but we may yet see a modified version of it brought into operation. If so, this booster, said to be powerful enough to launch multi-man space stations, might even have a role in establishing bases on the Moon.

While man consolidates his position in near-Earth space during the rest of this century, flights to remote parts of the Solar System will continue. These may use new propulsion systems such as the electric rocket, or ion drive, which is more efficient than existing chemical propellants. In an electric rocket, atoms of a heavy element such as mercury or caesium are *ionized* (have their electrons stripped from their atoms) by electric fields or high temperatures. The resulting ions are accelerated by electric fields to produce a high-speed exhaust which provides the craft's thrust. Although the thrust of such a rocket is low, it can operate for long periods in space. Electric rockets have already been used in orbit to keep geostationary satellites in their correct position.

To provide the rocket's electricity, solar panels are necessary, although in more advanced designs an atomic power plant could be employed. Designs have been drawn up for electric rockets, powered by sunlight, that would allow a space probe to rendezvous with a comet. Other projected uses of an electric rocket are for a mission to an asteroid, or even a probe to orbit the ringed planet Saturn. In the far distant future, some form of nuclear-electric rocket might be employed for journeys to the stars.

Another, more romantic, suggestion for propulsion within the Solar System is the solar sail, which uses the pressure of sunlight to push it around. Such a device is particularly attractive, for it needs no fuel, and there are no moving parts to wear out. At its simplest, a solar sail consists of a thin plastic film coated with a highly reflective layer of silver or aluminium. Solar-sail propulsion has been considered for a rendezvous with a comet, and one suggested design is for a so-called heliogyro consisting of 12 long sails like the blades of a windmill. Sunlight bouncing off their reflective surfaces would provide the propulsive force. Unfortunately, once a solar sail has receded too far from the Sun it ceases to work efficiently, but a modified version might yet be used for long-distance missions, powered by an intense beam of laser light projected from Earth.

Of all the corners of the Solar System to be explored, the planet Mars remains the most fascinating – and potentially the most welcoming for humans. There are no definite plans for future exploration following the landing of the two Viking craft in 1976, but a variety of ideas have been suggested. One proposal is for a Mars rover, an automatic laboratory like those carried by the Vikings, but on caterpillar tracks which would allow it to move around. Some scientists believe that life might be found only in oases on Mars, and that the Vikings, which were naturally targeted for the safest landing areas, missed these life-bearing spots. A rover could be steered over the most favourable niches for life, making a far wider survey of the planet's surface than was possible from the static Vikings.

Widely favoured is the even more exciting suggestion of bringing back a sample of Mars soil by a robot probe, similar to the Soviet craft which have automatically returned samples from the Moon. But while a Mars rover might be possible in the late 1980's, a Mars sample return mission is unlikely to be flown until 1990 at the earliest. In view of the possibility that Mars might harbour some organisms missed by the Viking probes, it would seem safest to examine any returned samples from Mars in the quarantine of a space station. Suitable space laboratories should exist in orbit around the Earth by then.

By the turn of the century, humans should have been back to the Moon, this time to stay more permanently than the brief visits of the Apollo programme. Lunar bases, long predicted, will eventually be established – and so may bases on the planets.

As with the first space stations around the Earth, so the first bases on the Moon may be made from modified rocket stages. Empty cylinders delivered to the surface of the Moon can be insulated by a metre or so of lunar soil, which will both help keep a constant temperature and also act as protection against radiation and meteorites. Alternatively, some kind of prefabricated shelter or even an inflatable plastic dome like those now familiar on Earth could provide living accommodation for the first lunar campers. Eventually, as in the Antarctic, the lunar colonists may hollow out underground living chambers.

Energy on the Moon is no problem, for sunlight provides a plentiful source. Oxygen for breathing can be extracted from the lunar rocks; and this oxygen will also be combined with hydrogen brought from Earth to produce water. Plants will grow happily in lunar greenhouses, and any further supplies the colonists need can be delivered simply by robot ferry craft.

What will we do on the Moon? Apart from simply studying the Moon itself, there will be plenty to keep the lunar colonists busy, not least by mining the surface to supply the space industrial centres with lunar ores. Industrial processes requiring a high vacuum – as in the production of rare metals like tantalum, molybdenum and niobium, or the blending of metals

Rendez-vous with Halley's Comet.

In 1986, one of the best-known comets, that named after the second British Astronomer Royal, Edmund Halley, is due to sweep in – as it does every 76 years – from the outer Solar System. A close analysis of the comet could reveal what comets are made of and more about the origins of the Solar System.

But the comet will be badly placed for observation. NASA scientists have suggested that the occasion would be ideal for testing at least three deep-space devices. Two will use the pressure of sunlight to 'sail' in formation with the comet. Of these, one, called the heliogyro, would look like a giant asterisk, with sails that could extend out to 7.4 km. (4.4 miles). A third probe would have scoop-shaped panels to catch sunlight to power an ion-drive rocket.

This solar-sail spacecraft has been nick-named the Yankee Clipper.

The ion-drive probe accompanies Halley's comet, its tail created by the solar wind sweeping away from the Sun. The heliogyro.

with paper or plastic – may be better suited to the Moon than space stations nearer the Earth, where stray atoms from the outer atmosphere are still to be found.

Astronomers will be particularly keen to settle on the Moon, where they can observe the sky without the intervening blanket of an atmosphere to obscure their view. The Moon provides a more stable platform than satellites or space stations. Radio astronomers, in particular, will want to set up radio telescopes on the far side of the Moon, where they are insulated from the radio noise of Earth by over 3,000 km. (1,875 miles) of solid rock.

The Moon's low gravity will also allow the construction of much larger telescopes, both optical and radio, than is possible on Earth, where gravity distorts mirrors that must remain in perfect shape whatever their position. Bowl-shaped lunar craters will be particularly useful for setting up radio telescopes like that at Arecibo which hangs in a natural hollow in the mountains of Puerto Rico.

Once man learns to live on the Moon, and the medical effects of long-duration flight in space stations have been assessed, a manned trip to Mars beckons as the next major goal. A round-trip to Mars is likely to take at least 15 months, and could be much longer depending on the engines used; here, electric rockets – with low thrust, but long-lasting and efficient – are almost certain to be employed.

For safety, and to ease the boredom of a long-duration flight, a manned Mars mission could consist of two craft each containing from three to six crew members. To reflect the international co-operation necessary for such a major undertaking, one craft might be sponsored by Soviet bloc countries, and the other by the US and Western Europe. In the event of a disaster befalling one craft, the other could act as a lifeboat.

Once close to Mars, the two craft would go into orbit, and landers would be dispatched to the surface. The first small steps on Mars will be a more complex, and more expensive, repeat of the

first landing of men on the Moon. Even the most optimistic space visionaries do not expect manned Mars landings before the year 2000, and they may well not occur until much later than that.

Bases on Mars, if they are judged desirable, will be set up using the experience gained from colonization of the Moon. Still farther in the future, men may venture to the satellites of the outer planets, gaining spectacular close-up views of the swirling clouds of Jupiter and the beautiful rings of Saturn. But, by then, we will probably have developed a range of robot explorers with such computerized intelligence that humans themselves will not need to stray too deeply into the Solar System.

Power from Space
There are a number of space engineering projects which sound quite fantastic at first, but which could nevertheless be undertaken in the near future. Among these is the solar power satellite, which is seen in some quarters as an answer to the world's energy crisis.

One of our most pressing needs is for energy. Only by the continued expenditure of vast amounts of energy can the developed nations maintain their standards of living, and the under-developed nations raise their standards to Western levels. Fossil fuels – coal, oil and gas – cannot provide the solution, because known reserves will be exhausted during the next century. Even if there were sufficient stocks of fossil fuels, burning them would produce so much pollution that the ecological balance of the Earth might be ruined for ever.

Nuclear power plants, using the energy of atomic fission, are slowly coming into operation, but they are widely disliked because of the dangers from radioactive nuclear wastes and the spectre of a major power plant disaster that would contaminate large areas. One projection is that the United States alone will need a million mega-watts of new electrical generating capacity by the turn of the century, and it seems unlikely that enough nuclear power stations can be built in time to satisfy this demand. Fusion power, which uses clean nuclear reactions like those inside the Sun, is being developed, but is unlikely to be available until well into the next century. A radical solution to the energy crisis is vital. And this is where solar power satellites enter the scene.

In essence, the solar power satellite collects solar energy in space and beams it down to Earth in the form of microwaves. Sunlight should in theory be our ideal energy source, for it is clean, endless (as far as mankind is concerned) and free (apart from the cost of harnessing it). But since sunlight is so spread out when it reaches the surface of the Earth, vast collecting areas are necessary to get sufficient power from it. What's more, the power is interrupted during cloudy weather and at night.

But above the blanket of the Earth's atmosphere there are no clouds and there is no night; sunlight in space is at least four times as plentiful

The Mars Rover, planned for landing in 1984, will be able to wander across the Martian desert landscape for about 100 km. (61 miles) and operate for at least one Martian year. Powered by a thermo-electric generator, it will carry about 100 kilos (220 lbs) of scientific instruments. Some devices are placed over each track, each of which can be positioned and left where required. Besides sending a continuous stream of pictures and information back to Earth, it would be able to build up a computer picture of its terrain and thus 'remember' any hazards.

as at the sunniest spot on Earth. Solar power satellites intercept solar energy where it is most abundant – in space. In one design, the solar power satellite collects sunlight on vast panels, each several miles square, studded with solar cells which turn sunlight into electricity. An alternative design envisages focusing the Sun's light with giant mirrors to heat helium gas which then turns turbines to generate electricity. Whichever way the electricity is generated aboard the satellite, it must then be converted into high-frequency radio waves, known as microwaves, for transmission to Earth.

Microwaves of about 10 cm. wavelength would pass almost unhindered through the Earth's atmosphere, so no power would be lost. The power beam would be of low intensity – far less intense than that in a microwave oven. Birds could safely fly through the beam and aircraft would not remain in it long enough to be affected.

Like communications satellites, the power satellites could be stationed 35,900 km. (22,400 miles) out in space in what is termed geostationary orbit; at this distance they would orbit at the same speed as the Earth spins, and so would appear to hang stationary above a point on the

The Solar Polar Mission

In 1983, a probe is due to set off to explore the Sun's poles in a joint project – linking NASA and the European Space Agency (ESA) – known as the Solar Polar mission.

In February 1983, NASA's Space Shuttle is to launch two spacecrafts to Jupiter. The crafts will separate en route. Then Jupiter's immense gravity will accelerate them free of the Earth's orbital plane and send them arcing back to the Sun. From above the poles, they will probe gaping holes in the Sun's corona, record cosmic rays from other stars, and trace the tangled web of magnetic field and intense energy that bathe our planet.

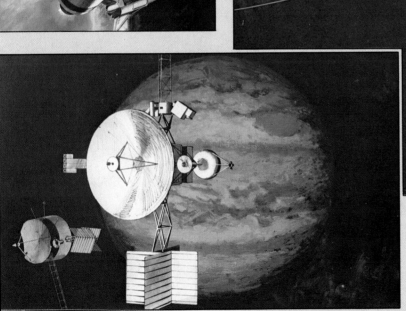

Above left: The linked spacecraft are launched from the Space Shuttle.

Left: The two craft, now separated, approach Jupiter.

Above: The double-slingshot manoeuvre throws them back towards the Sun's poles.

equator. Their microwave beams could therefore be directed accurately to receiver panels on the ground.

The receiver arrays, 10 km. (six miles) in diameter, could be placed anywhere on Earth within sight of the satellite, in remote areas or even off shore. They could turn the received microwaves into electricity to be fed into the local electric grid. This whole process of transmission, reception and conversion could be carried out with high efficiency, so that the advantage of putting the solar energy collectors into space was not lost.

Each satellite would deliver from 5,000 to 10,000 megawatts, depending on its size, enough to power a major city. Two of the larger-sized satellites could provide the entire electrical needs of India. And eight could supply the UK's current electricity demand, in contrast with the 200 or so conventional generating stations around the UK today. But in practice, solar power satellites are unlikely to supplant Earth-based generating stations completely. NASA and the US Department of Energy have considered using 100 or so satellites to provide 30% of the US electrical energy needs by the year 2025.

Experiments to prove the feasibility of the solar power satellite concept will be undertaken by the Space Shuttle. For one thing, supports will be needed for the solar energy collectors. During Shuttle flights, the construction of beams by remote-controlled robots will be tried out; if successful, this operation will be of value for building many other space structures, apart from power satellites.

A prototype power satellite could be set up in orbit by 1985, and a full-scale satellite solar power station could be operating by 1996, delivering power at a cost comparable to that of the cheapest energy sources on Earth. Income from sales of electricity on Earth could finance the solar power satellites, which would probably be under the control of an international organization like the Intelsat consortium that owns and operates the world's commercial communications satellites. With mass-production in orbit, each 5,000-megawatt solar power satellite is projected to cost $8,000 million, and to bring in an estimated $35,000 million in revenue over a 30-year working life.

It all sounds too good to be true, and perhaps it is. The catch is the enormous weight of each satellite – as much as 50,000 tons, or more than 600 times the weight of Skylab, the heaviest object ever launched. The current Shuttle can launch 29 tons at a time, so to build a complete power satellite in orbit would require at least one Shuttle launch a day for years on end. This is trivial in comparison with the number of flights from an airport, but is vastly in excess of any space activities to date.

What is needed is a larger and more economical launcher still than the Shuttle. There are already plans for one on the drawing board. One simple improvement is to replace the Shuttle's existing

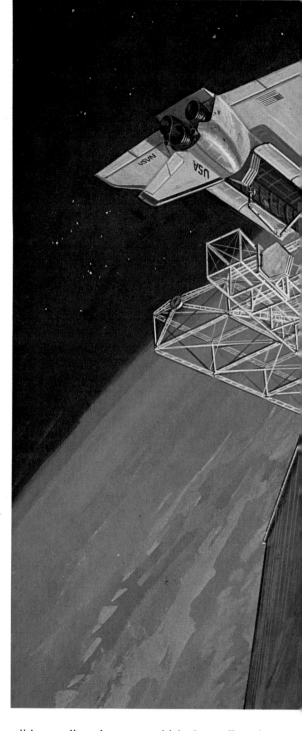

solid-propellant boosters, which drop off as the Shuttle ascends, with more powerful liquid-fuelled boosters, also reusable. These would increase the Shuttle payload to 45 tons. Another development is to replace the winged Shuttle orbiter with a payload. This increases the delivery capacity to more than 80 tons. And if four liquid-propellant boosters were clustered around the payload instead of just two, the system's launch capacity rises to 150 tons.

But even this is small compared with the advances that will eventually be possible. Designers are working towards a Heavy Lift Launch Vehicle (HLLV) capable of putting into orbit 500 tons – the equivalent of six or seven Skylabs. There are several different design ideas, all of which are astounding, and the last of which looks at first sight to be impossible.

Most HLLV plans centre on some kind of

massive two-stage booster, larger even than the Saturn V. Both stages would be recovered and reused to keep costs to a minimum; they might even be winged to fly back to Earth like larger versions of the Space Shuttle orbiter, or they could simply parachute back through the atmosphere. A suggested alternative is to cluster six winged boosters, each larger than the Space Shuttle orbiter, around the rocket's central core. These boosters would detach themselves and fly back to Earth as the central stage ascended. Once the payload is safely delivered to orbit, the central core parachutes down to be recovered and re-fuelled for the next mission.

But the ultimate goal of space designers is a rocket that will reach orbit with a single stage. Impossible? It seems so in a generation that has been brought up to believe in the inevitability of multi-stage rockets. Yet it need not always be

so, and foreseeable improvements in engine thrust combined with reductions in structural weight show that a single-stage-to-orbit (SSTO) rocket is on the cards.

A SSTO rocket would be stubbier than the pencil-thin rockets we are used to. One design study envisages a large SSTO as a gigantic Apollo capsule. The rocket's engines are arranged in a ring around its base, which incorporates a heat shield like that used in manned spacecraft to protect it during re-entry. Its payload is carried in a large forward bay. According to the plan, the rocket will splash down in a water-filled basin next to its launch site at Cape Canaveral, so that it can be immediately readied for the next launch.

Even though a fleet of such vehicles would dramatically cut the cost per pound of putting payloads into orbit, exhaust gases from a con-

For astronomers, one of the most exciting projects planned for the last two decades of this century is the space telescope. Freed from the distorting effects of Earth's gravity, the telescope could be scores of times bigger than the biggest Earth-based optical telescope. Beyond the atmospheric haziness that limits visual astronomy on Earth, the telescope will be able to resolve galaxies and stars in great detail. With it, astronomers hope to see to the edge of the Universe and spot individual planets round our stellar neighbours.

tinual stream of rocket launches would produce a major pollution hazard in the upper atmosphere. This, and the undoubtedly high development costs of the heavy-lift launch vehicles, is a major stumbling block of the solar power satellite project.

The Visions of Gerard O'Neill

But there is an alternative, and it is a startling one: Don't bring up the materials from Earth; instead, get them from the Moon, or even further afield. Although this suggestion may sound at first like mere fantasy, it has major practical advantages, at least in theory.

For one thing, there is no lack of suitable construction materials on the Moon. The samples brought back by the Apollo astronauts showed that as much as one-third by weight of the Moon's surface consists of metals, such as titanium, aluminium and iron. Silicon is abundant in the rocks of the Moon, as it is in rocks everywhere, so there is no shortage of raw material for making the silicon solar cells of the power satellite's giant collecting panels.

But the real beauty of mining the Moon is that the major expense of any large project – the launch costs – can be reduced to an absolute minimum. A rocket launched from Earth has to expend vast amounts of energy fighting against the Earth's gravity. As any aspiring young astronaut knows, the Moon has one-sixth the Earth's gravity, which makes launching from its surface much easier. And the Moon's lack of atmosphere is another advantage, for there is no air to resist the movement of a high-speed projectile. Payloads could be literally catapulted into space from the Moon, given a sufficiently powerful catapult.

Just such a device has already been suggested. A magnetic catapult could well be a suitable launcher for firing packages of lunar material into space. These packages would be picked up in orbit and transported to space industrial centres where the rocks would be processed for building power satellites.

This suggestion is one of many to come from an American physics professor, Gerard O'Neill, now considered the doyen of those who believe that mankind will be moving into space in a big way in the next century.

O'Neill's vision of how man could live in space is a staggering one. He sees the solar power satellite as the economic justification for a far greater project – nothing less than the establishment of large colonies in space. O'Neill, once a candidate to be an astronaut, has argued that we should build new land areas in space as a relief from the environmental evils that threaten the Earth's future. These land areas would be on the inside surfaces of giant spheres or cylinders, which would spin to provide gravity. Within these hermetically-sealed colonies, humans could walk around as freely as on a summer's day on Earth. The colonies would obtain their heat and light from the Sun, and grow their own crops. New land area could be built in the form of space colonies at a rate faster than the increase in population on Earth.

Says O'Neill: 'It is important to realize the enormous power of the space colonization technique. If we begin to use it soon enough, and if we employ it wisely, at least five of the most serious problems now facing the world can be solved without recourse to repression: bringing every human being up to a living standard now enjoyed by only the most fortunate; protecting the biosphere from damage caused by transportation and industrial pollution; finding high-quality living space for a world population that is doubling every 35 years; finding clean, practical energy sources; preventing overload of Earth's heat balance.'

Fantastic? It is as feasible now as landing on the Moon was 30 years ago. By the turn of the century, O'Neill speculates, we may see the first colonies dotting the night sky like faint stars. Probably they will not come quite that soon, but there is no doubting that we will have the technical ability to build such colonies next century.

How could such a radical change in our civilization come about? To begin building the colonies, we would need to set up a small mining base on the Moon, probably in the southeastern Mare Tranquillitatis, staffed by a few dozen people. These would be employed to scoop up Moon soil, parcel it up, and feed it into buckets on the magnetic catapult. This catapult works on the same principles as devised for magnetically levitated monorail trains on Earth.

Magnetic fields accelerate the buckets along a ramp to the Moon's escape velocity, 2.4 km. (1.5 miles) per second. The buckets are then rapidly decelerated, but the parcel of moonstuff flies onwards and outwards into space where it is collected in orbit. Back on the Moon, the buckets return along rails to be filled with more lunar soil for flinging into space.

According to O'Neill, two payloads would be launched each second. A million tons of lunar topsoil would be flung into space each year (but even at this rate a generation of lunar mining would not produce a crater big enough to be seen from Earth). Once the catcher is full, it moves off to the orbiting colony site where the ore processing plant is waiting to reduce the precious lunar material into metal and glass for construction.

O'Neill envisages that the first colony, which he terms Island One, will be a sphere 460 metres in diameter. It will be capable of housing 10,000 people on its inner surface. This is huge in comparison with the crews of traditional space stations, but is still quite small in comparison with the potential that the space colonization scheme offers.

Island One will rotate twice a minute to provide Earth-normal gravity at the equator. As the sphere curves towards its poles, artificial gravity on its inner surface will decline. At the poles, there will be no gravity at all. Here, human-powered flight will be possible.

Imagine living inside such a colony. The sphere would curve around you on all sides. Directly opposite you, a quarter of a mile away, the heads of other colonists would be pointed towards you, apparently hanging upside down from the inner surface of the spinning sphere. Lunar soil would coat the inner surface of the sphere, and with suitable nutrients plants and trees would be able to grow in it. Around the equator of the sphere a river could meander.

Sunlight would be reflected into the sphere through a ring of windows around the sphere's axis. By varying the amount of sunlight admitted, the colonists could vary their climate, and simulate night and day. O'Neill imagines that the climate of Island One will be lush and tropical, like that of Hawaii, with exotic plants, and even vines for wine-making. Life will certainly be more pleasant in the first colonies than on, say, an offshore oil rig. There will be dangers from radiation and micrometeorites, but these could be safely shielded out by surrounding the colony with a two-metre thick layer of slag left over after processing the lunar soil.

Houses in the colonies will be built from bricks made of compressed lunar slag, with plentiful glass composed of silicon from the lunar soil. Agriculture will be carried on in areas at either end of the sphere, so that the colony will be completely self-sufficient in food. Recycling of all air, water and food will be vital. Every day, a workforce will leave the colonies to construct more power satellites from incoming cargoes of lunar materials.

Although the Moon is rich in metals, it lacks important substances such as carbon, nitrogen and hydrogen. These are abundant in other bodies in space, the rocky asteroids. A number of asteroids, each a few hundred metres in diameter, pass close to the Earth, and could be harnessed for their raw materials, including hydrocarbons for making plastics.

As more colonies are built, some of them may assume special purposes. Different countries may have colonies built to order in which to carry out experiments such as social organization or genetic engineering that requires strict isolation from Earth. Astronomers will be fighting to get aboard the colonies. Eventually, whole industries will be forced to move into space where raw materials from the Moon and asteroids and energy from the Sun are more plentiful than on Earth.

A second generation of colonies, the Island Two design, could be cylinders 3.2 km. (two miles) long and 640 metres in diameter capable of housing 100,000 people. Well into next century, the time may be right to start building the largest cylinders of O'Neill's grand vision – the Island Three design, 32 km. (20 miles) long, 6.4 km. (four miles) in diameter, and housing millions of people. The cylindrical colonies would be split into alternating strips of land area and window. Each day, mirrors would peel back from the windows to reflect in the Sun's light. In the largest colonies, the atmosphere would be so deep that the sky would be blue, and clouds would form. Families will flock to the colonies in search of new lives and new opportunities in the vast ranges of space, in the same way that foot-loose immigrants crossed the Atlantic to America last century.

Initially the colonies will be established near the Earth and Moon, but in theory there is no reason why they should not be located anywhere within the Solar System. With suitably large mirrors the colonies could exist as far away as the orbit of Pluto, and with on-board nuclear power plants to supply energy they could exist in the even more remote and inhospitable depths of space, perhaps plying trade between the stars.

When we colonize space, we will do it not on the surfaces of planets, but in Earth-like habitats of our own devising. And once self-sufficient colonies like these have been set up – and it can happen within the next century – mankind as a

In this, one of several possible space colony designs, about 10,000 people can live in suburban comfort – complete with pools, rivers and lakes. The giant torus is a circular cylinder that spins on its own axis, thus creating Earth gravity at the rim. A normal sequence of nights and days is provided by the adjustable mirror that hangs above the colony. The hub is a docking area, the spokes for access. The Sun provides power. All materials (once the initial engineering and chemical ingredients are lifted from Earth) come from the Moon. Such colonies should be self-contained economically and ecologically, rapidly repaying invested capital by gathering and processing lunar and asteroidal materials for industrial use in space and on Earth.

An interplanetary 'barge' lies 'moored' off an asteroid, ready to be loaded with mineral-rich rock for processing by a space colony in Earth orbit. Ton for ton, freighting material from the asteroid belt would cost the same as rocketing it up from Earth. But in space, resources exceed those of Earth by many millions, the Earth is saved from pollution, and freighters can use low-thrust engines and take their time about travel, avoiding the need for rich fuels, crushing acceleration and split-second controls.

species is rendered safe from any disaster that may befall it on Earth. There will always be a pocket of humanity somewhere in space that will survive.

Journeys to the Stars

What of destinations further away? Will mankind ever reach the stars? In a way, interstellar travel has already begun, because the Pioneer 10 and 11 and the Voyager 1 and 2 probes to the outer planets will eventually leave the Solar System and drift out into the Galaxy, carrying messages for other civilizations. (See Chapter 6.) They are our fastest spacecraft to date, travelling at 17 km. (10.6 miles) per second, yet they would still take 80,000 years to reach the nearest star if they were aimed at it – and they are not. In practice, the Pioneer and Voyager probes will have stopped working by the time they leave the Solar System, but craft might one day be built that could survive an 80,000-year journey to the stars – were we willing to wait that long.

What we need is some way to cut down that yawning transit time to the stars – and that means new forms of rocket power. We have already discussed the possibility of electric rockets. These promise to bring down the transit time between stars to a matter of centuries – a considerable improvement on 80,000 years, but still not good enough.

Various other possible interstellar propulsion systems have been discussed, all of which sound like science fiction and most of which indeed may always remain so. One is the photon rocket, which uses a beam of light particles, or photons, for propulsion. This photon beam is produced by the annihilation of matter and anti-matter in the starship's engines – or that, at least, is the theory. In practice, no one has any idea how sufficient antimatter could be produced and stored without annihilating the entire spacecraft around it. Photon rockets are going to need some mighty advances in technology before they can be considered remotely possible.

A widely-favoured idea is the interstellar ramjet, a craft that releases energy by the fusion of hydrogen into helium and which scoops in its hydrogen fuel from interstellar space as it goes along. The ramscoop is not a solid affair, but is instead a gigantic magnetic cup which deflects atomic particles from space into the reactor. Again, this is the theory. And again, there are immense practical difficulties. One analysis suggests that more energy is needed for compressing the gas to suitable densities in the reactor than can ever be released by the subsequent fusion reactions. Therefore the ramjet will act as a brake, not as a propulsion unit. Passengers would have to get out and push.

Of all the proposals for interstellar propulsion, the nuclear pulse rocket looks like being the most promising. And nowhere has it received more detailed examination than in a British Interplanetary Society study called Project Daedalus. Daedalus is an unmanned two-stage probe in-

tended to reach Barnard's star inside a human lifetime, using technology likely to be available around the year 2000. The astronomers of the project chose Barnard's star because of its likelihood of having planets, but the probe could equally well be sent to any of the Sun's nearest stellar neighbours.

Nuclear fusion, as in a hydrogen bomb, releases sufficient power to make high-speed interstellar travel possible. The problem is harnessing that power. A nuclear pulse rocket works by firing a series of small nuclear explosions behind the craft, the energy from which pushes the craft along.

In Daedalus, the explosions occur in a 100-metre diameter reaction chamber. Small bombs, about the size of a table-tennis ball, are fired into the chamber where they are hit by energetic electron beams from electron guns ranged around the chamber's rim. Success with current experiments in fusion power show that this is a plausible mode of propulsion. The electron beams compress and heat the pellets so that fusion reactions occur, and the bomb explodes with a force of 90 tons of TNT. The starship takes a little jump forwards.

Why does Daedalus not blow itself to bits? Strong magnetic fields in the reaction chamber absorb the power of the blast and push the reaction products out of the back, acting like a magnetic spring. Bombs are injected at the rate of 250 every second. With such a rapid succession of explosions, the engine's thrust is virtually continuous.

The bombs of Daedalus are made of deuterium, a heavy isotope of hydrogen, and helium-3. But there is not enough helium-3 on Earth to make the 30,000 million bombs needed. Jupiter, though, has abundant deuterium and helium-3 in its atmosphere so the Daedalus team propose sifting the atmosphere of Jupiter for starship fuel. The whole craft will be so big and heavy – 230 metres long and 53,000 tons at ignition – that it will have to be built in space, probably by the same space industrial centres used to build solar power satellites and the O'Neill colonies. Only a civilization with extensive space industries can undertake starflight. We shall have such industries next century.

The starship's first stage would be under power for two years. Then it drops away and the second stage takes over. This is identical in operation, except that its reaction chamber is only 40 metres in diameter. After another 1.8 years of boost, the second-stage engines shut down. Daedalus then cruises towards Barnard's star at its top speed of 130 million km. (80 million miles) per hour or 12.2% the speed of light.

From engine ignition to encounter with Barnard's star would take a total of 49.2 years. During this time, Daedalus would be running its own affairs with a powerful on-board computer. Once on its way to the stars, a spacecraft will be too far away to rely on instructions from Earth. Daedalus could carry a payload of 450 tons, equivalent to five or six Skylab space stations, across the six-light-year gap to Barnard's star. This payload is mostly made up of 20 or so sub-probes designed to examine conditions in interstellar space and the Barnard's star system itself. As Daedalus approaches the star, large on-board telescopes begin to spy out whatever planets it may have – perhaps gas giants like Jupiter, plus smaller rocky bodies like the Earth. The sub-probes are then dispatched to examine the star and its planets.

According to the present proposal, Daedalus and its ensemble of sub-probes do not decelerate. They fly straight through the Barnard's star system collecting pictures and other data as quickly as they can. Then they radio back their results to be picked up by large radio telescopes on Earth. Several hundred photographs may be sent, showing anything from starspots on Barnard's star to cloud formations and surface features on its planets – our first close-up view of another solar system in space.

But what of sending manned spaceships to the stars? This is a prospect for the more distant future, particularly if we are talking about high-speed starships. But inhabited space arks could be made by fitting a small O'Neill colony with a Daedalus propulsion unit. This, like the interstellar arks of science fiction, would slowly ply its way between the stars, with whole generations being born, reproducing, and dying, knowing no home other than the ark. Eventually, their distant descendants step out at the destination after journeys that have taken 400 to 600 years, in the case of trips to Alpha Centauri and Barnard's star, or more than 1,000 years to Tau Ceti.

Once their destinations had been reached, the colonists would set up new habitats in orbit around their adopted star, using the same techniques as for building the original colonies in our Solar System. Some of these planetary systems may contain life (not necessarily intelligent), or they may be uninhabitable. With an O'Neill colony, the existence of habitable planets becomes irrelevant. The immigrants to the new planetary system will then scout ahead with robot probes like Daedalus to find more suitable target stars before sending off another ark to push the spread of mankind further into the Galaxy.

Even at these relatively slow speeds, the entire Galaxy could be colonized in a matter of a few million years – a long time in terms of our existence as a species, but a trivially short time compared to the lifetimes of stars, which are measured in thousands of millions of years. By then, or course, we may have come face to face with other life in space, some members of which may be engaged in similar interstellar colonization operations of their own.

Our first faltering steps into space today are a prelude to the possible spread of humanity throughout the Galaxy – a spread which can be completed in less time than it took for the ape-men of Africa to evolve into the spacefaring man of today.

FROM SPACE, POWER FOR EARTH

It is now generally acknowledged that the world's growth of industry and population cannot proceed indefinitely. Yet the alternative – enforced zero growth – is not pleasant. Space technology may offer a way out. In space, Earth could derive all its power from the Sun's almost infinite supply of clean energy.

The principles of tapping solar energy in space have already been researched. The plan foresees placing solar power stations in geosynchronous orbit. These would beam microwave energy to antennae which would feed electricity into the national grid.

The microwave beams – half the intensity of sunlight – would not damage birds or aircraft. The raised antennae would be transparent to light and rain, and the ground beneath could be used for agricultural purposes.

At first sight, the scale of the operation – even for one nation – seems forbidding. Each 'solar panel' and each antenna would extend for many square miles. And the cost of the first station would be perhaps a $100 billion.

But once the initial investment is made, the new technology could, in theory, pay its way. Gerard K. O'Neill, in *The High Frontier*, foresees power satellites planned in conjunction with space colonies. 'Within 13 years', he writes, 'the rate of construction could exceed the annual growth needs of the US.'

Other countries could then use the new technology. In theory, we could again see steady industrial growth without threatening Earth's delicate ecology.

Ferried into position by countless Space Shuttle journeys, the skeleton of the first 50,000-ton power satellite begins to take shape. In later power satellites, cost should be reduced many times by supplying materials from the Moon.

A completed power satellite,
20 km. (13 miles) long by
4.8 km. (3 miles) across, hangs
in geo-stationary orbit, beaming
its micro-wave energy back to
Earth.

Right: On the ground antennae
like this one, measuring 8 km.
(5 miles) by 12 km. (7.5 miles)
would transform the microwave beams into electricity.
Even directly beneath the
antennae, micro-wave levels
would be below the present
legal limits and the land could
be used for grazing and
farming. The antennae consist
of wire-mesh, transparent to
sunlight and rain and raised
several feet above the surface. If
such 'power-stations' supplied
all the US's electrical energy
needs in the year 2000, the land
area covered would be 0.2 per
cent of the continental United
States, that is, about one fifth of
the area already devoted to
roads. Moreover, the antennae
could be built in remote areas
away from population centres
and traditional fuel sources.

GLOSSARY/INDEX

Entries and cross-references in **bold**. Illustrations in *italic*.

A

Aberration. Fault that mars image in an optical instrument.
Aberration of starlight. Change caused by the Earth's motion in orbit in apparent positions of stars.
Absolute magnitude. Measurement of total light output of a star. Brightness at 10 parsecs/27, 32, 42.
Absolute zero. The coldest temperature (−273.16°C)/136.
Absorption lines. Wavelengths in an object's spectrum which have been absorbed/*28*, *136*, *155*.
Abt, Helmut/103.
Agena. Upper stage on American launch vehicles/182, 183.
Albedo. Measure of the proportion of light reflected by a non-shiny surface.
Aldebarar (star)/175.
Aldrin, Edwin Eugene (b. 1930). Lunar module pilot of Apollo 11 crew/183, 186–8, *187*, *188*.
Algol. Star in Perseus, also called Beta Persei/52.
Algonquin Radio Observatory/113.
Allende meteorite/73, 75.
Almagest/20.
Alpha Centauri (Rigil Kent). Triple-star system containing the nearest stars to our Sun/24, *25*, 25, 27, 31, 114, 216.
Alpha Cygni. See **Deneb.**
Alpha Virginis (Spica)/*24*, 24.
Altitude. The angle between a celestial object and the horizon.
Andromeda Galaxy (M31; NGC224). Separate spiral galaxy in Andromeda/10, *22*, 110, 112, 119, 120, *121*, 122, *123*, 123, 131, 141, 142.
Anglo-Australian Telescope/156.
Angstrom (A). Unit of measure of the wavelength of light, equal to one ten-billionth of a meter.

Apollo Flights

Unmanned

1/Feb. 26 1966
2/Jul. 5 1966
3/Aug. 25 1966
4/Nov. 9 1967
5/Jan. 22 1968
6/Apr. 4 1968

Manned

7/Oct. 11–22 1968
Walter Schirra, Donn Eisele, R. Walter Cunningham.

8/Dec. 21–27 1968
Frank Borman, James Lovell, William Anders.

9/Mar. 3–13 1969
James McDivitt, David Scott, Russel Schweickart.

10/May 18–26 1969
Thomas Stafford, John Young, Eugene Cernan.

11/Jul. 16–24 1969
Neil Armstrong, Michael Collins, Edwin Aldrin.

12/Nov. 14–24 1969
Charles Conrad, Richard Gordon, Alan Bean.

13/Apr. 11–17 1970
James Lovell, John Swigert, Fred Haise.

14/Jan. 31–Feb. 9 1970
Alan Shepard, Stuart Roosa, Edgar Mitchell.

15/July 26–Aug. 7 1971
David Scott, Alfred Worden, James Irwin.

16/Apr. 16–27 1972
John Young, Thomas Mattingly, Charles Duke.

17/Dec. 7–19 1972
Eugene Cernan, Ronald Evans, Harrison Schmitt.

Angular distance. The apparent separation of two objects.
Antarctic. See **Polar regions.**
Antares. Star/20.
Aperture synthesis. Technique which uses a number of small radio dishes to build a wide-angle view/160.
Apollo program. Space project that landed a total of 12 American astronauts on the Moon/11, 63–64, 166, 179, 182, 183–92, *185*, 193–5, 197, 207, 211, 212 (see also panel). Flights: 8/77; 10/187; 11/114, *184*; 14/190; 15/78, 168, 179, 191; 16/88, 191; 17/192.
Apollo telescope mount/195.
Apollonius of Alexandria/149.
Apollo-Soyuz. Joint US-Russian space project/197, *197*.
Apparent magnitude. The brightness with which a star appears as seen from Earth. (See also **Absolute magnitude)**/27, 32, 42.
Aratus. Greek philosopher/20.
Arctic. See **Polar regions.**
Arecibo Observatory. Site of the largest radio astronomy dish, in Puerto Rico/110, 111, 112, 113, 114, 159, 208.
Ariel satellites. A series of 5 British scientific satellites, launched by the US, 1962–74/161.
Aries/22.
Aries, 1st point of. See **Vernal equinox.**
Aristarchus of Samos (fl. 280–264 BC). Greek astronomer who calculated relative sizes of the Earth, Moon and Sun/62, 149.
Aristotle (384–322 BC). In astronomy, the major Greek philosopher. Developed **Eudoxus's** system of spheres. He rejected the idea that the Earth spins or that it orbits the Sun/*148*, 148, 149, 150, 151, 154.
Armstrong, Neil Alden (b. 1930). First man on the Moon/183, 186–8.
Asteroids. Thousands of small rocky bodies moving around the Sun/63, *74*, 74, 75, 76, *215*. See also individual names and types.
Asteroid belt/63, 70, 74.
Asthenosphere. Semi-molten rock layer in Earth/82, 85.
Astrolabe. An ancient instrument for observing stellar altitudes/149.
Astrology/19, 20, 60, 147.
Astrometry. The measurement of positions of objects in the sky/22, 24–32.
Astronauts. See individual names.
Astronomical unit (a.u.). The average distance between Earth and Sun; 149,597,910 km (92,955,832 miles).
Astronomy, history of/144–61.
Astrophotography/154–8.
Astrophysics. The application of physics to the study of the Universe/141, 155.
Atlantis. Mythical island in Atlantic/84–5.
Atlas rocket. Major American space launcher, used to launch America's first men into orbit in the **Mercury Project**/168.
Atmosphere of Earth/52, 63, 67, 82, 86–9, *88*, 91, 94, 100, 104, 110, 161; of Mars/69, 107.
Atomic time. A means of time-keeping based on oscillations of cesium atom.
Aubrey Circle/147.
Aurora. Phenomenon caused by impact on upper atmosphere (usually at poles) of solar particles/91; A. Borealis (Northern Lights)/91; A. Australis (Southern Lights)/91.
Azimuth. The bearing of an object around the observer's horizon.

B

Baade, Walter (1893–1960). German-born American astronomer who identified the two different populations of stars.
Babylonian astronomy/20, 147, 148.
Background radiation. Radiation believed to be energy left over from the **Big Bang**/10, 15, 136–139.
Bacon, Francis/85.
Baikonur. Official name for the Soviet space launching site at Tyuratam, near the Aral Sea.
Baily, Francis (1774–1844). English astronomer/87.
Baily's beads. Beads of light at solar eclipse caused by the Sun shining between mountain peaks along edge of Moon/*87*, 87.
Barnard, Edwin/102.
Barnard's star. A star 5.9 light-years away, named for Barnard. It has the fastest **proper motion** of any star/25, 27, 101, *102*, 104, *203*.
Barred spiral galaxies/122, 123–4.
Barycentre. Centre of orbiting system/*102*, *103*.
Bayer, Johann (1572–1625). German astronomer who in 1603 published *Uranometria*, charting over 2,000 stars/21, 24.
Bayeux Tapestry/75.
Bell Telephone Labs/136, 159.
Berenices/131.
Bessel, Friedrich Wilhelm (1784–1846). German astronomer who made the first measurement of a star's distance/10.
Beta Persei. See **Algol.**
Betelgeuse. A red giant, Alpha Orionis/50.
Big Bang. The supposed event which started the Universe's expansion 15–20 billion years ago/125, 126, 133, 135, 136, 138, 139, 143. Cyclical/143.
Big Dipper. See **Ursa Major.**
Binary Stars. See **Double stars.**
Black hole. A theoretical object whose gravitational pull is so strong that not even light can escape/15, 33, 39, 48, *50*, 50, *52*, 52, 134, *135*, 140, 141, 142.
Blue giant. A large, hot and bright star/51, *52*, 103.
Bode, Johann (1747–1826). German astronomer. Devised Bode's law, a number series that roughly coincides with the average distance of planets from the Sun/21.
Boeing jet. As carrier on **Space Shuttle** test/*201*, *202–3*.
Bondi, Sir Hermann (b. 1919). Austrian-born British mathematician, who with **Thomas Gold** proposed the **Steady State** theory of cosmology in 1948/125.
Borman, Frank (b. 1928). American astronaut, commander of the Apollo 8 mission/182, 185.
Bracewell, Roland/111, 114.
Brahe, Tycho (1546–1601). Danish astronomer, the greatest observer of the pretelescopic era/*150*, 151.
Bridle, Alan/113.
Brunhes, Bernard/90.
Bruno, Giodano (1548–1600). Italian supporter of **Copernicus'** theory that the Earth and other planets orbit the Sun.
Bunsen, Robert/*154*, 154–5.
Bykovsky, Valery (b. 1934). Soviet cosmonaut who made the longest solo space flight in history/181, 197.
Byurakan Astrophysical Observatory. Major Soviet observatory, near the city of Yerevan, Armenia/101, 103.

C

Calendars/146, 148, 149.
Callisto/71, 108.
Caloris Basin (on Mercury)/65.
Canals (of Mars)/67, 72–3, 104–6, *105*, 168.
Canis Major/27.
Canis Minor/24.
Cape Canaveral. Main launching site for US space missions, shared by NASA's **Kennedy Space. Center,** and the Cape Canaveral Air Force Base/11, 164, 166, 168, 180, 181, 182, 185, 191, 194, 200, 211.
Cape Kennedy. Previous name for **Cape Canaveral.**
Capricorn, Tropic of. The southernmost latitude at which the Sun appears directly overhead.
Carbon cycle. A chain of nuclear reactions by which energy is released in stars.
Carbonaceous chrondite. Type of stony meteorite containing carbon/73–4, 75, 76.
Carina/33.
Carina Nebula/54–5.
Carpenter, Malcolm Scott (b. 1925). Second American astronaut in orbit/181.
Carr, Gerald. US astronaut/196.
Cassegrain telescope/156.
Cassini, Giovanni (1625–1712). Italian-French astronomer who identified dark gap in Saturn's rings/71.
Cassiopeia/33.
Celestial sphere. The imaginary sphere of the heavens, with the Earth at its center, which appears to rotate once very day/24, 118, 148.
Centaur. American upper-stage rocket.
Centaurus A. (NGC 5128)/142, *142*.
Cepheid variable. Type of star which swells and contracts in size/28, 51, *51*, 120, 122, 123, 124.
Cepheus/51.
Ceres. Largest of the **asteroids**/74.
Cernan, Eugene (b. 1934). American astronaut, commander of the final Apollo mission to the Moon/183, 186, 191.
Cerro Tololo telescope/156.
Chaffee, Roger. US astronaut/185.
Chandrasekhar, Subrahmanyan (b. 1910). Indian-born American astrophysicist, who defined limits of **White dwarf** stars/50.
Chesaux, de/118.
Chondrite. The commonest form of stony meteorite.
Chromatic aberration. The failure of a lens to bring light of all wavelengths to the same focal point.
Chromosphere. The layer of gas about 10,000 miles (16,000 km) thick above the Sun's **Photosphere.**
Chryse Planitia (on Mars)/106, 107, 169.
Circumpolar. Term describing a celestial object that does not set when seen from a given latitude on Earth.
Coalsack. A large dark cloud of dust and gas.
Coccone, Guiseppe/112.
Collins, Michael (b. 1930). Command-module pilot on the Apollo 11 flight to the Moon in July 1969/183, 188.
Comet. A small icy body embedded in a cloud of gas and dust moving in a highly elliptical orbit around the Sun/59, *62*, 62, 63, 72, *75*, 101, 118, 120, 206.
Command Module. Crew's compartment in manned spacecraft during launch and landing/*184*, 188, *188*, 189.
Communications satellites. Relays in space for sending telephone, radio and television signals around the world/174, 175, 176, 209.
Comte, Auguste/154, 155.
Comsat. The US Communications Satellite Corporation.
Conrad, Charles. US astronaut/182, 183, 189, 194, *195–6*.
Constellations. 88 star patterns which provide a set of references for the recognition and identification of objects in the sky. See panel; also **Zodiac**; modern names/20, *20*, 21, 22, 29, 147, 149; also see under individual names.
Continental drift/67, 82, *84*, 84–6, *86*.
Cooper, Leroy Gordon (b. 1927). Astronaut who made the sixth and last flight in the **Mercury** series, May 1963. In 1965 he flew in Gemini 5/181, *181*, 182.
Copernicus, Nicolaus (1473–1543). Polish astronomer whose **Heliocentric** theory of the Universe defined the Earth as an ordinary planet/9, 10, 63, 67, 85, 122, *149*, 149, 150–1, 152.
Copernicus satellite/111.
Corona. The outermost layer of the Sun's atmosphere/*48*, *87*, 209.
Cosmic rays. Nuclei of atoms, stripped of all their electrons, shooting through space at speeds close to that of light/40, 135, 136.
Cosmology. The study of the origin and evolution of the Universe/118–26, 133–43.
Cosmological satellites. Continuing series of Russian Earth satellites – numbering over 800. Successors to the Sputnik series.
Cosmos, Russian spacecraft/193, 194.
Coudé focus telescope/156.
Crab nebula. The expanding cloud of gas ejected by a Supernova seen to explode in July 1054/21, 24, *49*, 153; Chinese account/49.
Craters on Moon/64, 77, *78–9*, 167; Alphonsus/166; Archimedes/166; Cone/190; Copernicus/*167*, Descartes/191; Le Monnier/168; Littrow/191; Plumb/191; St George's/78–9; Shorty/191; Tsiolkovsky/78, *79*.
Crystalline spheres. Outer realm of ancient's universe/24, 118, 148.
Curie point/185.
Curvature of space. A distortion caused by the presence of matter/135, 137.
Curwin, Joseph. US astronaut/195.
Cyclops. See **Project Cyclops.**
61 Cygni/10.
69 Cygni/*25*.
Cygnus X-1/24, 52, *52*.
Cygnus A/141, 142.
Cygnus/32, 52, 127, 141.
Cygnus Loop. See **Veil Nebula.**

D

Davis, R. Astrophysicist/45; Neutrino experiment/45, *45*.
Declination. The celestial equivalent of latitude on Earth/21, 22, 24.
Deimos. Smaller and more distant of the two moons of Mars.
Deneb. The brightest star in the constellation Cygnus, also called Alpha Cygni.
Dicke, Robert/138.
Diffraction grating. A device for diffracting light into a spectrum/158.
Discoverer satellites. Series of 38 US Air Force satellites, 1959–62.
Dispersion. The spreading out of light into its constituent wavelengths to form a spectrum.
Dixon, Robert/112.
DNA/114.
Dog Star. See **Sirius.**
Doppler, Christian/124.
Doppler effect. The change in frequency of waves emitted by an object as it moves toward or away from an observer/27, 104, 114, *124*, 124, 155, 161.
Double-lobe structure/141, 142.
Double stars/25; see also individual names.
Drake, Frank (b. 1930). American radio astronomer who made first attempt to detect signals from other civilizations in **Project. Ozma**/100–5, 111, 112, 113.
Draper, John/154.
Dreyer, Johan (1852–1926). Danish astronomer, compiler of the NGC, the New general catalogue of Nebulae and Clusters of Stars.
Dryden Flight Research Centre/201.
Duke, Charles. US astronaut/191.
Dwarf galaxies/123.

E

Eagle. Apollo lunar lander/187, 188.
Early Bird. The first communications satellite launched by the Intelsat organization, April 6, 1965/176.
Earth/12, 16, 31, 43, 60, *60*, 80–97; viewed from space/93–7; gravitational field/91; interior/82–4; crust/82, 84, 85, 86, 88; mantle/82, 85, 88; inner core/82–3; outer core/82–3; outer core/82–3, 91; age/84; mid-ocean ridges/85; composition/82–4, *83*; as centre of Solar System/9, *20*, 62–3, 148, 149;

size/149; flat/148, 149; rotation/21.
Earthquakes/82, 83, 84, 86.
Eccentricity. A measure of how an orbit deviates from being a true circle.
Echo satellites. Two giant plastic balloons used as reflector communications satellites, (1960–64)/136.
Eclipse. The passage of one astronomical body into the shadow of another; the term is also applied to the passage of the Moon in front of the Sun (a solar eclipse), more correctly an **Occultation** of the Sun by the Moon; lunar/149; annular/87; solar/87, 87.
Eclipsing binary. A system of two stars circling one another, with each periodically blocking off the other's light.
Ecliptic. The plane of the Earth's orbit around the Sun.
Eddington, Sir Arthur (1882–1944). British astrophysicist.
Einstein, Albert (1879–1955). German-born physicist, regarded as the greatest theoretical physicist of the century, and responsible for fundamental advances in a wide variety of fields/119, 125, 126, 134, 135, 137, 154. See **Relativity**.
ELDO. The European Launcher Development Organization. See **European Space Agency**/177.

Electromagnetic radiation. The range of radiation, from gamma rays through the spectrum of visible light, to radio waves/40, 41, 126, 135, 136.
Elliptical galaxies/122, 123, 139, 140.
Emission. The production of Electromagnetic Radiation, including visible light and radio waves/40.
Enterprise, part of **Space shuttle**/201, 201–3.
Epicycle. Device once used by astronomers to explain movement of planets by a combination of circles/63, 149, 150.
Epsilon Indi/111.
Equatorial Mounting. Mounting for telescopes that have one axis aligned with the Earth's axis.
Equinox. The moment when the Sun crosses the celestial equator, and day and night are of equal length anywhere in the world (equinox: 'equal night')/22. See **Vernal equinox**.
Eratosthenes of Cyrene (c.276–c.194 BC). Greek astronomer who first measured Earth's size/149.
ERTS. Abbreviation for Earth Resources Technology Satellite, the original name for **NASA**'s Landsat program.
Escape velocity. The speed an object must attain to escape from a

gravitational field.
ESRO. The European Space Research Organization of 10 countries.
ESSA satellites. Series of meteorological satellites launched by the Environmental Science Services Administration. They superseded the earlier Tiros series.
Eta Carinae/54.
Ether. The imaginary medium in which light was assumed to travel in the 19th century.
Eudoxus of Cnidus (c.400–c.350 BC). Greek astronomer who produced the first detailed system to account for the observed motions of the planets. His system of spheres was adopted by Aristotle/148, 149.
Europa/71.
European Southern Observatory/156.
European Space Agency 176, 200, 209.
Evening star (Hesperus). Popular term for the planet Venus in the evening sky shortly after sunset/61. See **Venus**.
Event horizon, in **Black Hole**/50, 156.
Exosphere/88.
Expanding universe. The apparent recession of distant galaxies with speeds which increase proportionally to their distance/10, 14, 15, 125, 126,

117–131, 133–6, 138, 143.
Exploding galaxies/141.
Explorer satellites. Series of over 50 American scientific satellites, begun in 1958 and transferred to NASA on its formation 164, 198.

F
Feldman, Paul/113.
Flare. A brilliant burst of light occurring near a **Sunspot**/46, 46.
Flying saucers. See **UFO's**.
Focal length. The distance between the lens or mirror of an optical instrument and the image it forms.
Fomalhaut. One of the brightest stars.
Fontenelle, Bernard le Bovier de/101.
Frau Mauro. Region on Moon/190.
Fraunhofer, Joseph (1787–1826). German optician and physicist, best known for charting dark **absorption lines** in the solar spectrum/28, 29, 154.
Frequency. The number of waves that pass a given point in a given period of time. In **electromagnetic radiation**, frequency is in cycles per second, or hertz.

G
G-1, Soviet rocket/172.
Gagarin, Yuri (1934–1968). Soviet cosmonaut; first man to fly in space,

1961/180, 180, 181, 192.
Galactic clusters. See **Open clusters**.
Galaxies. Systems of billions of stars/119, 140–2; distance from earth/123–4, 134; types/122, 123–4. See also under individual types, **Milky Way**, **Nebulae** and under individual names.
Galileo Galilei (1564–1642). Italian mathematician, astronomer and physicist and advocate of **Copernicus**'s heliocentric theory/17, 19, 151, 153.
Galilean satellites of Jupiter/10.
Gamma-ray astronomy. The study of celestial objects at wavelengths shorter than X-rays (i.e. less than 0.1 angstrom).
Gamow, George (1904–1968). Russian-born American astrophysicist, best known for his association with the **Big Bang** theory.
Gas Chromatic Mass Spectrometer/69, 108.
Ganymede. Moon of Jupiter/71.
Gemini project. American space program of 12 flights (1964–66) to practice rendezvous and docking techniques/181–3, 185; pictures from Gemini 6/182–3; 7/182–3; 9/183.
Geostationary. Term describing an orbit in which a satellite appears to hang stationary over a point on the Earth's equator. Also termed a synchronous orbit or Clarke orbit/174, 176, 206, 209, 217, 218.
Giant star. A large bright star.
Gilbert, William/89.
Gibbous. Term describing the phase of the Moon or a planet between half and full illumination.
Glenn, John (b. 1921). First American to orbit Earth/180, 181.
Globular cluster. Spherical-shaped cluster of 100,000–10 million old stars. Some 125 globular clusters form a halo around our Galaxy/31, 32, 33, 48, 114, 117.
Globular cluster in Hercules (M13)/114.
Goddard, Robert Hutchings (1882–1945). American rocket pioneer/165.
Goddard Space Flight Center. Facility of **NASA** in 1959 at Greenbelt, Maryland.
Gold, Thomas (b. 1920). Austrian-born American astronomer, who with Hermann **Bondi** proposed **Steady State** theory/125.
Gondwana. Early continent/85.
Granulation. A mottling effect on the Sun's **photosphere**/44.
Gravity. Fundamental property of matter, which produces mutual attraction/15, 31, 32, 40, 41, 43, 48, 72, 104, 135, 139. Gravitational theory/152, 155.
Great Bear/22.
Great circle. A line on the surface of a sphere which divides it into two equal hemispheres.
Great spiral. See **Andromeda Galaxy**.
Greek astronomy and early theories/20, 30, 148.
Greenhouse effect. The warming of a planet by its atmosphere/67.
Grissom, Virgil 'Gus' (1926–1967). American astronaut, first man to make two space flights/180, 182, 185.
'Guest stars' (Supernovae)/20, 49.
Gutenberg, Beno/82.

H
H regions. Areas of interstellar space containing hydrogen gas. In H1 regions the hydrogen is cool and un-ionized. In H2 regions it is hot and ionized.
Hadley Rille. Lunar feature/78, 168, 190.
Hale, George Ellery (1868–1938). American astronomer, pioneer of astrophysics. He founded the Yerkes, **Mount Wilson** and **Mount Palomar** observatories/156.
Hale Observatories. Name since 1970 of the Mount Wilson and Palomar Observatories, founded by **Hale** and operated by the Carnegie Foundation and the California Institute of Technology/156. See also under individual names.
Halley, Edmond or **Edmund**

Constellations

Constellation	Genitive case	English name	Abbreviation
Andromeda	Andromedae	Andromeda	And
Antlia	Antliae	Air pump	Ant
Apus	Apodis	Bird of Paradise	Aps
Aquarius	Aquarii	Water carrier	Aqr
Aquila	Aquilae	Eagle	Aql
Ara	Arae	Altar	Ara
Aries	Arietis	Ram	Ari
Auriga	Aurigae	Charioteer	Aur
Boötes	Boötis	Herdsman	Boo
Caelum	Caeli	Chisel	Cae
Camelopardalis	Camelopardalis	Giraffe	Cam
Cancer	Cancri	Crab	Cnc
Canes Venatici	Canum Venaticorum	Hunting dogs	CVn
Canis Major	Canis Majoris	Greater dog	CMa
Canis Minor	Canis Minoris	Lesser dog	CMi
Capricornus	Capricorni	Goat	Cap
Carina	Carinae	Keel	Car
Cassiopeia	Cassiopeiae	Cassiopeia	Cas
Centaurus	Centauri	Centaur	Cen
Cepheus	Cephei	Cepheus	Cep
Cetus	Ceti	Whale	Cet
Chamaeleon	Chamaeleontis	Chameleon	Cha
Circinus	Circini	Compasses	Cir
Columba	Columbae	Dove	Col
Coma Berenices	Comae Berenicis	Berenice's hair	Com
Corona Australis	Coronae Australis	Southern crown	CrA
Corona Borealis	Coronae Borealis	Northern crown	CrB
Corvus	Corvi	Crow	Crv
Crater	Crateris	Cup	Crt
Crux	Crucis	Southern cross	Cru
Cygnus	Cygni	Swan	Cyg
Delphinus	Delphini	Dolphin	Del
Dorado	Doradus	Swordfish	Dor
Draco	Draconis	Dragon	Dra
Equuleus	Equulei	Foal	Equ
Eridanus	Eridani	River	Eri
Fornax	Fornacis	Furnace	For
Gemini	Geminorum	Twins	Gem
Grus	Gruis	Crane	Gru
Hercules	Herculis	Hercules	Her
Horologium	Horologii	Pendulum clock	Hor
Hydra	Hydrae	Water snake	Hya
Hydrus	Hydri	Lesser water snake	Hyi
Indus	Indi	Indian	Ind
Lacerta	Lacertae	Lizard	Lac
Leo	Leonis	Lion	Leo
Leo Minor	Leonis Minoris	Lesser lion (Cub)	LMi
Lepus	Leporis	Hare	Lep
Libra	Librae	Scales	Lib
Lupus	Lupi	Wolf	Lup
Lynx	Lyncis	Lynx	Lyn
Lyra	Lyrae	Lyre	Lyr
Mensa	Mensae	Table Mountain	Men
Microscopium	Microscopii	Microscope	Mic
Monoceros	Monocerotis	Unicorn	Mon
Musca	Muscae	Fly	Mus
Norma	Normae	Level	Nor
Octans	Octantis	Octant	Oct
Ophiuchus	Ophiuchi	Serpent holder	Oph
Orion	Orionis	Orion	Ori
Pavo	Pavonis	Peacock	Pav
Pegasus	Pegasi	Pegasus	Peg
Perseus	Persei	Perseus	Per
Phoenix	Phoenicis	Phoenix	Phe
Pictor	Pictoris	Easel	Pic
Pisces	Piscium	Fishes	Psc
Piscis Austrinus	Piscis Austrini	Southern fish	PsA
Puppis	Puppis	Stern	Pup
Pyxis (= Malus)	Pyxidis	Compass	Pyx
Reticulum	Reticuli	Net	Ret
Sagitta	Sagittae	Arrow	Sge
Sagittarius	Sagittarii	Archer	Sgr
Scorpius (–o)	Scorpii	Scorpion	Sco
Sculptor	Sculptoris	Sculptor	Scl
Scutum	Scuti	Shield	Sct
Serpens	Serpentis	Serpent	Ser
Sextans	Sextantis	Sextant	Sex
Taurus	Tauri	Bull	Tau
Telescopium	Telescopii	Telescope	Tel
Triangulum	Trianguli	Triangle	Tri
Triangulum Australe	Trianguli Australis	Southern triangle	TrA
Tucana	Tucanae	Toucan	Tuc
Ursa Major	Ursae Majoris	Great bear	UMa
Ursa Minor	Ursae Minoris	Little bear	UMi
Vela	Velorum	Sail	Vel
Virgo	Virginis	Virgin	Vir
Volans	Volantis	Flying fish	Vol
Vulpecula	Vulpeculae	Fox	Vul

(1656–1742). Computer of the orbit of the comet that bears his name/207, 119.
Halley's Comet. The first periodic comet to be identified with a 76-year orbit/75, 119, 207.
Halo (of **Milky Way**)/33, 48. See **Globular clusters.**
Halo stars/33.
Harvard College Observatory/156.
Harvard Classification Scheme/156.
Hawking, Stephen. English astrophysicist/14.
Heavy Lift Launch Vehicle/210–11.
Heliocentric. Sun-centred. See **Copernicus.**
Heliogyro. Proposed space probe/206, *207*, 207.
Helios I. Spacecraft/173.
Helium. The second-lightest and second most abundant element in space.
Henry Draper Catalogue/156.
Hercules/24, 114.
Hermes. Apollo; ancient name for **Mercury**/62.
Herschel, Sir William (1738–1822), musician and avid amateur astronomer. recorded 2,500 nebulae. Suggested that the stars are arranged into a circular lens-shaped slab – our Galaxy/90–1, 32, 72, 120, 153, 154.
Herschel, Caroline (1750–1848), **William Herschel**'s sister/120.
Herschel, Sir John (1792–1871). William's son, made extensive surveys of the southern skies/*120*, 120.
Herschel's telescope/153, 154.
Hertzsprung, Ejnar (1873–1967). Danish astronomer who showed that color and luminosity of stars is related, and discovered the division of stars into giants and dwarfs, findings confirmed by Henry Norris **Russell.**
Hertzsprung-Russell Diagram. A graph on which the color or temperature of stars is plotted against their brightness/42–3, 48, 51.
Hesperus. See **Evening star.**
Hess, Professor Harry/85.
Hewish, Antony (b. 1924). British radio astronomer, discoverer of **pulsars.**
Hipparchus of Nicaea (fl. 146–127 BC). Greek astronomer, discovered that the Earth wobbles in **precession** over a period of about 26,000 years/20, 27, 149, 151.
Hoba Meteorite/73.
Homestake Goldmine/45.
Horsehead Nebula/35.
Hoyle, Sir Fred (b. 1915). British astronomer, best known for his support of the **Steady State theory** of cosmology/101, 125, *126*.
Hubble, Edwin Powell (1889–1953). American astronomer who proved that galaxies of stars exist outside our own Milky Way, and discovered that the Universe is expanding/10, 119, 120–2, 123, 125, 134, 136, 141.
Huggins, Sir William (1824–1910). English astronomer, a pioneer of stellar spectroscopy/155.
Humboldt, Alexander von/85.
Huygens, Christiaan (1629–1695). Dutch scientist, best known for his description of **Saturn**'s rings as a swarm of particles/72.

I

Icarus. Asteroid of the Apollo group/74.
Ice Ages on Earth/92.
Image Photon Counting System/161.
Inclination. The angle between the plane of a particular orbit and a reference plane.
Inferior planet. A planet with an orbit closer to the Sun than Earth, i.e. **Venus** and **Mercury.**
Infra-red astronomy/161.
Interference. An effect which occurs when two waves of the same wavelength are combined.
Interferometer. A pair of receiving devices linked to make the equivalent of a much larger receiver.
Interstellar absorption or **extinction.** The dimming of light by dust particles in space/140.
Interstellar communication. From Earth/114–5, *114*, 115; from space/110.
Interstellar dust and gas/32, 33, 40, 41, 101.
Interstellar molecules. Simple molecules found in clouds of gas and dust.
Io, moon of **Jupiter**/71.
Ions. Atoms which have either lost or gained electrons, and thus have acquired a positive or a negative charge.
Ionosphere. Region of the **Earth**'s atmosphere containing atoms and molecules that have been ionized/88, 114.
Irwin, James, US astronaut/*179*, 190, 191.

J

Jansky, Karl Guthe (1905–1950). American communications engineer, the founder of radio astronomy/158, 159.
Jastrow, Robert. US astronomer/15.
Jeans, Sir James (1877–1946). British astrophysicist/40.
Jeans mass/40.
Jefferies, Harold/85.
Jet Propulsion Laboratory (JPL). A division of **NASA**. in Pasadena. California/110.
Jodrell Bank. British radio-astronomy observatory near Macclesfield, Cheshire, operated by the University of Manchester; also known as Nuffield Radio Astronomy Laboratories. It was founded by Bernard **Lovell**/9, 15, 157.
Jupiter. The fifth planet from the **Sun**, and the largest in the Solar System/*60*, 60, 61–2, 63, *70*, 70–1, 72, 74, 75, 76, 100, 102, 103, 104, 108, 151; missions to/109, 174–5, 208, 209; life on/112; satellites/71, 151. See also under names of moons and panel. See also **Red Spot**.

K

K. Degrees Kelvin, the temperature above **absolute zero**/138.
Kahoutek, comet/75, 195.
Kant, Immanuel (1724–1804). German philosopher who suggested nebulae were other galaxies/119, 120.
Kardashev, Nikolai/112.
Kelvin, Lord/138.
Kennedy, President John F./166, 181, 186, 191.
Kennedy Space Center (KSC). NASA installation at Cape Canaveral, Florida/11, 169, 173.
Kepler, Johannes (1571–1630). German mathematician and astronomer who derived fundamental laws of planetary motion/63, 67, 150, 151, 152.
Kirchhoff, Gustav/*154*, 154, 155.
Kirkwood Gaps. Gaps in the asteroid belt.
Kitt peak National Observatory. Observatory located southwest of Tucson, Arizona/103, 104.
Komarov, Vladimir (1927–1967). First Soviet cosmonaut to make two space flights; first man killed during a space mission/181, 192, 193.
Kuiper Airborne Observatory/161.

L

Lacaille, Nicolas Louis de (1713–1762). French astronomer, pioneer in mapping southern skies.
Lagrange, Joseph (1736–1813). French mathematician who discovered Lagrangian points, five points at which a small body can remain in stable orbit with two massive bodies (see **Trojans**).
Laika, first creature in space/164.
Landsat. Two NASA satellites, originally called **ERTS**. Pictures from/*177*.
Large Magellenic Cloud. See **Magellenic Clouds.**
Large Space Telescope (LST). An astronomical telescope planned to be orbited by the **Space Shuttle.**
Laser. A device for producing an intense beam of light; *L*ight *A*mplification by *S*timulated *E*mission of *R*adiation.
Latitude. A co-ordinate for determining positions on Earth north or south of the equator.
Laurasia continent/85.
Leavitt, Henrietta/51.
Lehmann, Inge/82.
Leonov, Alexei (b. 1934). Soviet cosmonaut, first man to walk in space/181, 182.
Levy, Saul/103.
Libra/20.
Libration. The slow rocking motion of the visible face of the Moon.
Life in the Universe. Life zone/101; life elsewhere/13, 15, 16–17, 81, 88–9, 99–115, 139, 143; communications with/100, 102, 111. See also **Interstellar communications.** Hostile/114. See individual planets.
Light. The part of the spectrum of electromagnetic radiation to which the human eye is sensitive. Nature of/152, 154–5.
Light, velocity of. The speed at which electromagnetic radiation travels in a vacuum/137.
Light-year. The distance traveled by light in one year; 5.8786 trillion miles (9.4607 trillion km)/10, 25, 26, 27, 31, 100, 103, 104, 114, 125.
Lippersley, Hans/151.
Lithosphere/85.
Local Group. A cluster of about 20 known galaxies sincluding the Milky Way/*123*.
Longitude. Co-ordinate for determining the position of an object east or west of Greenwich meridian.
Lovell, Sir Alfred Charles Bernard (b. 1913). Radio astronomy pioneer, founder of **Jodrell Bank** Observatory/9.
Lovell, James (b. 1928). American astronaut, commander of the ill-fated Apollo 13 flight/182, 185, 189, 190.
Lowell, Percival (1855–1916). American astronomer, pioneered discovery of **Pluto**, and founder of the **Lowell Observatory**/67, 72–3, 104, 105.
Lowell Observatory/67, 72, 73.
Luna spacecraft. A series of 23 Soviet Moon probes, 1959–74; flights/9, 17, *166*, 167, 168.

Moons of Jupiter

Moon	*Distance (km)*	*Orbital period (days)*	*Diameter (km)*
1. Amalthea	181,000	0.498	240
2. Io	421,760	1.769	3,659
3 Europa	671,050	3.551	3,100
4. Ganymede	1,070,400	7.155	5,270
5. Callisto	1,882,600	16.689	5,000
6. Leda	11,100,000	239	15
7. Himalia	11,477,600	250.566	100
8. Lysithea	11,720,250	259.219	20
9. Elara	11,736,700	259.653	30
10. Ananke	21,200,000	631*	20
11. Carme	22,600,000	692*	20
12. Pasiphae	23,500,000	744*	20
13. Sinope	23,600,000	758*	20

(Details of a 14th moon, discovered in 1975, are not yet known.)
* Retrograde.

Moons of Mars

Moon	*Size (length)*	*Orbit radius*	*Period (hr/min)*
Phobos	27km	9350km	7.39
Deimos	14km	2350km	30.18

Lunar Module (LM). The two-stage craft in which Apollo astronauts landed on the Moon/*179*, 184, 186, *188*, 190.
Lunar orbiter. A series of five American Moon-orbiting craft, 1966–7/166; View from/*167*.
Lunar Roving Vehicle. Electronically powered Moon car used by **Apollo** astronauts for exploring the lunar surface/*179*, *190*, *191*, 191.
Lunar soil/63.
Lunokhod. Soviet automatic Moon car/167–8, *168*, 173.
Lyman alpha. Absorption/*136*.

M

M13. See **Globular cluster in Hercules.**
M31 (NGC 224). See **Andromeda Galaxy.**
M42 (NGC 1976). See **Orion Nebula.**
M104 (NGC 4594). See **Sombrero Hat Galaxy.**
Magellanic Clouds. Two small galaxies that orbit our Galaxy. Large/23, *122*, *123*; Small/*23*, *123*, 51.
Magnetic fields, of **Earth**/66, 81, 85, 89–91; of **Mercury**/66; of **Venus**/67.
Magnetic storms/46.
Magnetopause/91.
Magnetosphere. The outermost region of the Earth's atmosphere/91.
Main Sequence. Stars in **Hertzsprung-Russell Diagram**/42, 44–7, 51, 52.
Man. Evolution of/104–5; in space/9, 16, 17. See individual projects and astronauts.
Maria. Lunar seas/64, 65, 77, 78. Sea of Clouds/166; of Crises (Mare Crisium)/77, 167, 168; of Fertility/167; Ocean of Storms/166, 189; of Rains/167; of Serenity/168, 191; of Tranquility (Mare Tranquillitatis)/166, *187*, 212.
Mariner spacecraft. A series of American planetary probes/4, 6, 7, 9, 10, 65, 66, 69, 105, 106, 168, 169, 173, 175.
Mars. Fourth planet from the **Sun**/60, *60*, 62, 63, *67*, 67–70, 74, 76, 101, 151, 206; age/69; craters/69; exploration of/168, 169, 206–7, 208; possible life on/67–9, 105, 106–7, 112, 169–72; sunset on/*171*. See also **Canals** (of Mars).
Mars probes. A series of seven Russian space probes to investigate Mars, 1962–73/169.
Mars Lander see **Viking spacecraft.**
Mars rover/206, 207, *208*.
Marshall Space Flight Center (MSFC). NASA facility at Huntsville, Alabama.
Mass-luminosity relation. the relationship between a star's mass and its brightness/41, 42, 43.
Matthews, Drummond/95.
Mercury. The closest planet to the Sun/*60*, 60, 62, 64–6, *65*, 67, 70, 71, 76, 105; flights to/180, 181; craters/65.
Mercury project. the American program to launch a man into space, 1961–3/180, 181, 185, 190.
Messier, Charles (1730–1817). French astronomer best known for his list of nebulae and star clusters/120.
Meteor. The streak of light produced when a meteoroid burns out in the atmosphere/62, 113.
Meteor Crater, Arizona/*73*, 73.
Meteorite. A lump of rock or metal from space that crashes to Earth/63, 64, 73–4, 75, 76, 88, 100, 101, 108, 192. See types and individual names.
Michelson, Albert (1852–1931). Polish-born American physicist who measured the speed of light.
Micrometeorite. A particle from space which is small enough to be decelerated in the Earth's atmosphere without being vaporized.
Milankovich, Milutin. Geologist/91; model/91, 92.
Milky Way. A spiral galaxy of some 100,000 billion stars, of which the Sun is one/10, *13–14*, 16, *23*, 29–33, *31*, 40, *41*, 43, *116*, 117, 118, 119, 120, 122, *123*, 123, 134, 135, 136, 139, 141, 143; halo/139; disc/139.
Miller, Stanley/100.
Minor planet. See **asteroid.**
Mining in space/206, 207, 211, 214.
Mira Variable. Stars/51.
Mitchell, Edgar. US astronaut/190.
Mohorovicic, Andrija/82; discontinuity (Moho)/82.
Molnya Satellites. A series of Soviet communications satellites.
Moon. The Earth's natural satellite/60, 63–4, *64*, 65–6, 67, 70, 71, 73, 75, 76, 77, *78–9*, 105, 134, 141, 147, 168, 206; age/64; colonisation of/207–8; composition/77; craters see **Craters on Moon**; effect on Earth/89, 103; exploration of/164, 166–8, 179, 187–90; mining on/206, 207, 212; rock/101, 167, 187, 191; soil/63, 192.
Morning star (Phosphorus). The planet **Venus**, when shining in the morning sky/61.
Morrison, Philip/112.
Mount Hadley. On Moon/*78–9*, *191*.
Mount Palomar. Observatory/156; Schmidt 48″/157; 200″/157. See **Hale Observatories.**
Mount Wilson. Observatory/156; 100″/120–2, 153, 156. See **Hale Observatories.**
Moving cluster method. A technique for finding the distance of a nearby star cluster whose stars are moving through space along parallel paths/28.
Mullard Radio Astronomy Observatory (MRAO). The radio astronomy observatory of the University of Cambridge/*144*, *159*.
Murchison Meteorite/*100*, 101.

N

N-galaxies/141, 142.
NASA. The US's National Aeronautics and Space Administration, established October 1, 1958, to administer nonmilitary aeronautical research and space programs/110, 164, 177, 180, 181, 182, 185, 186, 188, 189, 195, 200.
National Radio Astronomy Observatory (NRAO). The largest US radio astronomy observatory, at Green Bank, West Virginia/100, 112, 113.
Nebula. A region of gas and dust in the Galaxy, usually fuzzy in appearance. Latin: cloud. Obsolete term for galaxy/40, *53*, 53, *54–5*, 118, 119, 153, 155, 158. See under individual names.
Neptune. The eighth planet in average distance from the Sun/12, 60, *61*, 61, 63, 70, 72–3, 112; discovery/154.
Neutrinos/43, 44–5. See also **Davis'** neutrino experiment.
Neutron star. A very small star, so dense that its protons and electrons have been compressed into neutrons/48, 49–50, *50*, 52, 53. See **pulsar.**
New moon. The Moon at the instant it lines up between the Earth and Sun.
Newton, Sir Isaac (1642–1727). British mathematical physicist who developed the concept of universal gravitation/15, 63, 137, *152*, 152–4.
Newtonian reflecting telescope/152, *156*, 156.
NGC. Abbreviation for *New General Catalogue of Nebulae and Clusters of Stars.*
NGC 205, elliptical galaxy/122.
NGC 224 (M31). See **Andromeda Galaxy.**
NGC 253. Spiral galaxy/*130*.
NGC 1365. Barred spiral galaxy/*122*.
NGC 1976 (M42). See **Orion nebula.**
NGC 3034. Irregular galaxy/*140*.
NGC 4565/*131*.

Glossary 223

Moons of Neptune

Moon	*Diameter (km)*	*Orbit radius (km)*	*Period (days)*
Triton	6,000	355,000	5.887
Nereid	500	5,562,000	359.88

NGC 4594 (M104). See **Sombrero Hat Galaxy.**
NGC 5128. See **Centaurus A.**
Nilosyntis. 'Canal' of Mars/104.
Nimbus satellites. A series of six US weather satellites from which the **Landsat** design was developed/171.
Northern Lights. See **Aurora Borealis.**
Nova. A faint star that suddenly erupts in brightness/49, 52, 53; classical/52; dwarf/52.
Nut. Egyptian Goddess of the sky/*148*, 148.

O

OAO satellites. Orbiting Astronomical Observatory US satellites.
Oberth, Hermann Julius (b. 1894). German rocket pioneer/*165*, 165.
Observatories. See under individual names.
Occultation. The obscuring of one astronomical body by another. See also **Eclipse.**
OGO satellites. Orbiting Geophysical Observatory US satellites.
Ohio State Radio Observatory/112.
Olbers, Heinrich (1758–1840). German astronomer, co-author of a cosmological paradox/*118*, 118.
Olbers' Paradox. That in an unchanging Universe filled with an infinite number of stars the sky should shine with solar brightness/118–9, 136.
Olympus Mons. Martian volcano/*68*, *69*, 169.
Omega Centauri/33.
O'Neill, Gerard K. Author of *The High Frontier*/212, 213, 216, 217.
Oort, Jan Hendrik (b. 1900). Dutch astronomer, who proposed that comets exist in a vast cloud at the edge of the Solar System. Oort cloud/75.
Open cluster. Galactic cluster; loosely packed cluster of stars/28, 32, 41, 46.
Ophiuchus/102.
Opposition. The instant at which a planet farther from the Sun than Earth appears opposite the Sun in the sky.
Orbiter in **Space Shuttle**/*201*.
Orion/20, 21, 25, 33.
Orion Nebula. A cloud in the constellation Orion, also called M42 and NGC 1976/*35*, *54*.
Oscillating universe. A theory which holds that the Universe expands and contracts in cycles. See **Big Bang.**
OSO satellites. Orbiting Solar Observatory satellites.

P

P-waves/182.
Paleomagnetism/190.
Palmer, patrick/110, 112.
Pangaea/185.
Parallax. The change in position of an object when viewed from two different positions/25–7.
Parsec. The distance at which the Earth and Sun would appear to be 1 second of arc apart, 3.2616 light-years/26.
Parsons, William. See **Rosse, Lord.**
Pegasus/131.
Penumbra. The partially shaded area in an eclipse/87.
Penzias, Arno. co-discoverer of **Background radiation**/136 8. *138*.
Perfect Cosmological principle. That the Universe is always the same/125–6.
Period-luminosity relation. The way in which **Cepheid variables** of longer period are more luminous than shorter period Cepheids.
Perseus/33.
Perturbations. Slight disturbances in the motion of a body caused by the gravitational pull of another object.
Phase. The proportion of an illuminated body that is visible to an observer.
Phosphorus. See **Morning star.**
Photometry. The measurement of an object's brightness.
Photomultiplier tube/158.
Photons. Particles of electromagnetic radiation/43, 44, 49, 158, 159, 161.
Photon rocket/214.
Photosphere. The visible surface of the Sun/46.
Pic du Midi Observatory, in the French Pyrenees/*160*.
Pioneer spacecraft. A continuing series of American space probes to explore the Solar Systems/77, 114, 115, 164, 166, 174–5, 214.
Pisces/22.
Planets. Details/61. See individual names.
Plate tectonics/85–6. See **continental drift.**
Planetary nebula. A gaseous shell surrounding a hot central star/50–1, 153.
Plato, Greek philosopher/148.
Pleiades. ('Seven sisters'). A cluster of stars in the constellation Taurus, also known as M45/*36*.
Plow, Plough. See **Ursa Minor.**
Plumb Crater on Moon/*191*.
Pluto. The farthest known planet from the Sun/10, *11*, 60, *61*, 61, 63, 66, 70; discovery/*72*, 72–3, 75.
Pogue, William/*196*.
Polaris. The north pole star lying in Ursa Minor/21, 23.
Polar regions on Earth/82, 89, 91, 92.
Pole star. See **Polaris.**
Posionius/10.
Power satellites/209–10, 216, *217–18*.
Precession. A slow wobbling of the Earth on its axis/22.
Prime meridian/22.
Primeval fireball/10, 11, 139, 143.
Procyon/27.
Project Cyclops. Proposed radio telescope array/110–11, 111, 114.
Project Ozma. Search for inter-stellar messages/100, 112. See **Drake** and **Interstellar communication.**
Prominence. A hot, bright cloud of gas projecting from the Sun's **chromosphere** into the **corona**/46, *47*.
Proper motion. The motion of a star across the sky, expressed in arc seconds per year/27, 28, 102, 103, 104.
Proton-proton chain. A nuclear process, by which energy is produced in stars like the Sun.
Proton rocket/174, 192.
Protostar. The early stage in a star's formation/41.
Proxima Centauri. The nearest star to the Sun, 4.3 light-years away/25, 27.
Ptolemy (c.100–c.178). Alexandrian astronomer, geographer, and mathematician. And theories/20, *149*, 149, 151. See also **Almagest.**
Pulkovo Observatory. Central observatory of the USSR Academy of Science, near Leningrad.
Pulsar. Rotating **neutron star** that emits short regular pulses of radiation/16, 49, 51, 110, 115, 135.
Pythagoras (c.580 BC1c.500 BC). Greek philosopher and mathematician, who believed that the Earth was spherical and lay at the centre of a spherical Universe/148.

Q

Quasar. Quasi-stellar object (QSO). Object which appears as a star-like point but which emits more energy than an entire galaxy/16, 33, 134–5, 136, 142; OH 471/*136*; 3C 273/*136*, *132*, *134*; CTA-102/113.
QEB. Egyptian Earth God/*148*.

R

Radar astronomy. The investigation of bodies in the Solar System by reflecting radio waves off them/65, 66, 110.
Radial velocity. The velocity of a star along the line of sight/27, 28.
Radiation Belts. See **Van Allen belts.**
Radiation pressure. The tiny force exerted on bodies by a beam of light.
Radio astronomy. The observation of the Universe at radio wavelengths/16, 33, 100, 112–3, 134, 158, 160.
Radio galaxies. Distant galaxies that are very powerful sources of radio waves/16, 141, *142*, 142, 160. See also under individual names.
Radio telescopes. Instruments for receiving radio waves from space/16, *111*, 112–3, 126, 136, 145, *157*, 159. See also under individual names.
Radio window/159.
Raleigh waves/82.
Ranger probes. A series of American Moon probes, 1961–5.
Rasalgethi/24.
Ratan 600. Telescope/110.
Red dwarf. A faint star of low surface temperature (2,000–3,000°C) and a diameter about half that of the Sun/42, 101, 102, 104.
Red giant. A star with a low surface temperature (2,000–3,000°C) and a diameter between 10 and 100 times that of the Sun/31, 42, 48, *50*, 50, 51, 53, 124, 125.
Red shift. The amount by which the wavelengths of light and other forms of **electromagnetic radiation** from distant galaxies are increased because of the expansion of the Universe/10, 15, 134, 136.
Red Spot. Apparently permanent feature among the swirling cloud bands of **Jupiter**/*70*, 71, 112.
Redstone rocket. An American rocket for space launches/164, *180*.
Reflecting telescope. A telescope in which a concave mirror collects and focuses light. See also under individual names.
Refracting telescope. Telescope in which a large lens or object glass collects and focuses the light/*154*.
Regulus/27.
Relativity. Einstein's theory of the structure of space and time, and its relation to motion and gravity/119, 137; special theory/137; general theory/137, 154.
Retrograde motion. Motion from east to west, opposite to the normal west-to-east, direction of motion in the Solar System.
Right ascension. The celestial equivalent of longitude on Earth/21, 22, 24.
Rigil Kent. See **Alpha Centauri.**
Ring Nebula/*53*.
Rockets. Comparative sizes/172. See individual names and types.
Rosse, Lord (1800–1867). Irish astronomer, William Parsons, Third Earl of Rosse/49, 120, *153*; 72-inch (183 cm) telescope/120, *153*.
Royal Astronomical Society/120.
Royal Greenwich Observatory. Britain's national astronomical observatory, now located at Herstmonceux Castle near Hailsham, Sussex.
Royal Society/152.
RR Lyrae variables. Pulsating stars (see **variable stars**) of standard luminosity used to calibrate distances in Galaxy/51.
Russell, Henry Norris (1877–1957). American astronomer, co-discoverer of the **Hertzsprung-Russel Diagram.**
Ryle, Sir Martin (b. 1918). English radio astronomer who developed the technique of **aperture synthesis**/160.

Moons of Saturn

Moon	*Diameter (km)*	*Orbit radius (10^3km)*
1. Janus	300	159
2. Mimas	500	186
3. Enceladus	600	238
4. Tethys	1,000	295
5. Dione	800	378
6. Rhea	1,600	527
7. Titan	5,800	1,222
8. Hyperion	500	1,483
9. Iapetus	1,600	3,560
10. Phoebe	200	12,951

S

S-waves/82.
Sagan, Carl Edward (b. 1934). American astronomer and biologist; pioneer in the possibility of life elsewhere in the Universe/113, 115.
Sagittarius/32, 33, 54.
Salyut. Soviet space station.
Sandage, Alan. US astronomer/10.
Satellites of the planets. See panels under separate planets and individual names of moons.
Saturn. The sixth planet from the Sun, renowned for its ring system/10, *61*, 61, 62, 63, 70, *71*, 71–2, 104, 135; possible life on/112.
Saturn rockets. Family of large space launchers developed for manned applications by a team under Wernher **von Braun**/164, 172, 184, 185, 186, 188, 192, 195, 197, 206, 211.
Scattering. The deflection of light and other forms of radiation.
Schiaparelli, Giovanni (1835–1910). Italian astronomer who first reported **canals** on Mars/67, 104, 105.
Schiller, Julius/21.
Schirra, Walter (b. 1923). Only astronaut to fly in all three types of American spacecraft: Mercury, Gemini and Apollo/181, 182, 185.
Schmidt, Maarten (b. 1929). Dutch-born astronomer, the first to interpret the spectrum of a **quasar.**
Schmidt telescope. A wide-angle photographic telescope developed by Berhard Schmidt (1879–1935). 48"/156, *157*.
Schwarzschild, Karl (1873–1916). German astronomer.
Schwarzschild radius. The distance from a **black hole** at which the **escape velocity** equals that of light.
SCO-X1/161.
Scorpius (also **Scorpio**, the form usual in astrology)/20, 161.
Scott, David Randolph (b. 1932). Commander of Apollo 15 mission.
Sculptor/131.
Seas on Moon. See **Maria.**
Seasons/72, 82, *89*.
Secchi, Father Pietro/155.
Service Module. Unmanned section of a spacecraft containing engines for course corrections, air and water supplies, and electrical power/184.
Seti. Search for Extra-Terrestrial Intelligence/110–1, 112.
Seven Sisters. See **Pleiades.**
Seyfert, Carl/140.
Seyfert galaxy. A galaxy with a very small, bright nucleus/141, 142.
Shapley, Harlow (1885–1972). American astronomer, who discovered that our Galaxy is much larger than was previously supposed.
Shepard, Alan Bartlett (b. 1923). First American in space/*180*, 190.
Shooting stars. See **meteorites.**
Sidereal day. The time the Earth takes to rotate on its axis with respect to a fixed point in space.
Siderite. An iron meteorite.
Siding Spring Observatory. Astronomical observatory, New South Wales/156. See also **Anglo-American Telescope.**
Sirius. The brightest star in the sky/24, *25*, 25, 27, 101, 147, 155.
Sitter, Willem de (1872–1934). Cosmologist/119, 124.
Skylab. First American space station/*93*, 195, 196; interior views/*196*; information from/3, 4, 47, 48.
Small Magellanic Cloud. See **Magellanic Clouds.**
Smithsonian Space Museum, Washington/197.
Solar apex/32.
Solar cycle. The period of about 11 years over which changes in the Sun's activity appear to go through a cycle/47.
Solar energy/16, 203, 206, 207–10, 213, 216–18.
Solar flares. See **flares.**
Solar granulation. See **granulation.**
Solar polar mission/209.

The Brightest Stars

Name		*Apparent Magnitude*	*Absolute Magnitude*	*Brightness (Sun = 1)*	*Distance (l.y.)*
(Sun		−26.5	+4.8	1	0.00002)
1. Sirius*	αCMa	−1.45	+1.41	23	8.8
2. Canopus	αCar	−0.73	−4.7	1,400	110
3. Alpha Centauri (Rigil Kent)*	αCen	−0.1	+4.3	1.7	4.3
4. Arcturus	αBoo	−0.06	−0.2	100	36
5. Vega	αLyr	+0.04	+0.5	53	27
6. Capella*	αAur	+0.08	−0.6	140	46
7. Rigel*	βOri	+0.11	−7.0	50,000	c.850
8. Procyon*	αCMi	+0.35	+2.65	6.9	11.4
9. Achernar	αEri	+0.48	−2.2	700	c.100
10. Hadar	βCen	+0.60	−5.0	10,000	c.380
11. Altair	αAql	+0.77	+2.3	7	16.3
12. Betelgeuse	αOri	+0.8	−6	14,000	c.650

* Multiple systems.

Solar radiation. See **solar wind.**
Solar sail spacecraft. See **Yankee Clipper.**
Solar System. The group of planets, comets and asteroids orbiting our Sun/10, 16, 31, 46, 49, 58–79, *60–1*, 60–1, 82, 115, 116, 117, 135, 149; origins of/75–6; old ideas of/*21*.
Solar wind. A continuous stream of particles emitted by the Sun/46, 75, 91, 188.
Solstice. The moment when the Earth's axis is inclined at its maximum (23°) toward the Sun.
Sombrero Hat galaxy/139.
Southern Cross/21, *23*.
Southern Lights. See **Aurora Australis.**
Soyuz. Soviet manned spacecraft for long-duration flights and rendez-vous and docking missions/192–3, 194, *194*, 195, *197*.
Space colonies/213–16.
Space curvature. See **curvature of space.**
Space Shuttle. Reusable winged space transporter/16, *17*, 175, 179, 197–202, *198–9*, *200*, *201*/3, *209*, 209, 217. See also **Orbiter, Enterprise.**
Space suits/*193*, 193.
Space telescope/16, 104, 161, *211*.
Spectracon/159.
Spectral lines. Narrow lines observed in a **spectrum**. Spectral lines may either be bright (emission lines) or dark **absorption lines**/Solar/*29*, 155; stellar/27, 28–9, *29*, 42, 155; nebular/28, 124, *155*.
Spectrographs/158.
Spectroscope. A device for observing the **spectrum**/120, 124, 155.
Spectrum. The entire range of **Electromagnetic radiation** from gamma rays to radio waves/120, 145, 156, 158, 159.
Spica. See **Alpha Virginis.**
Spicules. Vertical, jetlike features of the solar **chromosphere.**
Spiral galaxy. A type of galaxy in which many of the stars and nebulae lie along spiral arms/13, 14, 32, 41, 43, 120, *122*, 122, 123, *127*, 127, *130–1*, *139*, 140, 141, 153; nuclei/141; spiral arms/141.
Sproule observatory/102, 104.
Sputnik. A series of Soviet satellites, 1957–61/164, 166, 175, 180, 195.
Spy satellites/110.
Stafford, Thomas (b. 1930). US Astronaut/182, 183.
Star. A self-luminous ball of gas. Some 7,000 are visible to the naked eye. Nearest to sun/25; star map/22–3; measuring position of, see **astrometry**: composition, formation/13, 38–53, *43*, *44*, 53, 59; size/50; birth/39–41, 139; middle age/41–5; old/45–8; death/48–50; rate of formation/102; pulsating see **pulsars**; cannibalism/51–2; O-type/41, 42; B-type/41, 47; degenerate/103; brightest visible, see panel. See **red dwarfs, neutron stars, black holes** and individual names.

Moons of Uranus

Moon	*Diameter*	*Orbit radius*	*Period*
Miranda	550	130,000	1.41
Ariel	1,500	192,000	2.52
Umbriel	1,000	267,000	4.14
Titania	1,800	438,000	8.74
Oberon	1,600	586,000	13.46

Steady State theory. Theory of **cosmology**, now little supported, which regards the Universe as essentially unchanging over time/10, 125–6, 134, 136, 138.
Stonehenge. Prehistoric circle of giant stones near Salisbury, southern England, once used as an astronomical observatory/146–7, *147*.
Stones, in asteroids and meteorites/73–4.
Stratosphere/185.
Suess, Eduard/185.
Sun. Central body of the **Solar System**/30–1, *45*, 45, 46, 48, *50*, 59, 60, 147; as a star/*12*, *13*, 43, 44, 46–7; motion/32, *48*; composition, formation, see **stars**; surface, see **photosphere**; size/149; effect on earth/89, 92; distance from earth/10, 24.
Sunspot. A relatively cool, dark area on the solar **photosphere**/44, 46, *47*, 151.
Supergiant stars. The very brightest, largest stars.
Superior planet. A planet whose orbit is farther from the Sun than the orbit of the Earth.
Supernova. A star that explodes and ejects most of its mass/43, 49, 50, 53, 54, 76.
Surveyor probes. A series of seven US lunar soft-landers, 1966–68/166–7, 189.
Synchronous orbit. See **geostationary orbit.**
Synchrotron radiation. Electromagnetic radiation emitted by charged particles in a magnetic field.

T
Tau Ceti/100, 111, 112, 216.
Tektites. Small, glassy objects, probably formed by impacts of meteorites.
Telescopes. Invention and development/151, 152, 156. See under individual names, types and observatories.
Telstar. Two American communications satellites/174, 176.
Tereshkova, Valentina (b. 1937). Soviet cosmonaut, the first woman to fly in space/181.
Thermonuclear fusion/41.
Tides/*89*.
Tiros satellites. 10 US weather satellites; Television and Infra-Red Observation Satellite.
Titan (satellite of Saturn)/112.
Titan rocket. American space launcher/169, *173*.
Titov, German (b. 1935). Soviet cosmonaut, the first man to spend a full day in space/180.
Tombaugh, Clyde (b. 1906). US astronomer, discoverer of **Pluto**/72.
Trifid Nebula/*54*.
Troitsky, Vsevolod/112.
Trojans. Two groups of asteroids moving in the same orbit as Jupiter/74.
Troposphere/85.
Tsiolkovsky, Konstantin (1857–1935). Russian spaceflight pioneer/*165*, 165.
47 Tuchanae/33.
Twins paradox/137.
Tyuratam. The main Russian space launch site, near the town of the same name northeast of the Aral Sea/174, 180, 192, 195, 196, 197.

U
UFO. Unidentified Flying Object/61, 100, *113*, 113, 114.
Uhuru. X-ray satellite/161.
UK Meteorological Office/16, 92.
Ultraviolet astronomy. The study of the Universe in the ultraviolet part of the electromagnetic spectrum, between visible light and X-rays (3,000 Å down to 300 Å).
Umbra. In the Solar System that part of an object's shadow in which light from the Sun is totally cut off/87.
Universe. Size/10, *11–14*, 136; origin and evolution/10–15, 16, 138–40, 143.
Uranus. The seventh planet from the Sun. It has faint rings/*61*, 61, 63, 70, 71, 72, 76, 112; axis, orbit/72.
Ursa Major/20, 21, *22*.
Ursa Minor/
Utopia Planitia (on Mars)/106, 172.

V
V-2 rocket. German forerunner of modern ballistic missiles and space rockets, developed by Wernher **von Braun**/164, 165, 172, 174.
Van Allen, James (b. 1914). American physicist/91.
Van Allen belts. Two zones of charged particles surrounding the Earth/91, 164.
Van de Kamp, Peter (b. 1901). Dutch-born American astronomer who discovered the nearest planetary system to our own, orbiting **Barnard's star**/102, 104.
Vanguard project. A US Navy project begun in 1955, intended to launch the first US satellite/164.
Variable stars. Stars whose light reaching the Earth varies in brightness. See under individual names.
Vega/32, 154.
Vehicle Assembly Building/183.
Veil Nebula or Cygnus loop/54, *56–7*.
Venus. The second planet from the Sun/49, *60*, 60, *66*, 66–7; surface/16, 61, 101, 105, 151; life on/66.
Venus (or **Venera**) **probes.** A series of 10 Soviet space probes. Venus 9/173; 10/173.
Vernal equinox. The moment when the Sun, moving north lies on the celestial equator; also termed the spring equinox or First Point of Aries. See **equinox.**
Verschuur, Gerrit/112.
Very Large Array (VLA). The world's largest and most sensitive radiotelescope, about 40 miles (64 km) west of Socorro, New Mexico.
Very Long Baseline Interferometry/160.
Viking spacecraft. Two American probes designed to look for life on **Mars**/104, 106, 107, 108; biology experiments/69, 106–7, *107*, 107, *169*; **Mars lander**/69, *170*.
Virgo/*24*, 24, 139.
Volcanoes. On Earth/64, 82, 83, 84, 86, 88; on Moon/64, 66; on Venus/67; on Mars/*68*, *69*, 69, 106.
Von Braun, Wernher (b. 1912). German-born American rocket engineer, designer of the Saturn V Moon Launcher/164.
Voskhod. A modified Soviet **Vostok** spacecraft for carrying two or three cosmonauts/181.
Vostok. The first Soviet manned spacecraft/194.
Voyager probes/71, 114, 175, 214.

W
Weather satellites. Earth satellites which monitor the Earth's atmosphere and surface/175–7.
Wegener, Alfred/85.
Weightlessness/196.
Weinberg, Steven/138.
Whipple, Fred (b. 1906). American astronomer, who in 1949, suggested that a comet's nucleus can be compared to a dirty snowball.
Whirlpool Galaxy/122, 128.
White, Edward (1930–1967). First American to walk in space/*182*, 185.
White dwarf. A star with diameter 1 percent that of the Sun, and 10,000 times less luminous/31, 42, 48, *50*, 50, 51, 53.
White rock, in Mars crater/*68*.
Wickramasingle, Chandra/101.
Wilkinson, D. T./138.
Wilson, Robert/136–8, *138*.
Wright, Thomas/30.

X
X-ray astronomy. The study of the Universe at X-Ray wavelengths comprising **electromagnetic radiation** between 0.1 and 300 Å in wavelength.
X-ray sources/24, 49, *52*, 52, 135, 161.
X-ray satellites/161.

Y
Yankee Clipper, solar sail spacecraft/*207*.
Yeliseyev, Alexei (b. 1934). Soviet cosmonaut who flew on the Soyuz 5 flight in January 1969, which made the first docking between two manned spacecraft.
Yerkes Observatory/156.
Young, John Watts (b. 1930). Commander of the Apollo 16 mission/182, 183, 186, 191.

Z
Zelenchukshaya/156.
Zodiac. A belt of 12 constellations about 9° wide over which the Sun, Moon and planets (except Pluto) appear to move/20, *21*.
Zond spacecraft. A series of Soviet space probes that act as tests for future missions/192.
Zuckerman, Ben/110, 112.

The publishers would like to thank the following individuals and organizations for their kind permission to reproduce the photographs in this book:

Anglo-Australian Observatory/161 right; The Arecibo Observatory/114. (The Arecibo message of November 1974, designed by Professor Drake and the staff of the Arecibo Observatory), 159 above (The Arecibo Observatory is part of the National Astronomy and Ionosphere Center which is operated by Cornell University under contract with the National Science Foundation.); Courtesy of Bell Laboratories/138; The Bettmann Archive/18–19, 20 below, 21, 150 above, 165 below; Curators of the Bodleian Library, Oxford/20 above, 24; Boeing Aerospace/210–211, 217, 218–219; Professor Alec Boksenberg/161 left; Laurence Bradbury/92, 122 above, 156, 158 above; Brookhaven National Laboratory/45 above; Peter Cannings/22, 172; Crown Copyright – reproduced with permission of the controller of Her Majesty's Stationery Office/147; Mary Evans Picture Library/62 above, 73 above, 105 above, 118, 120 below; Fortean Picture Library/113; Michael Freeman/2–3, 17, 38–39, 52, 58–59, 98–99, 108–109, 116–117, 162–163, 184 above right, 193 above and below, 196 above left and right, 197 right, 200 above, 204–205, 214–215; Grummon Aerospace/175; H. H. Heyn/87 centre; Gary Hincks/11–14, 84, 86, 88 right, 186–187, 188–189, 198–199; Jet Propulsion Laboratory, California/65, 68 centre and below right; Jodrell Bank (Lascelles)/8–9, 8 inset; Keystone Press Agency/164, 165 above right; Kitt Peak National Observatory/142; Lowell Observatory Photographs/72, 105 below; Jim Marks 50, 51; NASA/4–5, 7, 46–48, 63 below, 64, 66–67, 68 above, 68 centre and below left, 69, 70, 71 inset, 77, 78–79, 80–81, 87 below left and right, 88 left, 91, 93, 94–95, 96–97, 106, 111, 115, 122 centre left, 168 below, 169, 170 above, 171 above, 173, 177, 178–179, 180–183, 184 centre and below right, 185–191, 192 inset, 192–193, 195, 196 below left and centre, 197 left, 200 below, 201–203, 207–209, 213; Bob Norrington/146–147; Novosti Press Agency/165 above left, 166 above and below, 168 above, 172–173, 194; John Painter/25, 107, 123, 170–171, 176, 184 left; Picturepoint/173 below; C. Ponnamperuma, University of Maryland/100; Popperfoto/126, 137 above, 157 below, 180 above; Michael Robinson/23, 26, 29–31, 47 right, 60–61, 62–63, 74, 75 above, 83, 87 above, 89, 90, 102, 103, 124, 125, 135, 137 below, 140 below, 143, 155 centre; Dr. Ian Robson/160 inset, 144–145, 160 inset; Ann Ronan Picture Library/28, 119, 120 above, 148–149, 150–155, 158; Royal Observatory, Edinburgh/54–55, 122 below left, 157 above; Robin Scagell/159 below, 160–161; US Geological Survey/endpapers, 1, 34–35, 35 inset, 36–37, 49, 53, 56–57, 63 above, 71, 75 below, 121, 122 below right, 127, 130–133, 139, 140 above, 167; US Naval Observatory Photographs/32, 41, 122 centre right, 128–129; Roy Williams/32 inset, 102 inset, 136. We would also like to thank Bill Pounds and the Air and Space Museum, Washington for their help.

Picture research: Susan de la Plain